T0155809

The 1702 Chair of Chemistry at Cambridge

Transformation and Change

The University of Cambridge's 1702 Chair of Chemistry is the oldest continuously occupied chair of chemistry in Britain. The lives and work of the 1702 chairholders over the past three hundred years, described here, paint a vivid picture of chemistry as it slowly transformed from the handmaiden of alchemists and adjunct of medical men into a major academic discipline in its own right. The book has twelve chapters, covering all fifteen chairholders, from Giovanni Francesco Vigani, a contemporary and friend of Isaac Newton, through Smithson Tennant, discoverer of osmium and iridium, and Alexander Robertus Todd, Nobel Laureate and elucidator of the structure of key components of the double helix, to the current chairholder, master molecule maker Steven Victor Ley. Containing personal memoirs and historical essays by acknowledged experts, this book will engage all who are interested in the pivotal role chemistry has played in the making of the modern world.

MARY ARCHER is a former fellow and lecturer in chemistry at Newnham College, Cambridge. She currently sits on the Chemistry Advisory Board and chaired the Tercentenary Steering Group in the Department of Chemistry at the University of Cambridge.

CHRISTOPHER HALEY was formerly Archivist and Historian of the Department of Chemistry at the University of Cambridge.

The 1702 Chair of Chemistry at Cambridge

Transformation and Change

edited by

MARY D. ARCHER
CHRISTOPHER D. HALEY
Department of Chemistry, University of Cambridge

CAMBRIDGE UNIVERSITY PRESS
Cambridge, New York, Melbourne, Madrid, Cape Town, Singapore, São Paulo

Cambridge University Press
The Edinburgh Building, Cambridge CB2 2RU, UK

Published in the United States of America by Cambridge University Press, New York

www.cambridge.org
Information on this title: www.cambridge.org/9780521828734

© M. D. Archer and C. D. Haley 2005

This publication is in copyright. Subject to statutory exception
and to the provisions of relevant collective licensing agreements,
no reproduction of any part may take place without
the written permission of Cambridge University Press.

First published 2005
This digitally printed first paperback version 2006

A catalogue record for this publication is available from the British Library

Library of Congress Cataloguing in Publication data
Chem@300 (2002 : University of Cambridge)
The 1702 chair of chemistry at Cambridge : transformation and change /
edited by Mary D. Archer, Christopher D. Haley.
p.   cm.
Includes bibliographical references and index.
ISBN 0 521 82873 2 (hardback)
1. Chemistry – Great Britain – History – 18th century.   I. Archer, Mary D.
II. Haley, Christopher D.   III. Title.
QD18.G7C48   2005
540´.941 – dc22   2004051857

ISBN-13 978-0-521-82873-4 hardback
ISBN-10 0-521-82873-2 hardback

ISBN-13 978-0-521-03085-4 paperback
ISBN-10 0-521-03085-4 paperback

The publisher has used its best endeavours to ensure that URLs for external websites
referred to in this publication are correct and active at the time of going to press. However, the
publisher has no responsibility for the websites and can make no guarantee that a site will
remain live or that the content is or will remain appropriate.

Contents

List of contributors   *page* vii
Preface   xv
Holders of the 1702 Chair of Chemistry at Cambridge   xviii
Illustration acknowledgements   xix

1   **'The deplorable frenzy': the slow legitimisation of
    chemical practice at Cambridge University**   1
    Kevin C. Knox

2   **Vigani and after: chemical enterprise in
    Cambridge 1680–1780**   31
    Simon Schaffer and Larry Stewart

3   **Richard Watson: gaiters and gunpowder**   57
    Colin Russell

4   **Lavoisier's chemistry comes to Cambridge**   84
    Christopher Haley and Peter Wothers

5   **Smithson Tennant: the innovative and eccentric eighth
    Professor of Chemistry**   113
    Melvyn Usselman

6   **Coming and going: the fitful career of James Cumming**   138
    William Brock

7   **Chemistry at Cambridge under George Downing Liveing**   166
    John Shorter

8   **The rise and fall of the 'Papal State'**   189
    Arnold Thackray and Mary Ellen Bowden

9   **Alexander Todd: a new direction in organic chemistry**   210
     James Baddiley and Daniel M. Brown

10   **Ralph Alexander Raphael: organic synthesis – elegance,**
     efficiency and the unexpected   237
     Bill Nolan, Dudley Williams and Robert Ramage

11   **Discovering the wonders of how Nature builds its**
     molecules   257
     Alan Battersby

12   **Chemistry in a changing world: new tools for the modern**
     molecule maker   283
     Steven Ley

     Index   304

Contributors

James Baddiley James Baddiley read chemistry at Manchester University, where he did his Ph.D. with Alex Todd on nucleotide structure and synthesis. In 1944, he moved with Todd to Cambridge, obtaining an ICI Fellowship at Pembroke College to work on nucleotide and nucleoside synthesis, and achieving the first structurally definitive chemical synthesis of ATP. He then worked independently in Stockholm, London and Harvard before taking up the Chair of Organic Chemistry, and subsequently the Chair of Chemical Microbiology, at the University of Newcastle-upon-Tyne. He established the structures of several nucleotide co-enzymes, notably co-enzyme A, and discovered the teichoic acids, which are major polymers in many bacterial walls and membranes. In 1980, he returned to Cambridge to continue research in the Department of Biochemistry. He is an Emeritus Fellow of Pembroke College. He has received many awards, including Fellowship of the Royal Society of Edinburgh and of the Royal Society (Leeuwenhoek Lecturer, Davy Medal). He was knighted in 1977.

Alan Battersby Alan Battersby began his chemical research at the Universities of Manchester and St Andrews (Ph.D. 1949). He gained a Commonwealth Fund Fellowship (1950–52) for study in the United States at the Rockefeller Institute with Lyman Craig on peptides and at the University of Illinois with Herbert Carter on pyruvate oxidation factor. In 1954, he joined the staff at the University of Bristol and initiated his research on biosynthesis. He was invited to a chair at the University of Liverpool in 1962 and was elected FRS in 1966. In 1969 he joined Alex Todd as the holder of a second chair of chemistry at Cambridge before being elected to the 1702 Chair in succession to Todd in 1988. He is an Honorary Fellow of St Catharine's College. He became Emeritus Professor in 1992 and continued both experimental research and his writing. He has received numerous national and international awards including the Davy,

Royal and Copley Medals of the Royal Society, the Roger Adams Medal, the Antonio Feltrinelli International Prize, the Wolf Prize and the Alonso Welch Award. He was knighted in 1992 for services to science.

Mary Ellen Bowden Mary Ellen Bowden studied history and chemistry at Smith College, Massachusetts. She undertook graduate studies at Yale University, first as a master's student in history and education in preparation for a brief stint as a high school teacher in the Maryland suburbs of Washington, DC. She returned to Yale as a doctoral candidate in history of science and medicine, studying with Larry Holmes and Derek Price among others and writing her dissertation with Price on seventeenth-century astrology. After a decade of administration and teaching at Goucher College and Manhattanville College, she came to the Chemical Heritage Foundation, Philadelphia, where she has researched and written exhibits and publications on subjects ranging from alchemy to solid-state chemistry. At CHF she is Senior Research Historian and Curator of the Roy G. Neville Historical Chemical Library.

William Brock William Brock read chemistry at University College, London before taking his doctorate in history of science at the University of Leicester. He taught History of Science and Victorian Studies at Leicester from 1959 to 1998, and is now Emeritus. He has published extensively on the history of chemistry, scientific periodicals and the development of scientific education. His books include *From Protyle to Proton* (1985), *The Fontana History of Chemistry* (1992), *Science for All* (1996) and *Justus von Liebig* (1997), and he is currently writing a biography of Sir William Crookes. He was President of the British Society for History of Science, 1978–80, and edited *Ambix*, 1968–83. He has been Chairman of the Society for the History of Alchemy and Chemistry since 1993.

Dan Brown Dan Brown was educated at the Glasgow Academy and the University of Glasgow and began his chemical research at the Chester Beatty Research Institute in London on anti-tumour agents. He joined the research group of Basil Lythgoe and Alexander Todd in Cambridge in 1948, later becoming an Assistant Director of Research and then Reader in Organic Chemistry in the Department of Chemistry. His early research was directed towards understanding the chemistry of nucleotide derivatives that led to general structures of DNA and RNA. This was followed by extended studies of chemical mutagenesis and antiviral agents, subsequently carried out as an attached scientist at the MRC Laboratory of Molecular Biology in Cambridge. He is a Fellow (and sometime Vice-Provost) of King's College, Cambridge and has also been a Visiting

Professor at the University of California at Los Angeles and at Brandeis University. He was elected to the Royal Society in 1982, and together with Professor Sir Hans Kornberg FRS he wrote the biographical memoir of Alex Todd for the Royal Society.

Christopher Haley Christopher Haley read Natural Sciences at St. John's College, Cambridge. This was followed by postgraduate studies in the Department of History and Philosophy of Science at Cambridge, where he completed his doctorate on nineteenth century models of the æther under the supervision of Simon Schaffer. As the historian and archivist of the Department of Chemistry at the University of Cambridge throughout 2002, he was one of the organisers of Chem@300, the conference that celebrated the tercentenary of the 1702 Chair and led to the commissioning of this volume.

Kevin C. Knox Kevin C. Knox completed his doctoral dissertation on culture and scientific change in Georgian Cambridge and London at Cambridge University before returning to North America in 1996 as a visiting professor at the University of California, Los Angeles. In 1997, he became the Ahmanson Postdoctoral Instructor in the Humanities at the California Institute of Technology, where he currently works as the historian of the Institute Archives. He has been involved in a number of multimedia projects related to the history of science and has published numerous works on late-Georgian natural philosophy and mathematics. With Richard Noakes, he co-edited *From Newton to Hawking: A History of Cambridge University's Lucasian Professors of Mathematics*, published by Cambridge University Press in 2003.

Steven Ley Steve Ley is the current BP 1702 Professor of Organic Chemistry at the University of Cambridge, and a Fellow of Trinity College. He studied for his Ph.D. at Loughborough University with Harry Heaney and then did his post-doctoral work with Leo Paquette at Ohio State University. He returned to the UK in 1974 to continue post-doctoral studies with Sir Derek Barton at Imperial College, London. He was appointed to the staff of Imperial College in 1975, became a Professor in 1983 and served as Head of the Department of Chemistry from 1989. He was elected to the Royal Society in 1990 and moved to Cambridge to take up the BP 1702 Chair of Organic Chemistry in 1992. His work involves the discovery and development of new synthetic methods and their application to biologically active systems. So far his group has synthesised over 90 major natural products. His published work has been recognised by 17 major international awards, and he was appointed CBE in 2002. He holds honorary degrees from three universities. He is currently chairman of the Novartis

Foundation Executive Committee, and he stepped down as President of the Royal Society of Chemistry in July 2002.

Bill Nolan Bill Nolan is the Bristol-Myers Squibb Teaching Fellow in the Department of Chemistry at the University of Cambridge. He was an under-graduate at Imperial College, London, 1982–85, during which time his organic chemistry tutor and final year project supervisor was Steve Ley, the current 1702 Professor. He moved to Cambridge to work for his Ph.D. with Ralph Raphael – he was Ralph's last research student and his Ph.D. work was directed towards the synthesis of indolocarbazole natural products. He spent a year as a post-doctoral fellow at Parke-Davis Neuroscience Research Centre in Cambridge and then returned to Lensfield Road as a post-doctoral worker with Andrew Holmes. He was appointed to his current post in 1995. He is a Fellow of Robinson College where, amongst other things, he is Director of Studies in Natural Science.

Robert Ramage Robert Ramage completed his B.Sc. and Ph.D. under the supervision of Ralph Raphael at Glasgow University. He then undertook post-doctoral work with Bob Woodward at Harvard University, 1961–63, on the synthesis of vitamin B_{12}, and at the Woodward Research Institute, Basel, 1963–64, on the synthesis of cephalosporin C. In 1964, he was appointed Lecturer in Organic Chemistry at the University of Liverpool and there followed a five-year period of research on the synthesis and biosynthesis of alkaloids with Alan Battersby, 1964–69, and later with George Kenner on protein synthesis. In 1977, he was appointed Professor of Organic Chemistry at UMIST, becoming Head of Department in 1979. He went to Edinburgh University in 1984 as Forbes Professor of Organic Chemistry, serving two spells as Head of the Department of Chemistry before his retirement in 2001, when he became Scientific Director of Albachem Ltd. He is a past-president of the Perkin Division of the Royal Society of Chemistry and a Fellow of the Royal Societies of both Edinburgh and London.

Colin Russell Colin Russell started his career as an organic chemist working in polytechnic education and researching heterocyclic chemistry. He joined the Open University in 1970 to take up a post in the history of science, which he had recently entered via the history of chemistry (his doctoral thesis was on the rise and development of valency theory). He has since written a number of books, from an elementary text on bakery science to a major biography of Edward Frankland and (as co-author) a new history of the British chemical industry. He became Professor of History of Science and Technology at the Open University in 1981, and is now Emeritus Professor. He was until 2002 a member of Council of the Royal Society of Chemistry and he is also a former chairman of its Historical Group. In 1986–88, he was President of the British Society for the History of Science. Having been a Visiting Fellow at Wolfson College, Cambridge, he is now a senior member, and also an honorary visiting scholar in the Department of History and Philosophy of Science at the University of Cambridge.

Simon Schaffer Simon Schaffer is Reader in History and Philosophy of Science at the University of Cambridge. He read natural sciences at Cambridge, then history of science at Harvard. His Cambridge Ph.D. on Isaac Newton's cosmology was awarded in 1980. He has taught at Imperial College, London and since 1984 at the Department of History and Philosophy of Science. He co-wrote *Leviathan and the Air Pump* with Steven Shapin (Princeton University Press, 1985), and has since produced work on the history of experiment and the physical sciences. Most recently, he has co-edited *The Sciences in Enlightened Europe* for Chicago University Press.

John Shorter John Shorter read chemistry at Exeter College, Oxford and did his D.Phil. on organic solution kinetics with Sir Cyril Hinshelwood. He was a staff member of what is now the University of Hull from 1950 to 1982, and is now Emeritus Reader in Chemistry. He has been Secretary of the International Group for Correlation Analysis in Chemistry (formerly Organic Chemistry) since 1982, and was a member of the IUPAC Commission on Physical Organic Chemistry, 1990–97. He has authored or co-authored many papers on physical organic chemistry, particularly on linear free-energy relations, and he is the author or co-editor of several books in this field. He was formerly Secretary and then Chairman of the Royal Society of Chemistry's Historical Group, and in 'retirement' he writes on the history of physical organic chemistry.

Larry Stewart Larry Stewart was educated in the UK, France and Canada, and received his Ph.D. in the history and philosophy of science from the University of Toronto. He is Professor of History and former Head of the Department of History at the University of Saskatchewan and currently visiting scholar at the Max-Planck-Institüt für Wissenschaftsgeschichte in Berlin. He is the author of *The Rise of Public Science. Rhetoric, Technology and Natural Philosophy in Newtonian Britain, 1660–1750* (Cambridge University Press, 1992) and (with Margaret C. Jacob) *Practical Matter. The Impact of Newton's Science, 1687–1851* (Harvard University Press, forthcoming). He is currently writing a study of chemistry, medicine and social reform in the late eighteenth century.

Arnold Thackray Arnold Thackray read chemistry at Bristol University and worked in industry before taking a Ph.D. in the history of science at Cambridge, where he was the pupil of Professor Mary Hesse as well as being the first student of Churchill College to be elected to its fellowship. He has made his career in the USA for the past 35 years. He has variously served as editor of *Isis and Osiris*, President of the Society for Social Studies of Science, Chairman of the Department and Joseph Priestley Professor of the History and Sociology of Science at the University of Pennsylvania, and Founding President of the Chemical Heritage Foundation.

Melvyn Usselman Mel Usselman read chemistry at the University of Western Ontario, Canada, and did his Ph.D. there with Paul de Mayo in organic photochemistry. As a post-doctoral fellow, he collaborated with de Mayo in the construction of a history of chemistry course and simultaneously obtained an MA in History at UWO under the supervision of Larry Holmes. He joined the Chemistry Department at UWO as an assistant professor in 1975, became an associate professor in 1981, and served as associate Chair, 1992–95. He teaches courses in general, organic and historical chemistry and has received teaching awards at UWO, as well as an Ontario Confederation of University Faculty Associations teaching award. In recent years, he has reconstructed the critical experimental works purporting to establish the law of multiple combining proportions, and also Liebig's 1831 combustion apparatus and his early analyses of alkaloids. In 2001, he completed a project to modernise all the chemistry articles and biographies of *Encyclopaedia Britannica*. He has published several papers on the life and science of William Hyde Wollaston as preliminary work towards a full scientific biography.

Dudley Williams Dudley Williams received his bachelor and Ph.D. degrees, in chemistry and organic chemistry, respectively, from the University of Leeds in 1958 and 1961. He subsequently studied as a post-doctoral Fellow at Stanford University, 1961–64. Since 1964, he has worked at the University of Cambridge, where he is a Fellow of Churchill College and Professor of Biological Chemistry. He was elected a Fellow of the Royal Society in 1983. He knew Ralph Raphael for much of his career, but particularly well during Ralph's years in Cambridge.

Peter Wothers Peter Wothers read chemistry at Cambridge, graduating in 1991 before doing his Ph.D. on stereoelectronic effects and conformational analysis with Professor Tony Kirby. In 1996, he was appointed to the newly established post of Teaching Fellow in the Department of Chemistry at Cambridge. He lectures to undergraduates and also runs the physical chemistry practical courses. Beyond that, he is involved in teaching chemistry to all age groups, giving demonstration lectures to the general public and school children, running courses for schoolteachers and supporting the International Chemistry Olympiad. He has co-authored two textbooks, *Organic Chemistry* and

Why Chemical Reactions Happen, both published by Oxford University Press. Throughout his time at Cambridge, he has remained at St Catharine's College, where he is the Director of Studies in Chemistry. In 2002, he was awarded one of the University's Pilkington Teaching Prizes. He has a keen interest in the history of chemistry and has amassed a fine collection of chemistry books from the seventeenth and eighteenth centuries.

Preface

It was a summer's day in 2001. We were among a group of Cambridge chemists making an excursion to the Whipple Museum of the History of Science to view the recently reconstructed post-war photochemistry laboratory in Free School Lane, only yards from where the young George Porter had built the first flash photolysis apparatus some fifty years ago. It was there that we realised that another anniversary in the long history of chemistry at Cambridge was upon us: the University's 1702 Chair of Chemistry would turn 300 years old the following year.

We all felt that we could not ignore such a landmark, and in December 2002 the Department of Chemistry held a two-day symposium, entitled *Chem@300*, to mark the tercentenary. The papers given at that meeting, with one or two later additions, form the basis of this book. *The 1702 Chair of Chemistry at Cambridge: Transformation and Change* pays tribute not only to the chemist's core skill of transforming one set of molecules into another, but also to the subject of chemistry itself, and how it has changed over the centuries from the handmaiden of alchemists and adjunct of medical men into a major academic discipline in its own right. The book is a history of the 1702 Chair, rather than a general history of chemistry in the University, so many important developments in Cambridge chemistry (and biochemistry) have fallen outside its remit.

The 1702 Chair of Chemistry at Cambridge is (save for an interregnum in World War II) the oldest continuously occupied chair of chemistry in Great Britain, although it is not the oldest chair of chemistry in the country: that honour goes to Oxford, where Robert Plot was appointed first curator of the Ashmolean Museum and Professor of Chemistry in 1683, but the chair lapsed for some time after his death: evidently Oxford thought the subject had no future.

This book presents a series of essays on the 1702 chairholders, each described by an expert or – in the case of Alan Battersby and Steven Ley – by the chairholder himself. The sweep is wide, from Giovanni Francesco Vigani, a contemporary and friend of Isaac Newton, through Richard Watson who sat in the House of Lords as Bishop of Llandaff, Smithson Tennant the discoverer of osmium and iridium, to George Downing Liveing, during whose long tenure of the chair the Natural Sciences Tripos was created and the college chemical laboratories waxed and waned. We enter modern times with Alexander Robertus Todd, Nobel Laureate and elucidator of the structure of key components of the double helix, the elegant synthetic and biosynthetic work of Ralph Raphael and Alan Battersby and end with the current chairholder, master molecule maker Steven Ley.

For simplicity, we use the title '1702 Chair of Chemistry' throughout, but this has never been the precise title of the chair, nor according to modern reckoning was it founded in 1702. True, the Grace of Senate that brought Vigani's chair into being was dated 10mo *Feb. 1702*, but the University, like most official bodies in England, was still using the old Julian calendar at the time. Under our present Gregorian calendar, the date of the chair's foundation was 21 February 1703. As to the name of the chair, Vigani held the simple title 'Professor of Chemistry', as did subsequent chairholders up to and including William Jackson Pope, even though four new chairs (Physical, Colloid Science, Metallurgy and Theoretical Chemistry) had been created in the Department by the time Pope died in 1939. Pope's death created a vacancy in the field of organic chemistry, so his post was renamed 'Professor of Organic Chemistry' by Grace of 26 February 1943 (*Reporter*, 2 February 1943, p. 358), and this was the title conferred on Pope's successor, Alexander Todd. In 1969, the University created a second Chair of Organic Chemistry *ad hominem* for Alan Battersby, and in the following year the date of 1702 was first attached to the earlier chair to distinguish the two. Thus Todd occupied the 1702 Chair of Organic Chemistry, as did his successor Ralph Raphael.

In 1990, the British Petroleum Company announced a generous benefaction of £1.5 million to re-endow the 1702 Chair, as one of a series of four endowments or re-endowments of British university chairs organised by Robert Horton, then chairman and chief executive of BP, and David Simon, then BP's Managing Director. In recognition of the benefaction to Cambridge, the General Board proposed that the chair be renamed the BP Professorship of Organic Chemistry (1702), and this was approved by Grace of 1 May 1991 (*Reporter*, 6 March 1991, p. 472). The structure of this formal title follows the University's 'subject, name (date)' convention for all dated chairs, but in normal usage the date is placed first. Thus it was when the current chairholder Steven Ley took up his appointment in 1992, he became the fifteenth holder of the '1702 Chair of Chemistry', but the first to take the title '1702 BP Professor of Organic Chemistry'.

We are indebted to Dr and Mrs Alfred Bader and to the Amberstone Trust for sustained financial support that enabled the Department to engage CH as archivist and historian for an eighteen-month period, and to Trinity College, Cambridge for their kind contribution towards the costs of publication of this book. We warmly thank our contributing authors for the dedication they have brought to their subjects and the patience with which they have dealt with our editorial queries. We also thank Jayne Aldhouse, Michelle Carey, Tim Fishlock, Andy Flower and Emily Yossarian of Cambridge University Press, and Liz Alan, Nick Bampos, Alan Battersby, David Buckingham, Brian Crysell, Peter Grice, Andrew Holmes, Jeremy Sanders, Jane Snaith, Brian Thrush, David Watson, Peter Wothers and Dudley Williams of the Department of Chemistry, Jacky Cox and Elisabeth Leedham-Green of the University Archives, Liba Taub of the Whipple Museum, Brian Callingham and John Eatwell of Queens' College, Richard Glauert and Denis Marrian of Trinity College, Nick Champion, Alison McFarquar and Mark Mniszko of the University Press and Publications Office, Deborah Easlick and Jane Crawford of the Development Office, Brian Emsley, David Giachardi and Cath O'Driscoll of the Royal Society of Chemistry, James Bamberg and Bernie Bulkin of BP, Peter Morris of the Science Museum, John Hudson of the Society for the History of Alchemy and Chemistry, Helen Brown, Hilary Todd and Sandy Todd, and Grant Buchanan, John Emsley, Roy MacLeod, Barbara Mann and Colin Russell. All these have commented on parts of the manuscript (as well as in some cases writing a chapter) or helped in other ways to redeem the sometimes fragile and dispersed history of the 1702 chairholders, thus making it possible to assemble the continuing story of chemistry in Cambridge.

Mary Archer
Christopher Haley

Holders of the 1702 Chair of Chemistry at Cambridge

	Date elected (n.s.)	Date vacated (n.s.)	
Vigani, John Francis (c. 1650–1713)	1703	1713	Died in office
Waller, John (c. 1673–1718)	1713	1718	Died in office
Mickleburgh, John (c. 1692–1756)	1718	1756	Died in office
Hadley, John (1731–1764)	1756	1764	Died in office
Watson, Richard (1737–1816)	1764	1771	Resigned
Pennington, Isaac (1745–1817)	1773	1793	Resigned
Farish, William (1759–1837)	1794	1813	Resigned
Tennant, Smithson (1761–1815)	1813	1815	Died in office
Cumming, James (1777–1861)	1815	1861	Died in office
Liveing, George Downing (1827–1924)	1861	1908	Resigned
Pope, William Jackson (1870–1939)	1908	1939	Died in office
Todd, Alexander Robertus (1907–1999)	1944	1971	Resigned
Raphael, Ralph Alexander (1921–1998)	1972	1988	Resigned
Battersby, Alan Rushton (1925–)	1988	1992	Resigned
Ley, Steven Victor (1945–)	1992	–	–

Illustration acknowledgements

Front cover: '*The Alchemist's Experiment Takes Fire*' by Hendrick Heerschop, by courtesy of the Fisher Collection, Chemical Heritage Foundation, Philadelphia.

Back cover: Vigani's arms on stained glass by Henry Gyles © Christopher Haley, with thanks to the Norwich Union.

Figure 1.1: Reproduced from Denis Dodart's *Mémoires pour servir à l'histoire des plantes*, Figure 1.4: Interior of the 'Publick Elaboratory' from William Combe's *History of the University of Cambridge*, and Figure 1.7: The Botanic Garden from William Combe's *History of the University of Cambridge*, by courtesy of the Institute Archives, Caltech.

Figure 1.2: Vigani's advertising flyer, by courtesy of the Bodleian Library, University of Oxford (John Johnson Collection, Patent Medicines 14).

Figure 1.3: Plan of the 'Publick Elaboratory' from Cambridge University Library (UL), Views.x.2[79a], Figure 1.5: Inventory of the 'Publick Elaboratory' from UL CUR 39.11.2(2), Figure 4.3: The Schools in the Botanic Garden from Cambridge University Library Views.x.2(73), Figure 5.2: Double distillation apparatus, from *Phil. Trans. Roy. Soc.* (1814), Figure 5.3: Ledger of Wollaston & Tennant's sales of platinum, from UL Wollaston Mss., Add MSS 7736, notebook I, Figure 5.5: Wollaston/Tennant financial agreement from UL Wollaston Mss., Add MSS 7736, notebook L1, Figure 5.7: Tennant's apparatus for potassium production, from *Phil. Trans. Roy. Soc.* (1814), Figure 6.3: Sketch of Cumming's laboratory from UL DAR.204.4 (14/11/1822), Figure 6.4: Development of the New Museums site from UL Archives UA.P.VIII.3, Figure 6.6: Cumming's electromagnetic instruments, from *Trans. Camb. Phil. Soc.* (1822), and Figure 8.5: Plan of Pembroke Street Laboratory from UL Archives, UA.P.VIII.10, by permission of the Syndics of Cambridge University Library.

Figure 1.6: Portrait of John Hadley © The British Museum.

Figure 2.1: Portrait said to be of John Francis Vigani, Figure 4.4: Wollaston's apparatus for the decomposition of water, Figure 4.5: Wollaston's apparatus for synthesis of water, and Figure 6.1: Reverend Professor James Cumming, by courtesy of the Master, Fellows and Scholar of Trinity College, Cambridge.

Figure 2.2: Vigani's *Medulla Chymiae*, by courtesy of the Master and Fellows of St Catharine's College, Cambridge.

Figure 2.3: Vigani's arms on stained glass by Henry Gyles © Christopher Haley, with thanks to Norwich Union.

Figure 2.4: Vigani's cabinet of *materia medica* © Department of Chemistry, with thanks to the President of Queens' College, Cambridge.

Figure 3.1: Wood engraving of Richard Watson, from a portrait by Romney, taken from Watson's *Anecdotes*.

Figure 3.2: '*Eloquence founded on Chemical Principles*', by courtesy of the National Portrait Gallery, London.

Figure 3.3: Calgarth Park, and Figure 3.4: Charcoal cylinders as gateposts in Gatebeck, Cumbria © Colin Russell.

Figure 3.5: Portrait of Richard Watson by Reynolds, reproduced with acknowledgements to the National Museum of Cuba.

Figure 4.1: Isaac Pennington, and Figure 4.2: Isaac Milner (LF 239.M5), reproduced by permission of the Master and Fellows of St John's College, Cambridge.

Figure 4.6: Engraving of William Farish, by courtesy of the Master and Fellows of Magdalene College, Cambridge.

Figure 6.5: All Saints' Church, North Runcton, by courtesy of Leicester University Library (English Local History Collection).

Figure 7.1: George Downing Liveing, Figure 7.3: Exterior of the Pembroke Street Chemical Laboratory, Figure 8.2: Pembroke Street Laboratory, Figure 9.4: Todd and HRH Princess Margaret, Figure 11.1: Alan Battersby, and Figure 12.1: Steven Ley, © Department of Chemistry, Cambridge.

Figure 7.2: St John's College laboratory, reproduced by permission of the Master and Fellows of St John's College, Cambridge, with acknowledgements to S. Muggleton.

Figure 8.1: William Jackson Pope, by courtesy of the Fisher Collection, Chemical Heritage Foundation.

Figure 8.3: '*What was it?*', reproduced by permission of Cambridge Newspapers.

Figure 9.1: Alexander Robertus Todd, and Figure 9.2: James Baddiley and his research group, from the private collection of Sir James Baddiley with permission.

Figure 9.3: Charles Dekker, Dan Brown and Hugh Forrest with Herchel Smith in front, from the private collection of Dan Brown with permission.

Figure 10.2: Ralph Raphael with Franz Sondheimer, Figure 10.6: Ralph Raphael with Ian Scott and Willie Porter, and Figure 10.13: Ralph and Prudence Raphael at Buckingham Palace, from the private collection of Prudence Raphael with permission.

Figure 11.6: Alan Battersby and Simon Bartholemew, from the private collection of Sir Alan Battersby with permission.

Figure 12.2: Ralph Raphael, Alexander Todd, Alan Battersby and Steven Ley, from the private collection of Steven Ley with permission.

Images of chemical structures are the copyright of the respective authors. Every attempt has been made to identify and contact the copyright holders where appropriate. However, if an image has been incorrectly attributed or unacknowledged, please contact the editors via Cambridge University Press, and they will endeavour to correct the error.

1

'The deplorable frenzy': the slow legitimisation of chemical practice at Cambridge University

Kevin C. Knox

Division of Humanities and Social Sciences, California Institute of Technology

On a mild autumnal evening in 1780, Cambridge's distinguished physician Robert Glynn rose from his chair in Somerset House to address the Royal Society. He announced that he wished to communicate a comical narrative concerning Cambridge's former professor of chemistry and the current Regius Professor of Divinity, Richard Watson. Dr Glynn reported how he had been enjoying a quiet evening reading in his rooms when a distraught messenger stormed in exclaiming, 'Dr. W[a]ts[o]n is absolutely ungovernable.' Immediately, Glynn speculated that his distemper could be attributed to 'chemical vertigo'. Upon arriving at Watson's college rooms, his 'auditories' were assaulted by 'tremendous growls'. Glynn soon came to realise that Watson had been immersed in chemical experiment and fiendishly penning 'two bulky volumes in quarto on chemistry'. His diagnosis was swift: Glynn supposed that Watson's 'kindly liking to the crucible' had resulted in 'a little too much of the phlogiston in his composition'. In response, Glynn's nurse declared, 'You perceive the natural easy transition from chemistry to divinity; you would almost think they were dependent on each other'. Glynn resolved upon an exorcism as the expedient remedy: 'Mr A[twood], almost the only *conjuror* in the Un-v-rs-ty', was summoned and in due course the malevolent spirits were coaxed from the body of the 'Philosophical Scriblerus'.[1]

This incident was recorded as *The Late Strange and Deplorable Frenzy of the Reverend Richard Watson*. No doubt this droll tale from the waggish Glynn amused the scholars, physicians and divines of the prestigious Royal Society. Yet this tale of Watson's 'distemper' is also representative of the problematic status of chemistry in Cambridge during the long-eighteenth century. Hitherto, because it has seemed self-evident that chemistry *is* one of the natural sciences,

The 1702 Chair of Chemistry at Cambridge: Transformation and Change, ed. Mary D. Archer and Christopher D. Haley. Published by Cambridge University Press. © Cambridge University Press 2005.

Figure 1.1. The cultivation, preparation and philosophical discussion of medicinal plants and herbs was a major part of eighteenth-century chemistry at Cambridge. Professors and students alike would be extremely familiar with all the steps required to dose patients, as is depicted in this French 1676 vignette from Denis Dodart's *Mémoires pour servir à l'histoire des plantes*.

historians have generally treated the foundation of chemical professorships as an inevitable component of the progression of universities. Two explanandæ stem from this assumption: why enlightened scholars such as Isaac Newton 'dabbled' in alchemy, and why dons who opposed the study of chemistry were misguided or 'irrational'.[2] (Lethargy, indifference, antiquarianism, dullness and Anglicanism have all been used as explanations for the 'lack-lustre' careers of Cambridge chemists.[3]) Some recent studies have challenged these assumptions; yet most historians still fail to elucidate the deep tensions embedded within chemical discourse and the convoluted topography of its practice and pedagogy during the *siècle de lumières*.[4] As late as 1800, Cantabrigian dons were uncertain if the 'chemical revolution' added or subtracted from the discipline's credibility. Meanwhile, to other detractors, chemistry was not a philosophical discipline but a 'sooty' and 'dingy' craft. Before it could pass muster in Cambridge, chemistry needed to be tamed and sanitised.

This domestication of chemistry has a protracted and complex history (Figure 1.1). With their exploration of the experimental techniques, local audiences and the global network in which the eighteenth-century chemists of the varsity were linked, Simon Schaffer and Larry Stewart skilfully describe several elements of this triumphant domestication in Chapter 2. Nevertheless, it is vital also to show the persistence of chemistry's awkwardness and the intense social negotiations that helped make chemistry a discipline fit for the university undergraduate. These negotiations also illuminate larger cultural and intellectual issues. At the turn of the eighteenth century, for example, chemistry found many proponents in forums such as coffee-houses, metropolitan waterworks,

Gresham College and provincial breweries. This 'rise of public science' contrasted with Cambridge, where the proponents of chemistry struggled to show that it was an appropriate activity for gentleman-scholars and future divines. In the eighteenth century, 'chemistry' was undoubtedly many different things to different people in different locales, but to many Cambridge scholars it remained an 'art', at best a non-essential accessory for future physicians. Even amongst the university's chemical professors, the uncertainty whether their lecture material was a branch of natural philosophy or a discrete enterprise persisted throughout the century. Nor were philosophers certain whether theoretical work such as Robert Greene's anti-Newtonian contractive forces or William Whiston's fluid dynamics of Creation fell within the category of chemistry.[5]

Though eminent philosopher-dons (such as Richard Watson) could march into its domains, chemistry remained tainted. What Glynn's nurse inferred from Watson's activities – that chemistry and divinity had become dependent on each other – was a common fear in the Hanoverian university. In his book on enlightenment Cambridge, John Gascoigne concedes that 'geology and biology' were subjects 'subject to particular scrutiny' because of their bearing on biblical exegesis. Yet he pronounces that 'a subject like chemistry . . . had no relevance to such religious issues', and 'was not viewed with quite the concern'.[6] If this is true, then it is difficult to explain why, for example, Glynn wrote his parody or why Watson felt himself compelled to torch his chemical manuscripts upon receiving his bishopric.

Throughout the eighteenth century, Cambridge dons realised that chemistry could have tremendous resonance in politico-theology. Although the dons exploited other scientific disciplines to comment on the nation's political and religious practices, particularly what Sir David Brewster dubbed the 'Holy Alliance' between Newtonianism and low Anglicanism, the chemical professors, wary of the subject's volatility, made little effort to forge such marriages.[7] They understood that chemistry threatened the status quo at Oxbridge. The anxiety of late-seventeenth century sceptics – especially their fear that chemistry was secretive, occult and inflammatory – persisted through the entire century.

This chapter focuses on the tactics that the champions of chemistry – from John Francis Vigani onwards – deployed to win over sceptics, especially the arguments that they used to convince their peers that chemistry had transformed from a craft to a philosophical discipline. As such, they made chemistry more gentlemanly and therefore suitable for a prominent place in the university curriculum. To show how extended and problematic this process was, I consider the context in which the professorship was founded, delving into the career of the foundation professor, John Francis Vigani, in relation to the political interests of Richard Bentley. The second section recounts some of the professors' minor triumphs of mid-century, as well as their enduring struggles to make the discipline a worthy subject for the varsity. In doing so I discuss the professors' departure from *materia medica*, their endeavours to import elements of Scottish 'philosophical chemistry' and their attempts to balance the discipline between a recondite activity and superficial entertainment. The final section examines the repercussions of the 'chemical revolution', as the professors tried to navigate through the minefield of French 'atheistic' nomenclature and nagging phlogistic vapours. Thanks to their sophisticated navigation, the new breed of chemical scholars were finally capable of purging the lingering doubts from even the most suspicious of dons, enabling chemical practice to flourish in Cambridge.

Diverting amusements

During the sixteenth and seventeenth centuries there was little reason why a scholar would come to Cambridge specifically to study chemistry. High Churchmen offered little encouragement to those who were interested in establishing a *laboratorium* within college cloisters. Nonetheless, the university was never an intellectual wasteland in terms of chemical investigations. Before the eighteenth century, undergraduate disputations often touched upon matter theory, usually in reference to Aristotelian forms, essences, affectations and qualities, as well as animal economies such as digestion and nourishment.[8] Such disputations could be very provocative. Richard Drake reminisced about his thesis on '*Pura Elementa non sunt Alimenta*' (Pure elements cannot provide sustenance), delivered in the Old Schools on Ash Wednesday, 1630: 'the speech . . . roused the hornets about my ears and so exited the anger of the Prochancellor, the Doctors, and I don't know whom else, that I was called to account before them.'[9]

Despite little support from college masters, late-Elizabethan Cambridge produced several scholars who practiced the chemical art. The exploits of John Dee are renowned; less celebrated is the work of gownsmen such as Samuel Norton, great-grandson of Thomas Norton, and Peterhouse's William Parys. Norton, for example, penned a *Key to Alchemie* and a treatise on mercurial preparations, while Parys published a *Booke of Secrets* which revealed 'diuers waies to make & prepare all sortes of Inke & Colours'.[10] William Harvey studied at Gonville and Caius College in the final decade of the sixteenth century, although the work for which he is remembered was completed in London and Oxford. While it is the 'Oxford physiologists' who are commemorated for their chemico-pneumatic researches, Cambridge's Regius Professor of Physic, Francis Glisson, also promoted a 'Harveian research tradition'.[11]

During the Interregnum and early years of the Restoration there is evidence that a 'Philosophical Club of Chymists', including Joseph Nidd, Isaac Barrow and John Beale, conducted chemical experiments.[12] Moreover, in their speculations about natural history and the Creation, John Ray and Tancred Robinson often investigated chemical subjects.[13]

Speculation about the Creation and the concealed processes of nature were regarded with suspicion by many Cambridge gownsmen. They fretted that such conjectures would evoke the nefarious contagion of the Civil War and Interregnum: Sectarianism. In general, this fear led them to agree with apologists for the Royal Society such as Thomas Sprat, who contended that the gentlemanly and semi-public space of Gresham College could alleviate the ravings of 'inspired' Britons.[14] However, while Robert Boyle assumed that Gresham College was a space that did not impinge upon theology, Henry More and the Cambridge neo-Platonists mobilised Boyle's experimental philosophy to comment on religious doctrines. Where Boyle claimed that his pneumatic engine produced value-free 'matters of fact' upon which everyone could agree, More and his Emmanuel College associates appropriated Boyle's pneumatic trials in order to comment upon the role of immaterial spirit and the existence of an hylarchic principle that governed the universe. Like Boyle, More was determined to evince that matter was 'brute and stupid' in order to demonstrate that an intelligent spirit 'umpired' inanimate and 'preposterous' matter. Yet, going further than Boyle, More reckoned that demonstrating the role of vital spirits must be the goal of a natural philosopher since this was a key 'antidote to atheism' and an efficacious remedy for sectarian distemper.[15]

Although Isaac Newton grumbled at 'the want of persons willing to try experiments', he undoubtedly inherited many of his (al)chemical interests from scholars such as More, Cudworth, Ray, Beale and Barrow. His own willingness to sully his hands in experiment and the protracted periods he sat in front of his furnace in Trinity College are well documented. As Newton's amanuensis remembered, 'ye fire in ye Elaboratory scarcely went out' during his indefatigable quest to regain the alchemical wisdom of the ancients.[16] It is from these experiments, and readings of Basil Valentine, George Ripley, Michael Maier and Sendivogius, that Newton came to envision the earth as a living entity: 'this Earth resembles a great animall or rather inanimate vegetable, draws in ætherial breath for its dayly refreshment & vitall ferment & transpires again wth grosse exhalations.'[17] For Newton, chemical analysis became crucial for comprehending the present state of the universe, man's place within the cosmos, as well as cosmogony and the millennium.

It remains uncertain to what extent John Francis Vigani acquiesced in Newton's conception of the earth as a breathing vegetable, but it has been recorded that the 'Great Man' enjoyed conversations with the Veronese apothecary. Humphrey Newton reminisced that 'Mr Vigani, a Chymist', was a regular guest of Newton, 'in whose Company he took much Delight and Pleasure at an Evening'.[18] This 'pleasure' ceased abruptly after the unfortunate Vigani recounted 'a loose story about a Nun', but before the unforgiving Newton broke with Vigani he wrote to Boyle that the Italian had 'been performing a course of Chymistry to several of o^r University much to their satisfaction'.[19] Vigani's first course probably took place in 1682, in an outdoor laboratory located in the cloisters of Queens' College. Almost certainly, students paid Vigani directly for his services, as they did for medical consultations, for it was Vigani's career as *practising* apothecary that proved the most enduring feature of his presence in the university town. Indeed, as late as 1709 Vigani was still peddling his medical wares, as the flyer shown in Figure 1.2 indicates.[20] Soon after his arrival in Cambridge Vigani's medicines became indispensable for Cambridge's student body:

> *Meus viridis Mercurius præcipitatus brevi momento conficitur, & Gonorrhæm*
> *infallibiliter, & radicaliter curat; non intelligo illum, de quo alii Authores*
> *mentionem fecerunt, sed tantum de eo, quem tanquam maxinuum arcanum*
> *conservo.*[21]

In effect, Vigani's trade secret perpetuated Cambridge's secret trade. Vigani's green precipitate of mercury, which he claimed was an infallible cure for venereal disease, undoubtedly was deployed frequently in an environment plentiful with 'women of doubtful virtue'.[22] It is perhaps ironic that an apothecary who often saw students in a state of distressed undress would later don sumptuous professorial robes. It is likely also that it was their mutual interest in medicines upon which the Vigani–Newton friendship was founded. Notwithstanding Newton's seeming indifference to his own body, he slaved over medicinal preparations, including his Lucatello Balsam, used 'ffor ye Measell Plague . . . & ye biting of a mad dog'. Among other preparations, he worked hard to perfect a '*primum ens*' of Balm, a restorative agent that was, among other things, capable of starting menstruation in seventy-year-old women.[23]

Vigani's expertise in *materia medica* also interested both members of the university and local practitioners of medicine. In 1704 the President of Queens' College agreed to the purchase of a handsome oak cabinet for Vigani, equipped with over 600 ingredients, ranging from *Sanguis Draconis* to opium (Figure 2.4 in the following chapter).[24] Vigani, however, did not lecture exclusively on pharmacopoeia. The year before he was awarded his professorship (1702/3), Vigani's lectures covered an array of subjects. Course notes from one attendee record that Vigani concentrated on processes related to the furnace, particularly

MR. VIGANI, Professor of Chymistry in the University of Cambridge, promises to sell unto his Old Customers the Medicines following, at the Prices annexed, from the First of August 1709. Which for their Satisfaction he hath thought fit to print. He only names the Medicines generally used; but if they think fit to send for any others, they shall have them at a proportionably low Price, and all the Preparations in as much Perfection as formerly.

NEWARK, August 1709.

	s.	d.		s.	d.
Sp. Corn. Cerv. ?			Antiheect. Poter. *per* ℥	2	9
Sp. Sal. Armon. } *per lib.*	5	4	Antimon. Diaphor. *per* ℥	1	0
Sp. Sal. Armon. Succinat. ?			Sulph. Antimon. *per* ℥	0	6
—Cum Galb.			Croc. Metall. *per lib.*	5	0
—Cum Gum. Ammon. } *per* ℥	0	9	Aurum Mofaic. *per* ℥	2	6
—Cum Aff. Fœt. }			Tart. Vitriolat. *per* ℥	1	0
Sp. Cran. Hum. *per* ℥	2	0	——Emet.	1	6
Sp. Sang. Hum. *per* ℥	1	0	Anima Mart. *per* ℥	1	0
Sal. Vol. Oleof. Sylv. *per* ℥	0	9	Mercur. Dulc. ? *per* ℥		
——Noftr. *per* ℥	1	6	Calomel. }	1	3
Sal volat. Salis Armon. *per* ℥	0	6	Refin. Jalap. *per* ℥	4	6
——Armon. Aromat. *per* ℥	1	2	——Scammon. *per* ℥	4	0.
Sal vol. C. C. *per* ℥	0	9	——Caftor. *per* ℥	1	3
Sp. Millep. ? *per* ℥			—Croci. *per* ℥	0	9
—Lumbric. }	0	9	— Succin. *per* ℥	0	9
Anodynum Diaphoret. *per* ℥	1	6	Ol. Succin. *per* ℥.	1	0
Elix. Vitriol. Mynf. ? *per* ℥			Ol. Succin. commun. *per* ℥	0	3
—Pæon. Mynf. }	1	3	—Sulph. per Camp. *per* ℥	1	3
Proprietat. *per* ℥	1	0	Sal vol. Succin. *per* ℥	5	0
Mart. *per* ℥	0	6	Sp. Caftor. *per* ℥	1	3
Tinct. Antimon. *per* ℥	1	3	——Theriac. *per* ℥	0	9
——Metallor. *per* ℥	1	0	Miftura Simpl. *per* ℥	1	0
Corall. *per* ℥	1	3	Pil. Matthæi. *per* ℥	2	6
Archealis. *per* ℥	1	0	Sp. Nitr. dulc. *per* ℥	0	9
Stomach. *per* ℥	0	10	—Sal dulc. *per* ℥	0	6
Martis Mynf. *per* ℥	1	0	Ens Ven. *per* ℥	1	0
Laud. liq. Cydon. *per* ℥	1	3	Sal Mart. River. *per* ℥	1	2
——Willif. *per* ℥	1	0	Balf. Sulph. Anifat. *per* ℥	1	4
——Tartariz. *per* ℥	1	2	——Succinat. *per* ℥	1	0
Londinenf. *per* ℥	5	0	——Terebinth. *per* ℥	0	4
Extract. Rud. *per* ℥	2	0	Crem. Tartar. *per lib.*	6	0
—Gentian. ?			Mars Willif. *per lib.*	2	6
Trifol. Paluft. (			Sp. Lavend. Comp. *per lib.*	8	0
Sabin. } *per* ℥		9	Sal. Abfinth. Chryftal. *per* ℥	0	6
Rut.			——Non Chryftal. *per* ℥	0	6
Centaur. Min.)			—Cochlear. *per* ℥	0	8
Lacryma Mart. *per* ℥	1	8	—Genift. *per* ℥	0	6
Panacæa Antimon. *per* ℥	1	6	—Artemif. *per unc.*	0	6

Printed at the University-Press in CAMBRIDGE, 1709.

Figure 1.2. Flyer advertising Vigani's preparations, published by the University Press in 1709.

distillations, cohobations (repeated distillation with the distillate returned to the residue), fermentations and sublimations. In doing so he touched upon metallurgical processes, vegetable products and animal distillations. Besides providing his students with a variety of recipes for elixirs, Vigani used his great skill with the unmortared brick furnace to impart the prevailing continental view of chemical principles: 'The Chemical Principles are commonly reckoned Five, whereof three are called active principles viz. The Spirit of Mercury, Sulphur & Salt, the other two are called passive principles viz. The Phlegm or Water, & Earth.'[25]

These lectures echo Vigani's only monograph, his 1683 *Medulla Chymiæ*, and initially seem to evince that Vigani's focus was not theoretical philosophy but rather practical chemistry.'[26] If Vigani seemed to display 'little concern for underlying goings-on,'[27] his contemporaries nevertheless grasped the deeper import of his practice. Although Hermann Boerhaave dismissed *Medulla Chymiæ* as a 'confused medley of experiments,' influential members of the Royal Society imagined that Vigani's work complemented the over-arching millennial aims of the Society. Prefacing the *Medulla Chymiæ* was an epistolary letter, possibly written by Sir Tancred Robinson, secretary of the Royal Society, cohort of John Ray and Sir Hans Sloane, and, later, physician to George I.[28] In the Latin epistle, the author hitched Vigani's chemical work to the visionary, Edenic and corpuscular mission of Gresham College. Condemning 'cowardly, idle sheep', 'old superstitions' and the '*deleria Jesuitica*', he praised the 'worthy Italian' for deciphering the cryptic 'hieroglyphs' of the 'omnipotent Creator.' Like 'Adam in Paradise calmly surveying God's creatures,' Vigani 'followed living nature herself'.[29]

Despite little discussion of theoretical matters, Vigani's work seemed to conform to the aims of latitudinarian natural philosophers. Had his work been contradictory or even irrelevant to these aims, it is doubtful whether he would have found support from Newton or two famous Boyle lecturers, Richard Bentley and William Whiston. Whiston, perhaps the university's most active philosopher from the 1690s until his banishment in 1710, understood that the biblical exegete needed to be well-versed in the new chemistry. In his *New Theory of the Earth*, he devoted a number of sections to the equivalence of cometary atmospheres and the 'Ancient chaos' in order to salvage the Mosaic account of Creation. He argued also that 'the Constitution of the *Antediluvian* Air was Thin, Pure, Subtile and Homogeneous, without such gross Streams, Exhalations, Nitrosulphureous, or other Heterogeneous Mixtures, as occasion Coruscations, Meteors, Thunder, Lightning, with Contagious and pestilential Infections in our present Air; and have so many pernicious and fatal (tho' almost insensible) Effects in the World since the Deluge'.[30] William Stukeley, who like Whiston focused much attention upon antiquarian subjects, offered a vivid description of the experimental regimen in the early 1700s university which evinces this desire to comprehend the state of postdiluvian animals and humans through chemical experiment:

> I went frequently a simpling, & began to steal dogs & dissect them & all sorts of animals that came our way. We saw too, many Philosophical Experiments in Pneumatic Hydrostatic Engines & Instruments performed at that time by Mr. Waller, after parson of Grantchester, where he dy'd last year beeing professor of chymistry, & the doctrine of Optics and Telescopes & Microscopes, & some Chymical Experiments, with Mr. Stephen Hales then Fellow of the College, now of the Royal Society. . . . We hunted after Butterflys, dissected frogs, usd to have sett meetings at our chambers, to confer about our studys, try Chymical experiments, cut up Dogs, Cats, & the like. I went to Chymical Lectures with Seignor Vigani at his Laboratory in Queens' College. . . . In my own Elaboratory I made large quantitys of sal volatile oleosum, Tintura Metallorum, Elixir Proprietatis, & such matters as would serve to put into our Drink. I usd to distribute it with a plentiful hand to my Tutors . . .[31]

Stukeley saw great value in the work of Hales, Waller and Vigani. Likewise, the formidable master of Trinity College, Richard Bentley, envisioned Vigani as an important ally. He assumed that the Italian apothecary would help him wrest the university from the grip of High Churchmen and aid in his bid to 'Newtonianise' Cambridge. It was in the midst of his fierce controversy with Tory Churchmen in 1702/3 that Bentley helped to secure the first professorship of chemistry for Vigani. The University Senate created the chair – albeit without emoluments or duties – in recognition of Vigani's two decades of 'laudable' service. Soon afterward, at the same time that he ordered the erection of an astronomical observatory for Roger Cotes, Bentley wooed Vigani to Trinity by converting a college shed to a chemical laboratory, described by Schaffer and Stewart in the following chapter.[32]

Shortly afterwards, other buildings in Cambridge were to be converted to laboratories. Following the appointment of John Waller (*c.* 1673–1718) as Vigani's successor to the chemistry chair in June 1713, the Senate announced that along with the Professor of Anatomy, Waller could arrange and make use of a new 'Publick Elaboratory' (Figures 1.3 and 1.4).[33] The wording of the Grace is revealing. While confirming that the Chair was conferring honour upon the institution it also implied that chemistry remained the handmaid to physick (*'Cum ad honorem academiæ et medicæ artis incrementum pertineat ut lections chemicæ in loco publico habeantur'*). Revealed also in the Grace is the fact that the Elaboratory was not to be purpose-built, but converted from an 'otherwise useless printing house.' To what extent these rooms, situated off

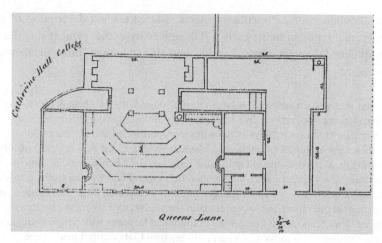

Figure 1.3. Plan of the 'Publick Elaboratory' in Queens' Lane.

Queens' Lane, were employed to promote the chemical arts is not well documented (there are no extant records of Waller's lectures), though the inventory (Figure 1.5) carried out after Waller's death in office in 1718 lists an abundance of chemical apparatus – such as furnaces, retorts and receivers – that would have been used for experiments and demonstrations. Importantly also, by acknowledging that the space was, in principle, public, the university was lending the discipline some new credibility within the town.

Yet, although the professorship and experimental spaces had been created, smuggling chemistry onto the curriculum proper as an examinable subject was no mean task, even with Bentley's fervent troop of Newtonians.[34] Bentley's detractors suggested that if Cambridge was to be a haven of right reason and safe politics, then any activity smelling of the furnace clearly had no place at the varsity. While Oxbridge humanists believed the new experimental philosophy threatened traditional learning, Sir William Temple, statesman, essayist, author of *Of Ancient and Modern Learning* (1690) and arch-enemy of Bentley, fired several volleys towards Trinity College. Temple, mastermind of the classicists' offensive on the new philosophers in the 'Battle of the Books', groused that he could not 'conceive well how [chymistry] can be brought into the number of the sciences'.[35] The danger to which Temple alluded was twofold: though he applauded the pragmatic endeavours of apothecaries and metallurgists, he argued that chemical practices should remain the domain of artisans. For Temple however, the real hazard lay in the 'wild visions' of occult alchemy in contrast to the 'diverting amusements' of artisans. Even Robert Boyle confessed that alchemy could jeopardise salvation: 'tis very dangerous to . . . procure the

Figure 1.4. The interior of the 'Publick Elaboratory'. From William Combe, *A History of the University of Cambridge* (London: R. Ackermann, 1815).

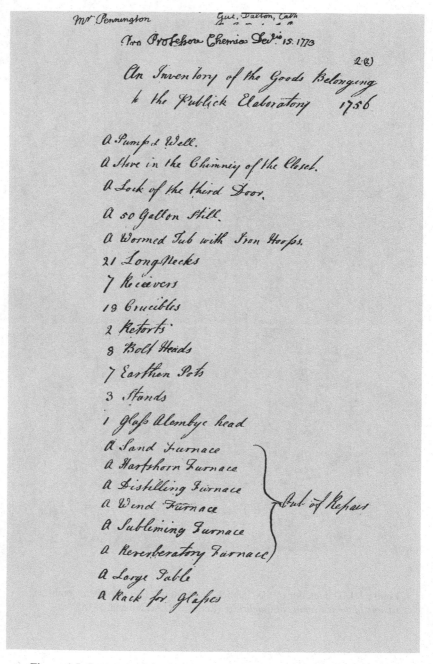

Figure 1.5. Inventory of the apparatus housed in the 'Publick Elaboratory' in Queens' Lane, from 1756, the year that Hadley succeeded Mickleburgh. Amongst the equipment are a variety of receivers, retorts and crucibles, several kinds of furnaces and a 'wormed tub with iron hoops'.

conversation of Angles [*sic*] for fear wee should by deluding arts of spirits farr more knowing than men be trapand [*sic*] into the fatal mistake of a divill for an Angle of light.'[36] But while Boyle numbered himself among the adepts, Temple railed against the 'shattered heads' who practised the art: 'For my part I confess I have always looked on alchemy in natural philosophy to be like enthusiasm in divinity and to have troubled the world to much the same purpose.'[37]

Even Bentley realised that it would be difficult to neutralise Temple's caustic remarks. As Steven Shapin has remarked, 'the solitary alchemist in his 'dark and smoky' laboratory was not a fit actor in a proper setting to produce objective knowledge'.[38] Although fashioning Vigani as a banal apothecary solved one problem about which Temple complained, it simultaneously created others. Critics like Temple might concede that as an apothecary Vigani was useful, but that he was providing little else than 'diverting amusement' and simple physick. It implied also that the lectures and experiments of Vigani and Waller were devoid of philosophy (i.e. theoretical knowledge). For the duration of the eighteenth century, Cambridge's professors struggled to dissociate themselves from these depictions.

The persistent problem

Isaac Newton recognised that underlying the mechanical philosophy was a more profound system of chemical principles. For Newton, the enterprise of natural philosophers could not be fulfilled without a complete understanding of the subtle activity of matter which accounted for such phenomena as the growth of metals, the communication of forces and human perception. In Cambridge, many of Newton's disciples – and some of his detractors – realised this also. In his 1707 Εγκνκλοπαιδεια, or *A Scheme of Study*, Robert Greene endorsed a heavy dose of chemistry for the undergraduate. Greene suggested that in Michaelmas Term the sophomore should concentrate upon the corpuscular philosophies of Descartes, Boyle and 'Le Clerk' (probably Jean Le Clerc, 1657–1736); during his third year the student should engage the 'Experimental Philosophy and Chymistry' through study of the *Philosophical Transactions*, Boyle's works, as well as the *Leipsick Acts*, Lemery (presumably James Keill's translation of *Cours de Chymie*) and the *Collegium Curiosum*.[39]

Yet the supposed profundity of chemical knowledge, and fear of its misapprehension, was itself very problematic. With the expulsion of the 'honest Newtonian', William Whiston, in 1710, the dons realised that the 'misapplication' of chemical ideas could foster heterodox doctrines unfit for an Anglican centre of learning.[40] Cambridge's dons could even point to their prize poet, Christopher

Smart, to show the dangers which the misuse of chemistry could occasion. Smart, a disciple of the anti-Newtonian Robert Greene, had versified the connections between biblical stories and chemical philosophy to help him show that 'Newton was ignorant'. For example, by postulating that 'QUICK SILVER is spiritual/and so is the AIR to all intents and purposes,' Smart simultaneously undermined the university's construction of Newtonian mechanical philosophy and the Anglican values that this philosophy was held to sustain.[41] Cambridge's professors of chemistry therefore needed to demonstrate that chemistry could glorify the Deity's Creation and yet still exclude non-Newtonian philosophies such as Hutchinsonianism.

The work of John Mickleburgh (also Mickleborough) (*c.* 1692–1756) indicates his sensitivity towards the heavenly expectations and dangerous pitfalls to which Smart's poetry pointed. Mickleburgh had first matriculated as a sizar at Gonville and Caius College, but soon migrated to Corpus Christi, perhaps at that time the centre for experimental philosophy in Cambridge. Combining this environment, his aptitude and the patronage of the Regius Professor of Physic enabled him to secure the Chair of Chemistry in 1718. Committed by university Grace to impart knowledge to 'Young Students,' Mickleburgh lectured at least five times before 1741 (though apparently not between 1742 and 1756). He made money lecturing: a guinea per head, with usually one to two dozen attendees. Mickleburgh was also on a mission: he worked hard to 'Newtonianise' his lecture series. In particular, his brief was to reduce chemical phenomena to axiomatic principles 'wch are either strict Mathematical truths or else propositions wch may be sufficiently evinced by clear and undeniable Experiments'. Exploiting the efforts of his friend, Stephen Hales, and the Oxonian John Freind, Mickleburgh's lectures first exploded Vigani's assumption that phlegmatic, alkaline, mercurial and sulphuric principles were the key to understanding the constitution of all material substances. Instead, to leave behind the 'miserably lost and bewildered' alchemists and 'discover the reason & philosophy of the . . . Art', Mickleburgh turned to John Freind's work, from which he derived nine central postulations and axioms.[42]

Of these axioms, it was '*viz repellens*', with its roots in Newton's 31st Query, that was the most significant. Mickleburgh likened this axiom to negative numbers in algebra, and, as will be discussed in the following chapter, looked to the 'ingenious Mr Hales' to ratify the existence of repulsive forces. Hales, after asking the question why oak trees 'do not render a vast explosion,' realised that air could lose its elasticity and be 'fixt' within solid substances.[43] And without this 'amphibious' air, Mickleburgh surmised, it was 'almost impossible to have true or right apprehension of any of the operations in Chymistry.' For the third professor of chemistry, the 'invisible fluid' gave his attendees – ranging from apothecaries and surgeons to the Woodwardian Professor and Lord Blanes – the opportunity to behold the Creator: 'by the all wise providence of the great Author of Nature we may now see w^t immense treasures of this noble & important Element are lodged in all material substances.'[44]

The manifestation of these godly, aerial forces and their affinity to algebraic manipulations was particularly important during a time in which Newtonian mathematics was increasingly coming under assault from High Churchmen such as Bishop Berkeley, who in 1734 disparaged the 'science's' unintelligibility. While Mickleburgh's colleague James Jurin affirmed that university mathematics was 'no friend of infidelity', Mickleburgh himself, by describing fermentation, digestion, calcination and putrefactions with reference to quasi-Newtonian laws, supposed that he demonstrated the presence of the Deity's handiwork in chemical phenomena.[45] Thus, for the chemical professor, to philosophise chemistry according to axioms was to buttress the marriage of Newtonianism and low Anglicanism. Moreover, since 'the surprizing propertys of the Air' showed how 'the beautiful frame of things is maintained & kept up in a continual round of production & dissolution,' Mickleburgh's lectures also pointed to the existence of a benevolent and economical circulation, an important theme that would dominate later analyses of airs.[46]

If nothing else, the career of Cambridge's next professor, John Hadley (professor of chemistry from 1756–64), evinces that there was little continuity between holders of the chair. Hadley (Figure 1.6) boasted that chemistry was 'rising in its reputation, & is becoming a very usefull as well as entertaining branch of natural philosophy'; yet he did little to consolidate the gains made by Mickleburgh in 'Newtonianising' the discipline so that it meshed gears with the increasingly mathematical regime of undergraduates.[47] He did, however, condemn the 'Vain pursuits', 'affectation of Mystery', and 'unintelligible jargon' of alchemists, something which both his predecessors and successors all did. These denunciations were undoubtedly ironic: since they did not have access to his manuscripts, the professors unwittingly condemned what Newton had practised – that is, the study of alchemy – in order to explain what Newton had apparently succeeded in doing – the eradication of unintelligibility as exemplified by the demise of alchemy.[48]

What sort of chemical practice did Hadley endorse? Interestingly, he did not strongly endorse iatrochemistry as one might expect, especially considering his background and career in London after leaving the cloisters and 'Publick Elaboratory' of Cambridge: his father had left Oxford in order to study medicine in Leyden while his younger brother became a surgeon. In Cambridge, at Queens' College, Hadley was taken under the wing of the Regius Professor of Physic, Russell Plumptre. While still Professor of Chemistry, Hadley would

Figure 1.6. John Hadley (1731–1764).

leave Cambridge to become Assistant Physician to St Thomas's Hospital and after being awarded his M.D. in 1763 would become the Physician to the Charterhouse. Shortly before his death he was elected a Fellow of the Royal College of Physicians.

Yet Hadley also maintained many connections to mathematics, natural history and natural philosophy. Both his uncles were prominent members of the Royal Society and helped him secure connection with mover-shakers like Sir Hans Sloane. Hadley himself was no stranger to Somerset House; he became a F.R.S. in 1758, and the Society's *Philosophical Transactions* of 1764 includes his 'Account of a Mummy', which he had inspected with the Hunter brothers and Dr Wollaston (almost certainly the distinguished doctor Charlton Wollaston, physician to the King and uncle to the William and Francis Hyde Wollaston discussed in Chapters 4 and 5).[49] Using his chemical skills Hadley analysed the process of mummification to comprehend the precise methods that Egyptians utilised. His skills in analysis were such that he impressed Ben Franklin, whom he received as a guest in 1758 and with whom he performed various experiments on the cooling effects of evaporation. Unfortunately, Hadley was unable to enjoy his rising reputation and financial security, for he died of a fever in 1764, only 33 years of age.[50]

In lieu of Newtonian, axiomatic chemistry and iatrochemistry, Hadley acknowledged he was 'much beholden to Becher, Stahl, Boerhaave, Henckel, Cramer, Pott, Margraave [and] Macquer'.[51] From them he gleaned a practical way to define the subject to his novice students: 'Chemistry is the art of separating Bodies into their simplest constituent parts, or uniting those parts again and reproducing the original Compound, and also by different mixtures and Various methods of Combining Bodies to produce new substances which exist not in Nature'. In many ways the knowledge he imparted resembled Vigani's chemistry of principles. While Hadley's lectures clearly evinced less concern with medicinal preparations than those of Vigani and even Mickleburgh, he often concentrated upon manufactures, including the production of sulphuric acid in Twickenham and green vitriol in Deptford.[52] This interest in industrial processes reflected, although perhaps unwittingly, the changing concerns of academic chemists in Scotland. The Scots, and in particular William Cullen, instilled civic virtue in chemistry by strengthening its bonds with the gentleman's farm, the bleaching field, the distillery and even the town meeting. When Cullen was appointed to Edinburgh's chair of chemistry in 1756, he secured patronage by pitching the discipline as a subject fit for the gentleman-scholar. Stripping chemistry of its mystique, he 'philosophised' it by using Adam Smith's argument in the *History of Astronomy* – that philosophers differ from artisans since seemingly mundane phenomena for the artisan often provokes wonder in the philosopher.[53]

Hadley, along with both his own successor, Richard Watson and Watson's deputy, Isaac Milner, learned valuable lessons from their continental counterparts. From their colleagues in Scotland, Watson and Milner also learned how to sell their discipline. When he procured the chair of chemistry, Watson sought, and obtained, patronage from Parliament: he did so by describing the value of chemistry to the government and descanting upon its suitability for the young gentleman. 'Chemistry', Watson enthused to the Duke of Rutland in the dedication of his 1781 *Chemical Essays*,

> is cultivated abroad by persons of the first Rank, Fortune and Ability: they find in it a never failing source of honourable amusement for their private hours; and as public men, they consider its cultivation as one of the most certain means of bringing to their utmost perfection, the manufactures of their country.[54]

No longer practised merely to derive medicaments or to 'astonish the vulgar', chemistry, according to Watson, had 'shaken off the opprobrium which has been thrown upon it'. Although the subject might be cultivated in private, it was nonetheless both honourable and fit for public men.[55] Natural theology and biblical exegesis underpinned this utilitarian bent, however. Watson and Milner both delighted in the benevolent systems of fiery, phlegmatic and aerial economies that the Creator had devised. Later, their fellow divine, William Paley, would point to exactly these circulatory systems in his *Natural Theology*. In his chapter on 'Elements', Paley averred that the 'beautiful and wonderful œconomy' of these elementary systems revealed to Christians the existence and attributes of the Deity.[56]

Nevertheless, chemistry's unrelenting, nagging links to irreligion and the occult persisted at the University. Watson's actions upon being elevated to the See of Llandaff are detailed in Chapter 3: in order to comply with '*Episcopal Decorum*', he burned his papers on chemistry 'with the zeal of the Idolators of old'.[57] In addition a new, but related, problem emerged mid-century. As problematic to Anglican dons like Hadley, Watson and Milner was the fact that in both metropolitan and provincial locales, scientific entertainers were exploiting the same chemical phenomena as the professors used in their lectures in order to astonish less learned (but nonetheless well-paying) audiences. Both showmen and their critics realised that the wonderful displays of chemical and electrical phenomena invoked weighty metaphysical problems. With astonishing shows like the 'electrical beatification' and *Creation of the World and Noah's Deluge* to be beheld at Bartholomew Fair, it was clear that rational diversion related to biblical speculation.[58] Yet, while showmen such as Benjamin Martin claimed that they provided 'entertainment for angels', Cambridge's natural philosophers worried that without sober tuition to accompany these performances, these shows could be theologically damaging.[59]

Although the chemistry professors had tried to differentiate their lectures from the scientific shows that passed through Cambridge – shows such as 'Mr. Gyngell and the anti-Combustible Man' – they were not always successful. Reminiscences of Isaac Milner's lectures are revealing. As Chapter 4 will explain, after serving as the deputy professor of chemistry for a number of years, Milner was elected the foundation Jacksonian Professor of Natural Philosophy in 1782. Despite the fact that chemistry was still not an examinable subject, the university also erected a purpose-built lecture hall for the Jacksonian Professor, a decision that must have pleased Milner since his income was derived principally from lecture fees. Although he counted himself among Britain's best natural philosophers, Milner was remembered as a 'first-rate showman', his lessons more 'contrived to amuse' than enlighten, and his optical lectures 'nothing more than Magic Lantern shows'.[60] After making his fortune in London, William Hyde Wollaston reminisced that 'Milner had a singular power of engaging the attention of the class, by appearing to view with wonder and delight the results of his own experiments'.[61]

Before the French revolutionary era, then, Cambridge's professors of chemistry had solved a number of problems to make their discipline a palatable one for the impressionable undergraduate. For instance, through the course of the century they proved that they were not merely handmaids to physicians. They proved also – including to Parliament, who had to vote each year to continue the professor's annual stipend – that chemistry's utilitarian aims should not be associated automatically with 'sootiness'. Finally, the number of students who voluntarily attended their lectures corroborates the professors' affirmations that their subject was indeed entertaining. Yet in overcoming these difficulties they encountered other obstacles that the next generation of dons would have to tackle. If it was admitted that chemistry was entertaining, it was not universally conceded that chemistry was natural philosophy proper. Though it is conceivable that chemistry could have been an integral part of determining each year's Senior Wrangler, the professors had yet to convince the university that chemistry was essential for determining the rank of the B.A. candidate. Moreover, after 1789 future chemists in Cambridge had to contend with the increasing associations of chemistry with democratic politics.

Despite delighting in astonishing spectacles, both Richard Watson and Isaac Milner were determined to remain at the forefront of their discipline. The shelf-life of Watson's works testifies to this: Humphry Davy remarked that he 'could

Figure 1.7. The Botanic Garden. One of the main purposes of the garden was to cultivate plants for medicinal use. From William Combe, *A History of the University of Cambridge* (London: R. Ackermann, 1815).

scarcely imagine a time or a condition of the science in which the Bishop's chemical *Essays* would be superannuated'.[62] As President of Queens' College, Milner could often reflect upon the state of Cambridge chemistry at the turn of the century since Vigani's *materia medica* cabinet was housed in his college (possibly in his private laboratory on the Backs). Yet Milner too was particularly forward looking in his researches. He joined the Jacksonian Gardener (Figure 1.7) in an attempt to 'inoculate plants', developed a new technique for producing gunpowder, looked for a cure for gout and scrutinised the debate between pro- and anti-phlogistonists.[63] Milner even paid close attention to the pneumatic researches of the Oxford dissident, Thomas Beddoes, telling William Wilberforce in 1793 that he thought it 'most improbable that Dr Beddoes reddened his face, & almost threw himself into Consumption by breathing Dephlog^d Air'.[64]

Seducing young men

The next generation of chemists proved to be equally reflective and innovative: Smithson Tennant, one-time student of Black, earned the Royal Society's prestigious Copley Medal. While William Wollaston earned fame and fortune in

London, his brother Francis John Hyde Wollaston became a greatly respected lecturer and practitioner. The naturalist and antiquarian Edward Daniel Clarke (for whom the Cambridge chair of mineralogy was later founded) and James Cumming each made substantial contributions to their discipline.[65] Nevertheless, the status of chemistry in Cambridge remained uncertain because of the subject's increasing affiliations with atheism and democracy. In his 1776 *Letter on Priestley's Experiments on Air,* Joseph Berington asked the chemico-philosopher: 'why do you not, as a *metaphysician,* aim to rise above the *visible world of matter,* where you may discover the existence and reality of other beings, whose ethereal forms cannot be confined in a tub of water, or a basin of quick-silver; nor be extracted by friction from a globe of glass; nor infine be analysed by all the powers of chymistry?'[66] A decade later similar concerns issued from Oxbridge presses: 'You have given the world much fixed air: let it have some fixed principles,' groaned Newton's great champion, George Horne.[67] Apprehension over Priestley's 'theological laboratory' escalated after Priestley chided the Prime Minister that his 'Universities resemble pools of stagnant water secured by dams and moulds, and [are] offensive to the neighbourhood'.[68] Famously, Edmund Burke was particularly scathing towards 'Gunpowder Joe' and 'alchemical legislators'.[69] Alluding to the destruction of property that had occurred in the production of gunpowder, Burke surmised that calculating chemists would level society: 'Churches, play-houses, coffee-houses, all alike, are destined to be mingled . . . into true democratic, explosive, insurrectionary nitre.'[70] Such antipathy led Priestley to flee to Pennsylvania in 1794. After the Home Office reported that Thomas Beddoes was a 'most violent *Democrate*' and apt to 'seduce Young Men to the same political principles', the outspoken professor was forced to quit Oxford.[71]

In Edinburgh, John Robison, the university's professor of natural philosophy, imagined that Masonic lodges had become the cauldron in which an atheistic distemper brewed 'projectors and fanatics . . . in science.' Numbering Priestley among these scientific fanatics, Robison blamed him for twisting the intentions of two Cambridge philosophers, namely Isaac Newton and David Hartley. In doing so, Robison contended that Priestley had reduced God to mere undulations in the '*vibratiuncloer*' and the mind to 'the quiverings of some fiery marsh *miasma*'.[72] Cambridge chemists, who had built up robust relations with Edinburgh, followed suit. In his 1799 pamphlet on *Receiving Jacobin Teachers of Sedition*, Francis Hyde Wollaston alerted his readers to the dangers of freemasonry, which enabled the 'insidious poison' of the Encyclopaedists to 'infect' England.[73] A decade later, during the Bible Society controversy, Isaac Milner compared his adversary to French calorimetry: like the French chemists, Milner surmised that his opponent's arguments contained much heat, but no 'real matter'.[74]

The dons who wished chemistry to remain a part of the scholar's life attempted to de-politicise and secularise it – at least in the lecture hall. In Cambridge, E. D. Clarke contrived a 'post-Enlightenment' space for the chemical lecturer. In many ways their attempts corresponded with the transition of Humphry Davy's career as he moved from Beddoes' Pneumatic Institute to London's fashionable Royal Institution in 1801. It was by this move, as Jan Golinski has shown, that 'Davy extricated himself from the confusion of the 'end of Enlightenment' and placed himself in a more central position from which he could address a wider public'.[75] Although some of his radical colleagues feared he surrendered his philosophical integrity since he no longer claimed that chemistry *must* be the engine of social progress, Davy pandered to the metropolis's elite by demonstrating to them how chemistry could be part of *their* progress in society. He did not entirely eradicate spectacle and natural theology from his lectures; but he did tame the subject for his affluent – and partially female – audiences by concentrating upon chemistry's value to the 'philosophical farmer' and the progressive manufacturer.[76]

Meanwhile, fifty miles north of London, E. D. Clarke was elected foundation professor of mineralogy in 1808. Clarke had returned from his extensive travels on the Continent with both an extravagant collection of ancient artefacts and unusual minerals, and also a firm conception of the ideal gentlemanly education. This enabled him to find success in the lecture room by identifying chemical analysis not with natural theology but rather with commerce, classical antiquities and the travelogue. Using his ingenious invention, the hydrostatic blowpipe, he offered to his students gentlemanly demonstrations, ranging from the decomposition of exotic marbles to the re-creation of Vesuvius's eruption. One student, for example, marvelled at the juxtaposition of 'blowpipes [and] the grotto of Antiparos with its beautiful stalactites'. Clarke gave his students a grand tour of chemical composition.[77]

Clarke's 'virtual' expedition to foreign climes chimed in nicely with William Farish's domestic tour of English manufacturing sites. Indeed, Farish's lectures on manufacturing demonstrate the degree to which the knowledge deemed appropriate for the university undergraduate had transformed since the Glorious Revolution. The content of the lectures by the seventh Professor of Chemistry did not hinge upon active principles, aerial economies or Jesuit delirium, nor even upon medical simples; instead Farish's students were treated to explanations of smelting, aquatinting, sugar refining, distilling, as well as the production of soap and gunpowder.[78] Neither Farish nor the university expected that Farish's students would ever make soap; but they did anticipate that in their future careers, these young gents could have substantial involvement in trade and manufactures, something increasingly accepted in late-Georgian England.

Victorian Cambridge would delineate precisely the role of chemistry in the life of the 'scholar and gentleman'.[79] Six years after he coined the word 'scientist' in order to differentiate mundane practitioners and genuine philosophers, William Whewell testified that those who engaged in chemical subjects did not necessarily fall into the former category.[80] In his *History of the Inductive Sciences*, he insisted that the discipline had reached a desirable stage of maturity, and no longer warranted mental links to wantonness, sootiness or insurgency. In contrast to earlier critics who equated French democracy and Lavoisierian chemistry, Whewell remarkably claimed that revolutionary principles were in fact antithetical to the praiseworthy 'Systematic Nomenclature' of the French: 'The democracy which overthrew the ancient political institutions of France, and swept away the nobles of the land', the Master of Trinity College exclaimed when treating Lavoisier's rendezvous with the guillotine, 'was not . . . enthusiastic in its admiration of a great revolution in science'. For Whewell, this entailed that modern chemistry could sidestep its democratic associations. Since it could 'avoid the embarrassments and contradictions of casual and unreflective classification' also, Whewell figured it was now fit for the Tripos proper.[81] During the major university reforms of 1848, he secured for chemistry a place on the Natural Sciences Tripos, and students were first examined on the subject in 1851. Whewell's aim to produce the liberally educated chemist was a success.[82]

Conclusion

Jeremy Bentham declared that his Oxford tutors summarily dismissed chemistry as 'fit only to make a man an atheist or an apothecary'.[83] Scholars, divines and gentlemen – what Oxbridge supposed they churned out – were neither of these things. Other detractors commonly pointed to the subject's confusion and even the professors themselves were apt to qualify their lectures with cautionary remarks. As Isaac Milner warned at the beginning of his lectures, 'the Subject is intricate & mysterious & that whenever we meddle with it w[ith]o[ut] the utmost care & circumspection, we are more likely to involve ourselves in Error & Absurdity, than arrive at any just conception of these hidden Powers'.[84]

Through the course of the long-eighteenth century, the dons developed a set of strategies to warrant meddling in the mysterious and intricate subject. Like William Cullen and Joseph Black to the north, the university professors emphasised the role that chemistry could play in promoting civic virtue. Waller, Mickleburgh and Hadley were all able to capitalise on the Senate Grace that enabled them to transmute an idle printing-house into a vital space for swan-neck flasks and furnaces. As with the rest of enlightened Europe where philosophers no longer considered the *laboratorium* as a place of seclusion and withdrawal, the professors could consider their 'Publick Elaboratory' a space for the interested community. Related to this strategy was the ability of the professors to react to various political crises that could have had devastating effects. As did Humphry Davy by his lectures at London's Royal Institution, the professors successfully detached chemistry from its connections to the enlightened cauldron of democracy.

Other tactics were put in play throughout the century, though success was mixed. Like the Lucasian Professor and the Plumian Professor of Astronomy (who, for instance, were prominent members of the Board of Longitude), the chemistry professors increasingly attempted to convince King and Parliament that their discipline was important to the aims of Empire: both Milner's development of a process to produce a key ingredient in gunpowder and Watson's ability to secure from the Crown a yearly stipend demonstrate this, although the latter may say more about Watson's networking skill than Imperial design. Like Joseph Priestley, who struggled to differentiate the 'slowness and blunders of mechanics . . . with the ardour of persons engaged in philosophical enquiries', Cambridge's chemists worked hard to fashion themselves as philosophers.[85] As the prefatory letter in Vigani's *Medulla Chymiæ* urged, it was desirable to create an environment 'where philosophers are mechanics and mechanics philosophers' (*ubi Philosophi sunt Mechanici, vel Mechanici Philosophi*).[86] Whether this was desirable at an English university was another question altogether.

So, though Cambridge could capitalise on some of the strategies that were deployed successfully in Edinburgh, Birmingham and London, the university needed to forge its own, unique space for chemistry. As a consequence, overcoming the tainted image of chemistry at the varsity took nearly two centuries. Besides fundamental changes in the perception of what the chemical discipline entailed, the perceived role of the gentleman in the social order had to change also in order for chemistry to gain a stronghold at Cambridge University. While the well-read Christian landlord remained an emblem of gentility, Britons increasingly conceded the propriety of business and industry. It is therefore tempting to say that Mickleburgh, Hadley and Farish, who all directed their students' attention to manufactures, were ahead of their time in some respects. It is perhaps safer to claim that only with Charles Babbage's *Economy of Machinery and Manufactures* (which acknowledged that business would be part of the student's afterlife) and Whewell's formulation of the 'liberally-educated chemist' was this conception of university chemistry articulated cogently.[87] However articulate, Bentley and his favourite Veronese apothecary would have stared in wonder.

Acknowledgements

This article could not have been written without the aid of Mary Archer, Jan Golinski, John Gascoigne, Chris Haley, Simon Schaffer and Larry Stewart. The California Institute of Technology graciously helped to fund research, while I am also indebted to the William Andrews Clark Library for a summer fellowship. I am very grateful to Gonville and Caius College, Trinity College, and Queens' College for enabling me to consult manuscripts. I am also grateful to the archivists of the Science Museum Library, London, while the archivists of the *Eighteenth Century Short Title Catalogue* helped me to procure a photocopy of a missing Vigani advertisement. Lastly, I am indebted to the President of Queens' College for enabling me to view Vigani's cabinet of *materia medica*.

Notes and References

1. Glynn, Robert (1781), 'The narrative of R-b-rt Cl-b-ry Gl-nn, M.D. concerning the late strange and deplorable frenzy of the R-v-r-nd R-ch-rd W-ts-n, D.D. F.R.S. R-gius Pr-f-ss-r of D-v-n-ty, in the Un-v-rs-ty of C-mbr-dge, . . . Faithfully copied from the original manuscript, transmitted . . . to the Royal Society and Royal College of Physicians, Oct. 21, 1780. . . . with notes, by the author of the Heroic epistle, Heroic address, &c. to Dr. W-ts-n', London. Glynn lectured in medicine at Cambridge until 1752. Richard Watson, in his memoirs, does admit to an hallucinogenic fever at this time, but claimed he fully recovered.
2. For Newton, see, for example, Brewster, David (1831), *Memoirs of the Life, Writings and Discoveries of Sir Isaac Newton* 2 vols., London: Murray; Boas Hall, M. and Hall, A. R. (1958), 'Newton's chemical experiments', *Archives Internationales d'Histoire des Sciences* **11**, pp. 151–2. For universities, see Ashby, Eric (1958), *Technology and the Academics; an Essay on Universities and the Scientific Revolution*, London: Macmillan; see also Winstanley, D. A. (1935), *Unreformed Cambridge: A Study of Certain Aspects of the University in the Eighteenth Century*, Cambridge: Cambridge University Press.
3. See, for example, Winstanley, D. A. (1935); Golinski, Jan (1992), *Science as Public Culture*, Cambridge: Cambridge University Press, pp. 52–3; Gascoigne, John (1989), *Cambridge in the Age of Enlightenment*, Cambridge: Cambridge University Press.
4. For revisionist histories of natural philosophy in early modern universities, see, for example, Gascoigne, John, 'A reappraisal of the role of the universities in the Scientific Revolution', in Lindberg, D. and Westman, R. (eds.) (1990), *Reappraisals of the Scientific Revolution*, Cambridge: Cambridge University Press, pp. 207–60; Clark, William (1992), 'On the ironic specimen of the Doctor of Philosophy', *Science in Context* **5**, pp. 97–137.
5. For the rise of public chemistry, see Stewart, Larry (1993), *The Rise of Public Science*, Cambridge: Cambridge University Press, passim; see also Golinski (1992). For the indecision of Cambridge professors, see, for example, Hadley, John, *Cambridge Lectures*, Wren Library MS R1.50:2; see also Watson, Richard (1818), *Anecdotes of the Life of Richard Watson*, 2 vols., London: Cadell and Davies, vol. I, pp. 46 and 54. For Greene's contractive forces, see Gascoigne (1989), pp. 167–71. For Whiston's cometary theories, see Force, James (1985), *William Whiston: Honest Newtonian*, Cambridge: Cambridge University Press, pp. 43–61.

6. See Gascoigne (1989), pp. 295–6. Gascoigne does recognise, however, the importance of chemistry as a polemical resource for Burke.
7. The 'holy alliance' is discussed in Gascoigne (1989), pp. 2–3.
8. For Cambridge's natural philosophical curriculum and disputations, see Costello,
William (1958), *The Scholastic Curriculum in Early Seventeenth-Century Cambridge*, Cambridge, MA: Harvard University Press; see also Mullinger, J. Bass
(1867), *Cambridge Characteristics in the Seventeenth Century*, London: Macmillan.
9. Drake cited in Costello (1958), pp. 91–2.
10. Norton, Samuel, *The Key to Alchemie*, Ashmolean MS 1421, article 26; Norton,
Samuel (1630), *Mercurium Redivivum*, London. William Parys (1596), trans. *A
booke of secrets: shewing diuers waies to make and prepare all sorts of inke, and
colours: as blacke, white, blew, greene, red, yellow, and other colours. Also to write
with gold and siluer, or any kind of mettall*, London. For Dee's career, see Clulee,
Nicholas H. (1988), *John Dee's Natural Philosophy: Between Science and Religion*,
London: Routledge. The work of Norton and Parys is discussed in Gunther, R. T.
(1937), *Early Science in Cambridge*, Oxford: Oxford University Press, pp. 217–20.
11. See Frank, Robert (1980), *Harvey and the Oxford Physiologists*, Los Angeles: University of California Press. For Glisson's programme, see Gascoigne (1989), p. 60.
12. The 'Club' is mentioned in a letter from Beale to John Evelyn; see Christ Church
College, Evelyn Correspondence, no. 79: cited in Gascoigne (1989), p. 63. The Nidd
circle is discussed in Gascoigne (1989), pp. 63–4.
13. For the speculations of John Ray and Sir Tancred Robinson, see, for example, Ray,
John (1713), *Three Physico-Theological Discourses, Concerning I. The Primitive
Chaos, II. The General Deluge, III. The Dissolution of the World*, 3rd edn, London:
W. Innys.
14. Sprat, Thomas (1667), *The History of the Royal Society of London, for the improving
of Natural Knowledge*, London: Printed by T.R. for J. Martyn and J. Allestry.
15. For Boyle's conception of matter, 'matters of fact', and the demarcation between
natural philosophy and theology, see Schaffer, Simon and Shapin, Steven (1985),
Leviathan and the Air Pump, Princeton: Princeton University Press, pp. 155–224,
283–331. For More, see More (1662), *A Collection of Several Philosophical Writings
of Dr. Henry More*, London: Printed by James Flesher for William Morden. More
cited in Henry, John, 'Henry More versus Robert Boyle: the Spirit of Nature and the
Nature of Providence', in Hutton, Sarah (ed.) (1990), *Henry More: Tercentenary
Studies*, Dordrecht: Kluwer Academic, pp. 55–76, on p. 59.
16. Humphrey Newton to John Conduitt: cited in Golinski, Jan, 'The Secret Life of
an Alchemist', in Fauvel, Flood, Shortland, & Wilson (eds.) (1988), *Let Newton
be!* Oxford: Oxford University Press, pp. 147–68, 153. For Newton's chemical and
alchemical work, see Dobbs, Betty Joe Teeter (1975), *The Foundations of Newton's
Alchemy*, Cambridge: Cambridge University Press; Westfall, Richard (1980), *Never
at Rest: A Biography of Isaac Newton*, Cambridge: Cambridge University Press,
pp. 280–308, and passim; McGuire, J. E. and Rattansi, P. M. (1966), 'Newton and
the Pipes of Pan', *Notes and Records of the Royal Society* **21**, pp. 108–43.
17. Smithsonian Library, Dibner Institute Burndy MS 16, ff. 3–4: cited in Westfall
(1980), p. 306.
18. Keynes MS 135; cited in Westfall (1980), pp. 191–2.
19. The nun story cited in Westfall (1980), p. 192. Newton to Boyle (1679?) in Yahuda
MS.2.4, f.25v: cited in Westfall, p. 339.
20. Westfall indicates that Vigani's first lectures may have been given as early as 1679,
but this is incongruous with other calculations of Vigani's arrival in Cambridge.
21. Vigani, J. F. (1683), *Medulla Chymiae*, London: Henry Faithorne, p. 53.

22. For prostitution in early modern Cambridge, see, for example, Gunning, Henry (1854), *Reminiscences of Cambridge*, 2 vols., London: Bell, passim. For secrets in early modern Europe, see Eamon, William (1995), *Science and the Secrets of Nature*, Princeton: Princeton University Press.

23. Newton's interest in medicine is discussed in Westfall (1980), pp. 284–8. For the Lucatello Balsam, see Iliffe, Robert, 'Isaac Newton: Lucatello Professor of Medicine', in Shapin, Steven and Lawence, Christopher (eds.) (1998), *Science Incarnate: Historical Embodiments of Natural Knowledge*, Chicago: University of Chicago Press, pp. 121–55.

24. The purchase of the cabinet is discussed in Peck, E. S. (1934), 'John Francis Vigani', *Proc. Cambridge Antiquarian Soc.* **34**, pp. 34–49. The specimens of the cabinet are listed in Gunther, R. T. (1937), pp. 472–81.

25. Vigani's lectures are recorded in the 'Notebook of John Yardley, 1702', Caius MS 631/460.

26. Vigani, John Francis (1683), *Medulla Chymiæ*, London: Henry Faithorne (published in shorter form in Amsterdam, 1682).

27. Guerrini, Anita, 'Chemistry Teaching at Oxford and Cambridge, circa 1700', in Rattansi and Clericuzio (eds.) (1994), *Alchemy and Chemistry in the 16th and 17th Centuries*, Amsterdam: Kluwer Academic, pp. 183–99, on p. 187.

28. Boerhaave, Hermann (1751), *Methodus Studii Medici*, Amsterdam: Sumptibus Jacobi a Wetstein, vol. i, p. 139: cited in Coleby, L. J. M. (1952), 'John Francis Vigani', *Annals of Science* **8**, no. 1, pp. 46–60, p. 49. The *'Epistola quondam conscripta ab Amico Londinensi'* is signed 'T. R.': Tancred Robinson (c. 1655–1748) graduated M.B. from Cambridge in 1679 and lived in London before travelling with Hans Sloane to the Continent in 1683. He authored several papers on chemical subjects in the *Philosophical Transactions*. See Michael Hunter (1982), *The Royal Society and its Fellows*, Chalfont St Giles; BSHS monograph, passim. See also the following chapter by Schaffer and Stewart, note 9.

29. T[ancred] R[obinson], *'Epistola'*, unpaginated.

30. Whiston, William (1696), *New Theory of the Earth*, London: R. Roberts for Benj. Tooke, Book II, pp. 69–71, 292.

31. Stukeley, W. (1880–5) *Family Memoirs of the Rev. William Stukeley, M.D.* 3 vols., Durham: Surtees Society, vol. I, pp. 80–2.

32. For the observatory, see Gunther (1937), p. 161. For the Trinity laboratory, see *idem*, p. 222.

33. Cambridge University Archives, Original Grace Book θ, 613 (11 June 1713) for Waller's appointment; 667 (10 Oct. 1716) for designation of the laboratory.

34. For the creation of the chair, see Gunther (1937), p. 220; for Bentley's tangle with High Churchmen, see Gascoigne (1989), pp. 91–100.

35. Temple, William, 'Some thoughts upon reviewing the essay of ancient and modern learning', in Monk, S. H. (ed.) (1963), *Five Miscellaneous Essays by Sir William Temple*, Ann Arbor: University of Michigan Press, p. 87.

36. Boyle correspondence: cited in Hunter, M. (1990), 'Alchemy, magic and moralism in the thought of Robert Boyle', *Br. J. Hist. Sci.* **23**, pp. 388–410, on p. 397.

37. Temple (op. cit.), p. 177.

38. Shapin, Steven (1988), 'The house of experiment in 17th-century England', *Isis* **79**, pp. 373–404, on p. 377.

39. Greene, Robert, Εγκνκλοπαιδεια, or *A Scheme of Study*. Manuscript (dated 1707) in the Old Library, Queens' College, Cambridge.

40. For Whiston's Arianism and expulsion, see Force (1985), pp. 17–19.

41. Smart, Christopher (1990), 'Jubilate Agno', in Williamson and Walsh (eds.), *Selected Poems*, Harmondsworth: Penguin, lines 217–222, 'B' fragment.

42. Mickleborough, John, 'Lectures 1726', Gonville and Caius MS 619/342, day 1.
43. Hales, Stephen (1727), *Vegetable Staticks*, London: W. & J. Innys.
44. Mickleborough, John, 'Lectures 1728', Gonville and Caius MS 619/342, day 3: *Of Air*.
45. For Berkeley's attack upon the calculus, see Berkeley, George (1734), *The Analyst*, London: J. Tonson. For the rebuttal, see [James Jurin] (1734), *Geometry no Friend to Infidelity; or, a Defence of Sir Isaac Newton . . . by Philalethes Cantabrigiensis*, London: T. Cooper.
46. Mickleborough, John, 'Lectures 1728', Gonville and Caius MS 619/342. For the theme of benevolent aerial economies, see, for example, Schaffer, Simon (1987), 'Priestley and the politics of spirit', in Anderson *et al.* (eds.) (1987), *Science, Medicine and Dissent: Joseph Priestley*, London: Wellcome Trust.
47. Hadley, John, 'Lectures 1759', Trinity College Library, MS.R.1.50:2, ff. 3–4. For Hadley's life and career, see Knox, K. C., 'John Hadley', *New Dictionary of National Biography* (forthcoming).
48. See Hadley, John, 'Lectures 1759', Trinity College Library, MS.R.1.50:2, ff. 2–4.
49. Royal Society Sackler Archives; Edwards, G. M. (1899), *University of Cambridge College Histories: Sidney Sussex College*, London: F. E. Robinson & Co., pp. 160, 163, 182.
50. Curiously, Charlton Wollaston – another of the mummy's inspectors – also died that same year, aged only 31. (Royal Society Sackler archives).
51. The Margraave mentioned is possibly Christiaan Marggraf (1626–87).
52. Ibid., pp. 13, 63–63; 83–84.
53. For the social construction of Scottish chemistry, see Golinski, Jan (1992), pp. 11–49; for aristocratic farmers, see Lord Kames (1776), *The Gentleman Farmer*, Edinburgh: printed for W. Creech and T. Cadell; for Smith, see 'The Principles . . . of the History of Astronomy', in Wightman, W. and Bryce, J. (eds.) (1980), *Essays on Philosophical Subjects*, Oxford: Clarendon, pp. 31–105.
54. Watson, Richard (1781–7), *Chemical Essays*, 5 vols., Cambridge: Cambridge University Press, vol. I, p. ii.
55. Ibid., vol. I, preface, p. 38. In his attempt to show the extent to which Britons, and especially himself, had philosophised chemistry, Watson even penned a volume in Latin in order to attract foreign natural philosophers: see Watson (1768), *Institutionum chemicarum in pralectionibus academicis explicatarum, pars metallurgica*, Cambridge: J. Archdeacon.
56. Paley, William (1802), *Natural Theology: or, Evidences of the Existence and Attributes of the Deity, collected from the Appearances of Nature*, London: R. Faulder, p. 402.
57. Watson, *Essays* (1781–7), vol. IV, pp. ii–iii.
58. For the dons' desire to separate themselves from vulgar mechanics, see, for example, Milner, Isaac (1778), 'Reflections on the communication of motion by impact and gravity', *Phil. Trans. Royal Soc.* **68**, pp. 344–78. For the theological import of display, see Schaffer, Simon, 'The consuming flame', in Porter, R. and Brewer, J. (eds.) (1993), *Consumption and the World of Goods*, London: Routledge, pp. 489–536. For Noah's deluge, see Altick, Richard (1978), *The Shows of London*, London: Belknap / Harvard University Press, p. 80.
59. Martin, Benjamin (1749), *A Panegyrick on the Newtonian Philosophy*, London, pp. 14–15.
60. For Gunning's experiences with Milner, see Gunning (1854), vol. I, pp. 236–8. For Milner as 'showman' see Milner, Mary (1842), *Life of Isaac Milner*, Cambridge: J. & J. J. Deighton, pp. 29–31. For the scientific entertainments of the metropolis,

see Altick (1978), or Stafford, B. M. (1994), *Artful Science*, Cambridge, MA: MIT Press.

61. Gilbert, L. F. (1952), 'W. H. Wollaston manuscripts at Cambridge', *Notes and Records of the Royal Society* **9**, pp. 311–32, on p. 314.

62. De Quincey, Thomas (1862), 'An essay on Coleridge', in *Miscellaneous Essays*, Edinburgh: Adams and Black, 3rd edn, p. 143.

63. For the plant inoculation, see 'Notes of the Jacksonian Gardener', Trinity College Library MS R.8.42[15–17]. For gunpowder, see Milner, Isaac (1789), 'On the production of nitrous acid and nitrous air', *Phil. Trans. Royal Soc.* **79**, pp. 300–13. For anti-phlogistic arguments, see 'Essays on Heat: Jacksonian Lectures for 1784–88', Trinity College Library MS R.8.42[1–4], ff. 42–43. Milner's chemical lectures are housed in Queens' Old Library, MS 77.

64. Milner to Wilberforce, 11 September 1793: Wilberforce Correspondence, Bodleian Library MS C.47 f. 116.

65. See Gunther (1937), pp. 230–6.

66. Berington, Joseph (1776), 'Remarks on Dr. Priestley's *Experiments on Air*', in *Letters on Materialism and Hartley's Theory of the Human Mind, Addressed to Dr. Priestley, F.R.S.*, London: M. Swinney, pp. 220–21.

67. [Horne, George] (1787), *A Letter to the Revered Doctor Priestley by an Undergraduate*, Oxford, p. 3.

68. The 'theological laboratory' is from Horne (1787), p. 21. Priestley's letter to William Pitt cited in Kramnick, I. (1986) 'Eighteenth-century science and Radical Social Theory: the case of Joseph Priestley's scientific liberalism', *J. Br. Stud.* **25**, p. 13.

69. See, for example, Stewart (1993), pp. 166–86; Schaffer (1987); Crosland, Maurice (1987), 'The image of science as a threat: Burke versus Priestley and the 'Philosophic Revolution", *Br. J. Hist. Sci.* **20**, pp. 287–318.

70. Edmund Burke, 'Letter to a Noble Lord', in *Works*, London: G. Bells & Sons (1903), vol. V, p. 143.

71. Report cited in Levere, Trevor (1981), 'Dr. Thomas Beddoes at Oxford: radical politics in 1788–1793 and the fate of the Regius Chair in chemistry', *Ambix* **28**, pp. 61–9, on p. 65.

72. Robison, John (1798), *Proofs of a Conspiracy Against all the Religions and Governments of Europe, Carried on in the Secret Meetings of Free Masons, Illuminati, and Reading Societies*, London: T. Cadell, pp. 7, 340.

73. Wollaston, Francis (1799), *A Country Parson's Address to his Flock, to caution them against Being Misled by the Wolf in Sheep's Cloathing, or receiving Jacobin Teachers of Sedition*, London: G. Wilkie, p. 18.

74. Milner, Isaac (1813), *Strictures on . . . Richard Marsh*, London: T. Cadell, p. 6.

75. Golinski (1992), p. 188.

76. Ibid., pp. 190–203.

77. For Clarke's blow-pipe, see Oldroyd, D. R. (1972), 'Edward Daniel Clarke, 1769–1822, and his role in the history of the blow-pipe', *Ann. Sci.* **29**, pp. 213–35. For Clarke's conception of the liberal education and the use of the blow-pipe, see Dolan, Brian, *Governing Matters: The Values of English Education in the Earth Sciences, 1790–1830*, Cambridge University Ph.D. dissertation, 1995; for the memories of Clarke's lectures, see Anon. (1836), *Facetiæ Cantabrigiensis* (London 3rd edn, 1836), p. 150.

78. Farish, William (1813), *A plan of lectures on Arts and Manufactures, More Particularly as they relate to Chemistry*, Cambridge: J. Smith; for the substance of these lectures, see H. I. Sperling's manuscript notes interleaved with *idem.*, Science Museum Archives MS 433.

79. For the tension between pedantry and honour, see Shapin, Steven (1991), "A Scholar and a Gentleman': The problematic identity of the scientific practitioner in Early Modern England', *Hist. Sci.* **29**, pp. 279–327.

80. Whewell, William (1834), 'Review of Somerville's *On the Connexion of the Physical Sciences*', *Quarterly Review* **51**, pp. 58–60.

81. Whewell, William (1837), 'Transition from the chemical to the classificatory sciences', *History of the Inductive Sciences*, 3 vols., London: J. W. Parker; 3rd edn, (1857), vol. III, pp. 119–21, 153–4.

82. For the institution of chemistry on the NST, see Roberts, Gerrylynn K. (1980), 'The liberally-educated chemist: chemistry in the Cambridge Natural Sciences Tripos, 1851–1914', *Historical Studies in the Physical Sciences* **11**, pp. 157–83.

83. Bentham cited in Schaffer, S. (1990), 'States of mind: enlightenment and natural philosophy', in Rousseau, G. S. (ed.) (1990), *Languages of the Psyche*, Los Angeles: University of California Press, pp. 233–90, on p. 261.

84. Milner, Isaac, 'Jacksonian Lectures', Trinity College MS R.8.42, f. 2.

85. Priestley, Joseph (1775), *History and Present State of Electricity*, 2 vols., 3rd edn, London: Printed for C. Bathurst and T. Lowndes [et al.], vol. I, p. ix.

86. Vigani, J. F. (1683), p. iv.

87. Babbage, Charles (1832), *On the Economy of Machinery and Manufactures*, London: Charles Knight; Whewell, William (1845), *Of a Liberal Education*, London, J. W. Parker, pp. 40–1.

2

Vigani and after: chemical enterprise in Cambridge 1680–1780

Simon Schaffer

Department of History and Philosophy of Science, University of Cambridge

Larry Stewart

Department of History, University of Saskatchewan, Canada

The Prince of Chemists

On 10 February 1703, the Senate of the University of Cambridge resolved that since Giovanni Francesco Vigani, a native of Verona, 'has with great praise exercised the art of chemistry amongst us for twenty years', he should be 'honoured with the title of Professor of Chemistry'. This was not exactly the inauguration of a chemistry chair, since the Senate granted Vigani neither pay nor rooms.[1] But chemists honour past titles. Vigani traced his art from the metallurgist Tubal Cain, eighth man after Adam. Other seventeenth century chemists found similarly ancient ancestors. Nicolas Lefèvre, lecturer at the Paris botanical garden, then Huguenot refugee and manager of a London chemistry laboratory for King Charles II, complained that 'they who reckon Chymistry amongst the modern Arts and of but late invention, betray their knowledge both in the history of Nature and the reading of ancient Authors.' Like Vigani, Lefèvre recalled Tubal Cain, but added Moses' chemical exploits. Had not the patriarch made the golden calf potable? Moses must have had chemical understanding of mineral acids' solvent effect on gold.[2] The implication was that chemistry, though a mere *art*, as the Cambridge Senate carefully observed, was noble because it was old. Its sometimes dubious, often arrogant, practitioners should be respected as masters of hallowed skills.

Cambridge followed leads established elsewhere, as in Leiden, where students intent on medicine pursued practical chemistry under the steadying hand of Herman Boerhaave. But even this was the consequence of a trajectory which, in most of Europe at least, tied chemistry and experimental work to practical uses and medical treatment. Closer still, in Oxford, John Freind merged chemistry

The 1702 Chair of Chemistry at Cambridge: Transformation and Change, ed. Mary D. Archer and Christopher D. Haley. Published by Cambridge University Press. © Cambridge University Press 2005.

and medicine while exploring Isaac Newton's speculations on matter – hot off the press in 1704.[3] Even the ventures of obscure Fenland apothecaries were not leagues apart from the laboratory life of great philosophers. The wider development of natural and experimental philosophy at Cambridge laid a new foundation for the enterprises of chemistry. Medical context made the Cambridge chair. Chemistry was what the apothecary, surgeon and physician demanded. Similar initiatives were launched throughout the German lands and in France and northern Italy. Something like industrial-scale production of pharmaceuticals was set up in those major European cities that commanded long-range trade networks, especially Amsterdam and Venice. Travellers toured these workshops as students and adepts, gathering and touting recipes, secrets and carefully guarded boxes of potent substances.[4]

Vigani's patchy career fits this European landscape of pilgrimage, medical pharmacy and entrepreneurial patronage. One of his Cambridge students recalled him as 'a very learned chemist, and a great traveller, but a drunken fellow'.[5] A recently discovered portait carrying a label which bears Vigani's name (Figure 2.1) shows a prosperous and confident young man. He was born around the middle of the seventeenth century in Verona, a city almost ten times larger than Cambridge. His native city, under the rule of the Venetian Republic, was a centre of apothecaries' art. Verona boasted herbaria and museums, as well as an academy of local physicians, the 'Aletofili', keen to promote newfangled chemical remedies. The Veronese apothecary Francesco Calzolari publicised the herbs of nearby hills, and established a museum above his shop to show instruments, chemicals and preparations. Its learned catalogue by two local medics was read throughout Europe. Remains of Calzolari's collections stocked Verona museums during the seventeenth and early eighteenth centuries, especially those of the apothecaries Maffeo Cusano and Mario Salò.[6] In 1663, several Veronese 'cabinets or collections of natural and artificial rarities', including those of Cusano and Salò, were visited by the great naturalist John Ray while he toured botanic and medical sites in exile from Restoration Cambridge. Ray's curious travels match those of Vigani. Before 1682 the Italian reached England, which he reckoned 'the great Laboratory of Arts'.[7] He set up as apothecary in bustling Newark-on-Trent, and soon arranged the London republication of his brief chemistry handbook, *Medulla chymiae* (Marrow of Chemistry, Figure 2.2), already released in the German lands. It was reviewed in Rotterdam by Pierre Bayle, in Leipzig by the *Acta eruditorum* and in Paris by the *Journal des Savants*.[8]

The London edition carried a prefatory letter from a metropolitan friend, 'T.R.', praising Vigani's unusual clarity and lauding Newark as a retreat from metropolitan cares. However seemingly withdrawn, Vigani had useful contacts.

Figure 2.1. This portrait, labelled on the reverse 'Vigani . . . Professor of Chemistry in the University of Cambridge . .'; was purchased at auction by Trinity College, Cambridge in June 2004. The label gives a date of 1744–5, well after Vigani's death. No portrait of Vigani has previously been known to exist.

His book carried images of chemical hardware designed by John Troutbeck, a Cambridge medic turned royal surgeon, and was dedicated to the Yorkshire aristocrat Thomas Belasyse Earl Fauconberg, ambassador to Venice in 1679. Belasyse's north Italian journey may have been an early English contact for the Verona chemist. Vigani also dedicated his book to other royalist peers who had toured northern Italy, Philip Stanhope Earl of Chesterfield and William Cavendish Earl of Devonshire, Fellow of the Royal Society. T.R.

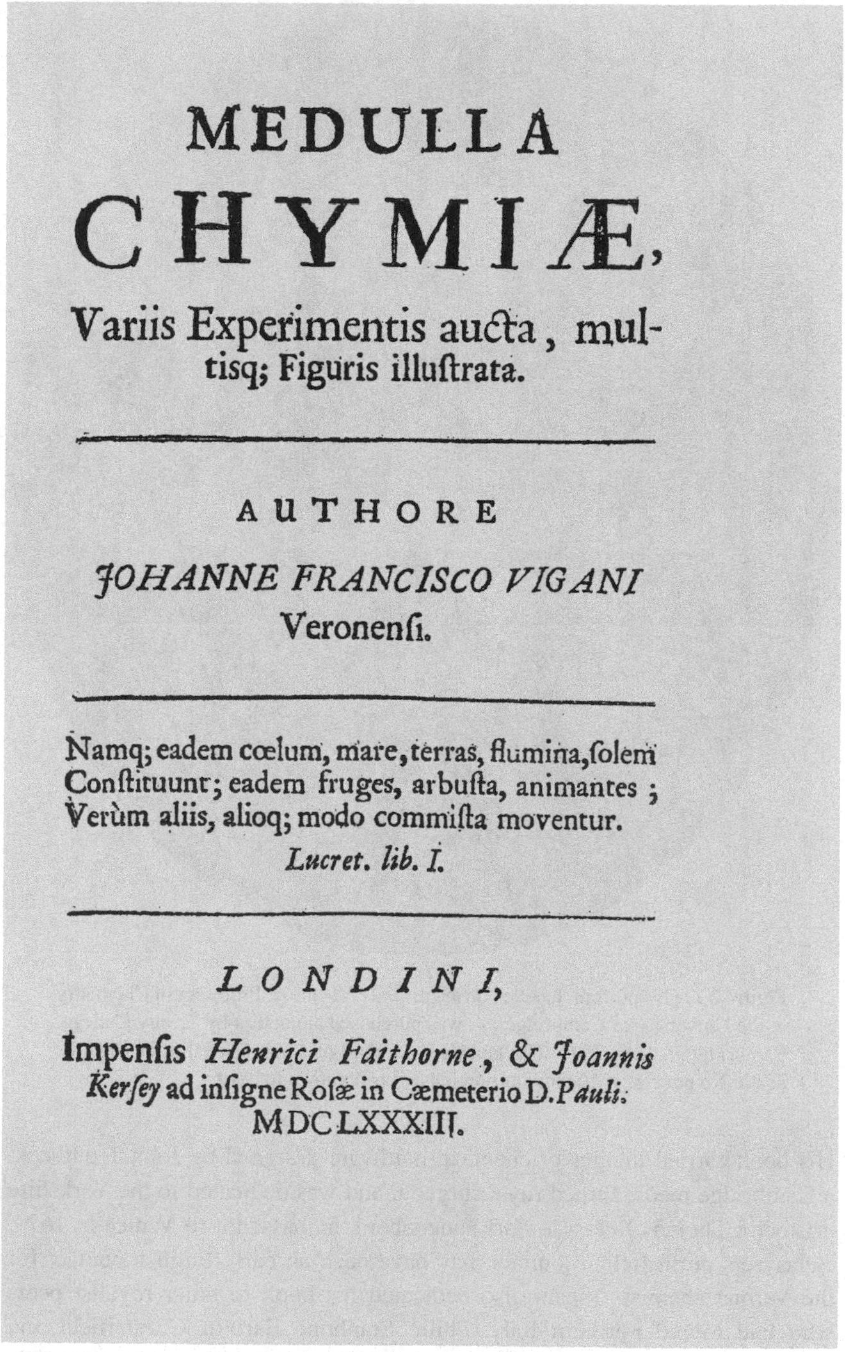

MEDULLA

CHYMIÆ,

Variis Experimentis aucta, mul-
tisq; Figuris illustrata.

AUTHORE

JOHANNE FRANCISCO VIGANI
Veronensi.

Namq; eadem cœlum, mare, terras, flumina, solem
Constituunt; eadem fruges, arbusta, animantes ;
Verùm aliis, alioq; modo commista moventur.
Lucret. lib. I.

LONDINI,

Impensis *Henrici Faithorne*, & *Joannis
Kersey* ad insigne Rosæ in Cæmeterio D. *Pauli.*
MDC LXXXIII.

Figure 2.2. Frontispiece of Vigani's *Medulla Chymiae.*

judged Vigani's chemical work a contribution to the Royal Society's new experimental programme, exalting, as Knox has commented earlier, 'that Republic, more utopian than Platonic, where philosophers are mechanics and mechanics philosophers'. It was easy to place Vigani in 'Boyle's camp', as T.R. put it.[9] In 1682 Vigani began corresponding with Robert Boyle, most famous of experimental philosophers in Britain. The same year the Italian translated, for these colleagues' benefit, the new polemical work of the Roman Jesuit Daniello Bartoli, who incautiously attacked Boyle's corpuscularean doctrines. T.R. told Vigani the Jesuit was a 'dwarf' in comparison with the 'Hercules' Robert Boyle. Vigani agreed. He sneered at Aristotle's four elements and was sceptical of chemists' favoured principles. Rather, so he told his readers, both ancients such as Lucretius and moderns such as Descartes, Gassendi and Boyle had it right in deriving chemical properties from the shape, size and motions of arrays of minute corpuscles. According to Vigani, Boyle 'came, saw and conquered'.[10]

As the Stuart monarchy foundered, Vigani's career and custom prospered, especially in Cambridge. The Regius professor of medicine, Robert Brady, and the medically interested Master of Catharine Hall, John Eachard, got recipes and chemicals from him. Students at Eachard's college bought medicines from the Italian. Another of Vigani's clients, John Covel, the new Master of Christ's, was active in chemistry and botany, using the college garden to cultivate rare herbs. Covel had also been a traveller, acting as chaplain to the English embassy in Constantinople; then he resided in Italy and made himself an expert on oriental medicaments and languages. Covel would proudly show visitors his 'museo' of paintings, coins, urns, herbs and Turkish and Sanskrit manuscripts. In summer 1692 Vigani sent Covel a detailed recipe for crystallising fruits, and told of his many new contacts in Newark and elsewhere: 'my affairs are going from good to better'.[11] During this decade Vigani extended his medical circle to several provincial physicians.[12] Important contacts, perhaps via Fauconberg, were the self-styled 'York virtuosi'. This group, which included the eminent naturalist Martin Lister, recruited chemists, physicians, antiquarians and painters. One was the ingenious glass painter Henry Gyles, who worked on the chemistry of furnaces, ruby glass and resilient enamels. Gyles installed glass windows in Fauconberg's mansion, and got his friend John Place, physician to the Florentine court, to seek glass and chemicals in Venetian glassworks and throughout Italy. These virtuosi had Cambridge contacts too, though dons were not always good patrons. 'All those old fellows are extremely covetous', Place told Gyles. In 1690 Gyles made a fine glass window for the hall of Trinity College and in late 1692 began negotiating to decorate the windows of its impressive new library.[13] Vigani visited York and prescribed medicine to the ailing Gyles.[14] He tried getting Gyles work via his natural history networks. The ambitious Vigani

Figure 2.3. Vigani's arms by Henry Gyles, from a stained glass window now situated in Surrey House, Norwich.

soon also commissioned an emblematic pair of painted glass windows from Gyles in 1697, with his arms and the initials of his wife Elizabeth prominently entwined (Figure 2.3). The proud Latin motto reads: 'Vigani of Verona, Prince of Chemists'.[15]

Vigani, like his York friends, established crucial links with metropolitan printers, such as Thomas Newborough. In the early 1690s the eminent London printer Samuel Smith, publisher of Boyle, John Ray and Thomas Sydenham, distributor of the Royal Society's *Philosophical Transactions* and of Isaac Newton's *Principia Mathematica*, commissioned a new chemistry text from Vigani. It was to be composed in Italian then translated into English by a London medical man, but it never appeared.[16] Vigani's Cambridge chemistry lectures did prove successful, pulling audiences from medical students and local practitioners apt for work on pharmacy and *materia medica*. The St John's College undergraduate Abraham de la Pryme was inspired to order chemical equipment from London 'to try and invent experiments and all the things that I shall do I intend to put down in a proper book'.[17] Such lecture notes survive from several of Vigani's courses, including that delivered after his elevation to professor at 'his laboratory at Queens' College' in winter 1705, when the young medical student William Stukeley was in attendance. Stukeley reported that he 'took down all his readings into writing, and have them in a book'. Medical students used their notebooks for recording the physicians' lore of signs of conception, or ointments for the itch, before opening new and lengthy sections where Vigani's potent recipes were recorded in great detail.[18]

By early 1707 the fiery Master of Trinity, Richard Bentley, ambitious to turn his college into a centre of natural philosophical instruction, had commissioned a new chemistry laboratory in the north range of Great Court, providing the professor with temporary rooms. It was Bentley's view that his college had 'grown an University within ourselves, having within our own walls better instruments, and lectures for Astronomy, Experimental Philosophy, Chymistry, and etc., than Leyden, Utrecht, or any University could shew'.[19] Stukeley attended courses at Vigani's new laboratory in company with his Corpus Christi colleague Stephen Hales, a keenly pious experimenter and naturalist. Hales and his fellows tried Boyle's experiments on the distillation of mercury at the Trinity laboratory during 1707.[20] Bentley saw the establishment of Vigani's 'elegant chemical laboratory' as but part of his schemes, which included the provision of an observatory and lecture room over the Great Gate for the brilliant young Plumian astronomy professor Roger Cotes, and the overhaul of the University Press. Vigani then used the Press to advertise his laboratory-made drugs (as previously shown in Figure 1.2), and exchanged news with Cotes. Another experimenter, Stephen Gray, described by Stukeley as 'the first eminent propagator of electricity,' first came to Cambridge to help Cotes at the new observatory, and there met Hales. Hales was briefly to consider the possibility that Gray's Cambridge electrical experiments could actually provide an explanation for muscular action.[21] This was one foundation, at least at Cambridge, for the appropriation of experimentation by chemistry. But Vigani's contribution to this Trinity programme was brief. Soon he was back with his wife and daughters in Newark, where in February 1713, a decade after his elevation to professor, he died.[22]

Lectures, laboratories and cabinets

What, then, did Vigani profess? The Paris review of his work in March 1685 summed up: 'while the majority of chemists only talk to us of a thousand uncertain preparations which they have never put into practice, Vigani abandons their principles to embrace the atomic system, and only gives us what he knows with the utmost certainty from his own experiments'. Much was made of a chemistry both mechanical and experimental.[23] These were pieties of seventeenth-century chemical discipline. Vigani paid due respect to the masters of his art, Helmont, Boyle and their colleagues. But, as Kevin Knox has described in the previous chapter, it was not obvious that such enterprises were proper to academic life. In a work addressed to 'the apothecaries of England', the King's chemist Lefèvre scolded scholars who imagined 'they should wrong their Gravity and

Doctoral State to defile and sully their hands with the blackness of Coals'.[24] Chemistry was advertised as a practical art which revealed the order innate in nature through assays, collection and writing. Vigani left traces of all three of these enterprises, in records of his laboratory, his cabinet and his teachings. His project highlights the challenges in making artful analysis of nature a fit enterprise for the colleges.

Cambridge in Vigani's time was a fenland town of around 5500 souls afflicted by ague, pox and marsh fevers. There was no systematic distinction between physicians, surgeons and apothecaries, despite futile attempts by licensers to police trade boundaries. Many apothecaries were paid by parish overseers for the sick poor. As Vigani's ambitions show, those who dealt in pharmacy were often men of substance. John Crane, a very wealthy Cambridge apothecary, left the profits of his own house for use by the Regius professor of physic. John Fage, an apothecary who died in Cambridge in 1694, also owned the town's main inn. The local apothecary Thomas Day, who died in 1681, left more than £1500 in his will. One Cambridge apothecary of the period, Peter Dent, trained a St John's student who then developed an effective recipe for the celebrated Jesuits' bark, cinchona, a sovereign remedy for ague. Cambridge medical students were often taught herb and pharmacological lore by these local apothecaries.[25]

Significantly for Vigani's career, he was expected to earn his income solely from lecture fees paid by his auditors, and the years 1680–1720 saw a comparative peak in numbers of medical students, especially in the colleges where Vigani was most active.[26] University numbers, housed in some fifteen colleges, were in slow decline from a peak in the early seventeenth century to less than 2000 after 1700. It was a university under the control of the masters of arts, not, as elsewhere in Europe, dominated by its students, and certainly not, as the careers of Vigani and his professorial colleagues testify, a university keen or able to fund organised collective enquiry in such arcana as experimental philosophy and chemistry. Such projects were almost always private affairs, voluntarily supported; hence the shock which greeted Bentley's remarkable initiatives at Trinity. Bentley himself recalled that the lumber-room he converted into a laboratory had previously been 'the thieving house of bursars of the old set, who in spite of frequent orders to prevent it would still embezzle there the college timber'. His enemies in the fellowship reckoned the cunning Master was aiming to turn the laboratory into a greenhouse for his expanding riverbank gardens. One moaned that though 'I am none of those who glory in despising and running down chemical observations and experiments, but yet with regard to this famous laboratory of ours, I have talked with those that have gone the course and they all seem to be of opinion that as those matters are managed, the learned world is not like to reap any mighty profit or advantage from anything that is there taught'.[27]

What mattered most to Vigani's appeal to the colleges was his characteristically chemical capacity to match the production of *materia medica* with the making of orderly texts. His book, and his auditors' lecture notes, focused on pharmaceuticals. Vigani's friend and student Stukeley went from the course in *materia medica* to 'the apothecary's shop to make myself perfect in the knowledge of drugs'. De la Pryme, slightly less enthused, heard an entire lecture from Vigani on 'star-shot jelly', which the Italian reckoned a chemically potent product of the stars themselves, though de la Pryme more cautiously judged it the residue of dead earthworms.[28] Though the 'atomic system' was canvassed at the head of Vigani's treatise, it was but rarely applied in the everyday laboratory course. Rather, each medicine was analysed and its effects charted. Emetics were the more violent, Vigani lectured, when residues were unwontedly left in the chemist's vessels and not properly separated. In speaking of vitriol of ammonia (ammonium sulphate), a common drug, Vigani first explained how best to prepare it. Green vitriol must first be purified of the copper it so often contained by plunging a hot iron sheet into the liquid until all the copper was precipitated.[29] Vigani backed up the recipe with lessons on standard hardware.[30] His performances were stocked with chemical recipes and laboratory lore designed to aid understanding of commercial remedies' effects. Consider his discussion of antimony, the regulus of which played a decisive role in Newton's work in the 1670s on the vegetation of metals. Chemists in London and Paris, such as Lefèvre, were wont to market antimony cups, selling for as much as 40 shillings each, in which wine left standing overnight would then marvellously become a potent emetic. They claimed their cups would never lose their power or their weight. Vigani riposted that antimony must indeed lose weight in the process, the effect hidden by the combination of antimony with tartrate in the liquor. One student dutifully noted the conclusion: 'It is certain that the powder receives some small alteration in its virtue, otherwise an apothecary that has but one ounce of it, it will serve both him and all his successors forever, but we observe that apothecaries do often buy fresh to renew it, and at the same time they will swear that they had it of their masters'.[31]

Though Vigani sometimes denounced apothecaries' mysteries for the benefit of their customers, he also instructed his scholars how to make seeming marvels. Many relied on the occult powers of spirits and odours, the key concern of much chemical work. Vigani's chemistry favoured fragrances, long associated with sanctity and therapeutic regeneration. A Paduan medical professor wrote that a specifically aromatic agent such as balm, proverbially linked to the miracles of Christ, was 'so praised among ancient and modern physicians that its oil or juice were considered and celebrated by the whole world as a divine gift'.[32] Vigani held the common view that gold grew as a vegetable animated by such steams underground in veins, and taught how to whiten copper so none but skilled assayers could tell it from true silver. He could manufacture crystals in winter which resembled artificial roses, and an aromatic salt which – oddly – would make Rhenish wine taste like that of Spain. He was master of a sulphurous balm that would strengthen any odour to fill the room, and devoted pages to the astonishing effects of one of the most powerful of imported aromatics, tobacco.[33]

Nor was publicity given to all recipes. Knox has mentioned a form of green mercury which could cure gonorrhoea, but which Vigani prudently kept as 'a great secret'. Some tricks were especially startling, such as Vigani's method for reducing amber oil to a volatile salt by sublimation after mixture with common salt, and then the reproduction of the original amber from its *caput mortuum*. These tactics for recovery of the original substance after analysis were the stock-in-trade of the jobbing chemist. Vigani made much of the fact that when he distilled verdigris the amount of volatile acid needed to reconstitute the original from the latent copper was roughly the same as the amount of acid separated during distillation. We might perhaps see here an experiment to demonstrate the law of constant proportions – Vigani, rather, saw this as an elegant performance of a dramatic chemical anatomy akin to his other laboratory shows.[34] We catch the sense of these performances in the enthusiasm of some of Vigani's students. The professor lectured on the 'liquid gunpowder' used to dissolve intestinal stones. His student William Stukeley recalled similarly explosive tricks, which 'often surprised the whole college'. Stukeley and Hales set up their own laboratory at Corpus Christi where, as the previous chapter described, they would share their table with dead dogs for anatomy and chemical apparatus for medical experiments. It is salutary to reflect on the irruption of such practices into the academic world.[35]

Vigani's command of a laboratory was crucial in the new geography of learning. The very term 'laboratory' was novel in seventeenth-century English. Laboratories typically hosted the preparation of medicines and the distillation of spirits. Above all, the chemist's workplace was to be stocked with large and unwieldy furnaces. Pride of place was given to reverberatory furnaces, with domed upper chambers to reflect the heat. Care was to be taken in discriminating between hot blasts to distil minerals and the slower, delicate warmth demanded by lengthier analyses.[36] Vigani thus touted his own skill at oven design: 'all the matter consists in managing the fire to keep it slow', and 'every small hole between the bricks is a regular and constant pair of bellows'. He learnt much from his colleague Newton, who designed his own ovens and kept painstaking records of furnace design. 'His brick furnaces he made and altered himself without troubling a bricklayer', recorded Newton's labourer. Understandably, too, it was Vigani's recipes for 'building furnaces of dry bricks without iron or mortar and his manner of regulating the fire to any degree of heat' which attracted Stukeley's attention.[37] The security of these workplaces was decisive. Laboratories had often been located in monasteries and colleges. In Venice, for example, laboratories' secrecy was assured since, so it was said, none could enter without promising to stay forever. Robert Boyle referred to his own laboratory as 'a kind of elysium'. The laboratory was often situated at the rear of a house, where access was controlled.[38] When public lecturers in Oxford, Cambridge and London began to forge a version of public, commercially viable, and, as we shall see, decisively Newtonian chemistry in the early eighteenth century, they had to shift their art's place from sequestered colleges and laboratories to urban coffee-houses, dyeworks, glass-houses and print-shops.[39] These puzzles of publicity made performances in ambiguous sites such as those of Vigani peculiarly troublesome. In many great European cities, the laboratories were crucial nodes of the global networks of goods and personnel on which the authority and practice of the chemists increasingly relied.

Thus Vigani had not merely to recruit audiences, publish handbooks and set up workplaces. Remote Cambridge workrooms had to be linked with centres of global commerce. Many enterprises in stationers' workshops, botanic gardens or cabinets of curiosities involved the accumulation in central depots of transient materials which could there be studied and turned into powerful tools. This was why Vigani's skill at fixing perishable substances, such as drugs, plants or minerals, counted. He distilled snakes, and human skulls, to extract their vital longer-lived essences. Pharmacists like Vigani were major customers of glass-makers and potters whose wares allowed them to store vulnerable goods. Dried herbaria, a northern Italian technique introduced in the 1540s, helped renovate medical botany. Vigani's students wrote down his instructions that herbal oils must be extracted from dried specimens of plants and roots, since fermented plants would generate a 'vinous spirit', while raw samples would yield nothing useful. It was apt that the Trinity fellowship guessed Bentley might use the new chemistry laboratory as a greenhouse: just such innovations in garden management secured exotic plants for European study and exploitation.[40]

Accumulation of fragile overseas merchandise was crucial for Vigani's enterprise. So during the year following his elevation to professor, orders flowed from the Italian chemist via the Queens' College proctor to London druggists such as Francis Porter and the Cheapside dealer Henry Colchester, and to glass suppliers and cabinet makers. The result was the oak cabinet of *materia medica* that still graces the College's rooms (Figure 2.4).[41] It embodies the reliance of such enterprises on long-range systems of trade in spices and resins, shells and woods, metals and earths. These systems were inextricably linked to new European colonial markets and their links with pharmacy assays.[42] The antimalarial

Figure 2.4. Vigani's cabinet of *materia medica*, currently housed in the President's Lodge, Queens' College, Cambridge.

cinchona supplied from Jesuits' bark was a prized example. Amongst its more than 600 specimens, Vigani's cabinet held this Peruvian bark, alongside spices, jewels and opium, his favoured aromatics such as benzoin gum from Siam and *tacamahaca*, balsamic poplar oil from New Spain, which when burnt on coal cured hysteria. One jar held a stock of Lucatello's balsam, a compound of

Venice turpentine, beeswax, olive oil and sandalwood, a fashionable panacaea much favoured and carefully analysed by Vigani's professorial colleague Isaac Newton. Newton took the balsam internally, mixed with cochineal. Cochineal reached Vigani from Henry Colchester, who also sent palm oil and spermaceti. There was *nux vomica* from Bengal and *terra sigillata* from the Aegean, a special earth (our aluminium silicate) extracted ceremonially on the former Venetian colony of Lesbos, then publicly sealed, as the name implied, by local officials. Porter supplied Vigani with Brazilian guaiacum, powerful lodestones, and oriental bezoars – stony antidotes for snakebite taken from the bellies of Persian goats, and one of the more expensive items on the Queens' College shopping list.[43] Snake venom itself also played an important role, both as opportunity to exercise therapies and source of invaluable chemicals. In a remarkable public ritual especially common in Vigani's native land, venom was used ceremonially to prepare theriac, a potent antidote to other poisons. It was significant for Vigani's projects that north Italian physicians and chemists were so expert at these rituals of public assaying and pharmacological trial.[44] Through Vigani's kind of enterprise, cloistered Cambridge began to gather worldly goods to match its impressive collections of print and paper. In his Cambridge teachings, his cabinet and his laboratory, the Veronese chemist professed ingenious ways of amassing chemical commodities and making them matter.

Towards public and useful experiment

While practically commercial medical chemistry may have been paramount in Vigani's enterprise, it was hardly devoid of larger theoretical issues in the world of natural philosophy. The publication of Isaac Newton's natural philosophy had considerable impact on how chemistry was received in Cambridge. Vigani early established atypically warm relations with the reclusive Newton. He reportedly often visited the mathematics professor, learning of improved oven designs and distillation instruments. Newton's laboratory projects involved a telling contrast between 'a more subtile secret and noble way of working' and the 'vulgar' or 'purely mechanical' actions of nature: this distinction was marked in the layout of his backstage chemical work. 'This may be done in your chamber as privately as you will, and it is a great secret', his alchemical masters taught him. But Newton's projects also needed help from travelling advisors. In spring 1696, just before he left Cambridge, Newton was visited by one of Boyle's former London acquaintances who advised him on making volatile spirits, a process which must last at least nine months: 'it was not necessary that the vitriol should be purified, but the oil or spirit might be taken as sold in shops without so much as rectifying it. When you draw off the spirit,' Newton's visitor told him, 'you must leave the soul not thick like honey or butter but thinner than oil so that you may pour clean out of your glass like a liquor.' Even the most cloistered alchemy relied on shopkeepers.[45]

Newton used these Cambridge laboratory studies as resources for his published chemical manifestos. He bound his annotated copy of Vigani's book with a prized text by the great sixteenth-century German humanist and metallurgical expert Georg Agricola. As mentioned, it was Newton who told Boyle that Vigani was already 'performing a course of chemistry for several of our University much to their satisfaction'.[46] Vigani in turn seems to have used a microscope to examine crystalline structure, even perhaps seeking a Newtonian explanation based on inter-particular forces. The attractive forces of matter which played an essential part in early eighteenth-century chemistry were described by Newton in all three editions of his *Opticks*. Newton's most notable speculations on matter theory appeared, as Knox has mentioned, as a brief chemical treatise in his famous thirty-first Query (1706) to this book. It is striking that Hales, one of Vigani's auditors, took up the challenge of exploring chemical reactions even after he had moved from Cambridge in 1709 to take up duties as minister of Teddington. Hales' debt to Newtonian speculations on attraction was explicit. He took note of this in 1727 at the start of a work heir to Cambridge laboratories, *Vegetable Staticks*, which founded the British exploration of pneumatic chemistry for much of the eighteenth century.[47] Hales was especially concerned with the air that emerged from the decomposition of animal, vegetable and mineral substances when subjected to heat or fermentation:

> Where it appears by many chymio-statical Experiments that there is diffused thro' all natural, mutually attracting bodies, a large proportion of particles, which, as the first great Author of this important discovery, Sir Isaac Newton, observes, are capable of being thrown off from dense bodies by heat or fermentation into a vigorously elastick and permanently repelling state: and also of returning by fermentation, and sometimes without it, into dense bodies: it is by this amphibious property of the air, that the main and principal operations of Nature are carried on; for a mass of mutually attracting particles, without being blended with a due proportion of elastick repelling ones, would in many cases soon coalesce into a sluggish lump. It is by these properties of the particles of matter that he solves the principal Phaenomena of Nature. And Dr. Freind has from the very same principles given a very ingenious Rationale of the chief operations in Chymistry.[48]

Like Hales and several Cambridge contemporaries, the virtues of chemical experiment and its medical import were discovered by the physician James Jurin, a Cambridge student who was elected a Fellow of Trinity in 1709 just after

Vigani's departure. Before his appointment as a Fellow of the Royal Society (1717) and physician to Guy's Hospital (1725), Jurin joined a small group of Newcastle philosophers who gave lectures there. In 1716 Jurin took his M.D. and left for London for good, turning his expertise to the crucial medical topic of the force of the heart.[49] Medicine's long experimental and chemical train was carried further by John Mickleburgh, the third Professor of chemistry at Cambridge, from 1718. Mickleburgh seems to have continued trade as a dispensing chemist, as well as lecturer, catering to his medical connections with useful knowledge. This may have been the result of established connections: it was the Professor of Physick who recommended Mickleburgh for the post.[50] Upon his appointment, he was to share the existing Queens' Lane laboratory with the Professor of Anatomy, a precedent established during the brief tenure of the second chemistry Professor, the Reverend John Waller (*c.* 1673–1718) – about whom little is known save the few details mentioned in the previous chapter and his connection with Corpus Christi (then Bene't Hall), where he was a tutor.[51] Only at that point did the University apparently begin to pay for the requisite apparatus. Even in such circumstances Mickleburgh performed his duties for more than a quarter of a century, delivering at least five courses of which there are surviving traces.[52] Mickleburgh's auditors came from many colleges, but included as well were the Cambridge surgeon Francis Sandys and his assistant Samuel Brigham, who took the courses twice. The local apothecaries John Roper and John Benwell also attended. It is remarkable that among Mickleburgh's students were three successive Professors of Anatomy: John Morgan (1728–34), George Cuthbert (1734–35), and Robert Banks (1735–40), along with Charles Mason, Professor of Geology (1734–62). Mickleburgh's lectures attracted many who were to proceed to take medical degrees, several of whom were to continue studies at Leiden under the influence of Boerhaave.[53]

Mickleburgh convincingly argued it was Trinity College's furnaces and utensils that had allowed Newton to find 'the first hints and notices of these Phenomena which very hints and notices have since been reduced by the reverend and ingenious Mr. Stephen Hales into plain facts and rendered even visible to our eyes by an almost infinite variety of experiments'.[54] Likewise, John Hadley, the fourth Professor, appointed in 1756 following the death in office of Mickleburgh, cited the views of Hales on the air which resulted from the application of acids to 'tarter, human calculi, chalk, and iron files', adopting Joseph Black's language, for instance, on the production of 'fixed air'. Hadley was nevertheless careful to wrap himself in the Newtonian dictum of *hypotheses non fingo*. He explained, for example, that 'I shall not attempt to account for the fluidity of water or think of deducing its properties from any Physical reasonings [as] the nature of this as of all other bodies is to be learnt only from Experiments'.[55] The appreciation of forces was grounded in fact-production, in an instrumental and experimental method which Newton and his disciples loudly proclaimed and elaborated.[56] Thus, for example, Hales used a burning glass to extract and assess the qualities of airs so liberated, in experiments similar to those which had been conducted in London by the Newtonian reverends John Harris and John Theophilus Desaguliers.[57] At Cambridge, John Hadley's lectures were very much in the same experimental and dramatic tradition of what he then described as 'a very usefull as well as entertaining branch of Natural Philosophy'.[58] The growing interest in chemical reactions also spawned a new industry of chemical apparatus, such as the design by Peter Shaw and Francis Hauksbee of a portable laboratory intended, most likely, for practical chemists as well as for those gentlemen increasingly enamoured of the rage for experimentation.[59] Chemistry then became one branch of the flood of experiment and instrumentation, to the point where such apparatus could be bought off the shelf in numerous shops in London and provincial towns. Newton's work provided much of the inspiration for chemists in Cambridge and beyond. It was also Newton's work which asserted that only experiment and an arsenal of apparatus could achieve what hypotheses could not.

Whatever the extent of interest in Newtonian speculations, whatever the degree of concern about theory, by mid-eighteenth century there is considerable evidence of a broad public culture of experiment and demonstration. This was as true among the coffee-house lecturers of London, in manufacturing sites like Manchester or industrial concerns around Newcastle, in spa towns like Bath or Scarborough, as it was in Mickleburgh's Cambridge lectures.[60] Lecturers promoted experiment before audiences whose motives included curiosity, practicality, even investment. It was in such a circumstance, for example, that Hales tried experiments on air in the sap of plants by using an air pump, a staple in the armoury of many of the lecturers who acquired devices, sometimes at immense expense.[61] Such apparatus led Hales to further experiments on pneumatics, especially attempts to eliminate any impurities in air (which he still regarded as an element) by passing the results through water and collecting them in a trough.[62] Airs' qualities, their abilities to sustain combustion or respiration, were the next stage, notably in Hales' *Statical Essays* published in 1733. Similarly, Hales was averse to hypotheses, as we might well expect. He was especially reluctant to adopt any general iatrophysical explanations of circulation, especially in the capillaries and any attendant relationship to muscular motion.

At the end of Hadley's tenure as Professor in 1760 he went on, predictably, to practise medicine, first at St Thomas's then at the Charterhouse. Only on his death in 1764 was the position filled by Richard Watson. As discussed in the following chapter, Watson – notoriously, and apparently cheerfully – claimed no knowledge of chemistry but would give it a try 'being tired of mathematics and natural philosophy'. Auditors demanded entertainment. Watson thus recruited from Paris a skilled assistant for the chemical laboratory who was able to aid him in lectures which attracted 'persons of all ages and degrees'.[63] This was the age of the dramatic electrical and instrumental shows in Paris of the Abbé Nollet and the chemical lectures of Etienne-François Geoffroy and Guillaume-François Rouelle, the last of which attracted large audiences and from whom Enlightenment *philosophes* Jean-Jacques Rousseau, Denis Diderot and even Lavoisier learned their chemistry. Popularisation was one route through which chemistry gained an audience. Rouelle's reputation made him an authority in Cambridge as much as in Paris. He was, typically, an apothecary who had learned the techniques of demonstration at the Jardin du Roi. More importantly, Rouelle too was an admitted heir to the innovations of Hales, which already had effect with both Hadley and Mickleburgh.[64] So to recruit a Parisian assistant from the school of French chemistry that adopted improved Hales' apparatus for the collection of airs was a natural and wise choice for those who sought to isolate the elemental characteristics of material bodies.[65] Watson's lectures proved an immense success and helped, at last, to secure from the Crown the regular patronage of £100 per annum.

Since the time of Hadley, there were many who felt that there was enough in nature for experimentalists – without privileging theory. Indeed, in his manuscript *Introduction to Chemistry*, Hadley defined his science, as had others, as

> the art of Separating Bodies into their simplest constituent parts, of uniting those parts again and reproducing the original Compound, and also by different mixtures and Various methods of Combining Bodies to produce new Substances which exist not in nature.[66]

This simple manifesto is rather more important as a reflection of the promotion of experiment in chemistry than might first be appreciated – precisely because it proposes even the artificial creation of compound bodies. Indeed, it is not at all surprising that, like many of his contemporaries in London and the provinces, Hadley was interested in practical chemistry, as in the manufacture of alum, green vitriol (ferrous sulphate) and sulphuric acid.[67] It is crucial to see his not as a sole voice suggesting an application of experimental knowledge in industrial production. Rather it mirrors the developments which were taking place widely in the British natural philosophical community. For example, in response to some vitriolic criticism of his motives in giving experimental courses in London, Benjamin Martin, an instrument-maker, author and popular demonstrator, envisaged a world of useful knowledge, where 'the Chemists, the Anatomists, the Physicians, and Divines, everywhere read lectures for money'.[68] This was precisely what Cambridge's Professors of Chemistry were doing, at least until they received sufficient patronage. There was a wide community of experimentalists who were largely concerned with the practical consequences of their analysis and discoveries.

Hadley had embraced this very attitude. So too did his successor Richard Watson, who came to chemistry, as had others, from that natural philosophy which, in an age much influenced by Newton's Queries, was a fount of chemical speculation. Watson's wide range of technical interests flies in the face of the Enlightenment character of a merely gentrified scientific community, as deliberately constructed by elements within the Royal Society and by subsequent historians. In fact, technical issues were of widespread interest as, for example, in the work of William Cullen and Joseph Black in Scotland.[69] Thus, Watson, according to one account, 'decided that the application of chemistry to the arts and manufactures was the most fitting theme at that time for a university course', and his chemical passions moved beyond the medical to the industrial. There is considerable evidence of this from his *Chemical Essays*, which, it would seem, were based on his lectures. Thus, he argued that

> The uses of chemistry, not only in the medicinal but in every economical art are too extensive to be enumerated, and too notorious to want illustrating . . . It cannot be questioned that the arts of dyeing, painting, brewing, distilling, tanning, of making glass, enamel, porcelane, artificial stone, common-salt, sal-ammoniac, salt-petre, potash, sugar, and a great variety of others, have received improvement from chemical inquiry and are capable of receiving much more.[70]

The issue of the end of knowledge dissemination merged Cambridge and Enlightenment. Cambridge chemistry long gave access to an interested public. Lectures, exhibitions, even books helped to make the industrial world approachable. Such dissemination brought chemistry and industry face-to-face much earlier and in a wide variety of ways.[71] *Materia medica* already provided a model. This indeed was the view from the early nineteenth century. For example, in 1815 Samuel Parkes, then of the Haggerston Chemical Works in London, published five volumes of *Chemical Essays* in imitation of Watson's tomes. In them Parkes asserted that the enterprise of Scheele, Bergman and Watson had 'contributed in no small degree to the information of the public mind, and to that growing taste for chemical pursuits which is one of the characteristics of the present age'.[72] This is the difference the promotion of chemical experiment had made by the beginning of the nineteenth century.

Conclusions

An image of torpor in Cambridge chemistry, in contrast to Midlands chemical entrepreneurs, is no longer sustainable.[73] The pantheon of natural philosophy included the chemistry professors as much as the freebooting adventurers of eighteenth-century industry. This was the crucial consequence of a culture of public science. When private societies were organised in the metropolis, as in various rapidly expanding provincial towns, they made it their task to promote the deliberation of new developments in chemistry, among many other subjects. In soot-laden Birmingham, the Lunar Society, with such luminaries as Erasmus Darwin and Joseph Priestley as members, would discuss at some length chemical discoveries and industrial implications. Less well known, perhaps, were the chemical obsessions of the engineer James Watt, worked out in frequently dangerous conditions in his own private laboratory. Or, likewise, the short-lived, but intensely dynamic Chapter Coffee House Society in London from 1780–1787, organised largely by the chemist Richard Kirwan, which debated the latest discoveries from the continent, the rise of hot air balloons, the passion for electricity and the making of bar iron, or ferromania. It is no surprise that many of the members of Kirwan's club of philosophers were also physicians. The link between practical chemistry and medicine continued. And it did so particularly in the intertwined careers of James Watt and Thomas Beddoes, who promoted the new pneumatic chemistry as an answer to the miasma of growing industrial towns and the endemic tuberculosis which consumed so many in increasingly squalid and crowded tenements and factories.[74] As Beddoes would put it, and as Cambridge professors well knew, chemistry had a heavy burden to bear.

At the climax of the long-eighteenth century, when the world seemed turned upside down in politics, Beddoes called upon chemistry for salvation. Being a reformer in politics, and despite the extremities of French constitutional experiments, he adopted a formula of medicine, chemistry and social transformation mixed with the radical alchemy of democracy and a public scientific culture. To him, as for chemists and republicans like Joseph Priestley, Enlightenment science was the way to escape kings and oligarchies. But principle had a price. Beddoes removed himself in 1793 from Oxford and any opportunity at a Regius chair in chemistry was effectively blocked by heightened political tensions.[75] He had plans when he then relocated near Bristol – including the application of the newly discovered gases, like oxygen and nitrous oxide, to intractable diseases such as endemic consumption – and, on a crucial parallel track, setting up a course of lectures in chemistry. In a few years, he was writing to his friend and collaborator, James Watt, that the public interest in chemical lectures was gratifying. This was especially so, he believed, because 'the effect of a number of people receiving agreeable ideas together may be to soften animosity & that there will be thus a chance of preventing some acts of barbarity in the times that I fear are coming'.[76] Beddoes was overly optimistic. Yet his conclusion was not unusual among physicians and the heirs to a popular experimental tradition that had emerged over the century since Vigani.

The development of early modern chemistry, both at Cambridge as elsewhere, was determined not exclusively by institutions but by the entrepreneurial designs of lecturers, like Vigani, Mickleburgh, Hadley and later Watson. Beddoes' chemical burden reflects the spectacular evolution of experimentation and useful knowledge. Our comprehension of a century is defined not simply by the trope of *progress* but by upheaval, in politics in America and in France, in industrialism and manufacturing, first of all in Britain, and amidst the broad intellectual force of Enlightenment to which experiment and chemistry were essential.[77] It is, therefore, impossible to accept the characterisation made almost sixty years ago by Herbert Butterfield that eighteenth-century chemistry failed to achieve anything of note, indeed that there was no real chemistry, until the work of Lavoisier and his *Traité Elémentaire de Chimie* in 1789. In fact, Lavoisier's creation of a chemical taxonomy owes much to the general attempt to sort out and to classify the promiscuous treasure house of nature that European empires and their natural philosophers encountered.[78] The laboratory was a treasure trove, as rich as the promise of the New Worlds of America or Tahiti. But there was also a passion in the quest for chemical order as much as for the secrets once attached to alchemy and those trades policed by guilds. Knowledge was transmuted into power. Hence, in 1815, the chemist and Fellow of the Linnean and Geological Societies, Samuel Parkes, recounted the Chinese 'story of nine virgin sisters, who passed their lives in celibacy, intent on chemical pursuits'. Whether this enterprise had something to recommend it or not, it does bespeak an image of the 'ardent and inquiring mind' which professors of chemistry could promote.[79]

Notes and References

1. Clark, John Willis (1904), *Endowments of the University of Cambridge*, Cambridge: Cambridge University Press, p. 181; Coleby, L. J. M. (1952), 'John Francis Vigani, first Professor of Chemistry in the University of Cambridge', *Ann. Sci.* **8**, pp. 46–60, p. 48, note 9.
2. Vigani, J. F. (1683), preface to *Medulla Chymiae*, London: Faithorne and Kersey; Lefèvre, Nicolas (1670), *A Compleat Body of Chymistry*, first part of 2 parts separately paginated, London: Pulleyn, p. 1.
3. Debus, Allen G. (2001), *Chemistry and Medical Debate. Van Helmont to Boerhaave*, Canton: Science History Publications; Simcock, A. V. (1984), *The Ashmolean Museum and Oxford Science*, pp. 1–10. Oxford: Oxford University Press;

Guerrini, Anita (1994), 'Chemistry teaching at Oxford and Cambridge circa 1700', in Rattansi, P. M. and Clericuzio, Antonio (eds.), *Alchemy and Chemistry in the 16th and 17th centuries*, Dordrecht: Kluwer, pp. 183–99, pp. 188–89.

4. Multhauf, Robert (1966), *The Origins of Chemistry*, London: Oldbourne, p. 266.
5. de la Pryme, Abraham (1870), in Jackson, Charles (ed.), *Diary*, Durham: Surtees Society, p. 24–25.
6. Maffei, Scipione (1731–2), *Verona Illustrata*, 2 vols., vol. 1, part 2, Verona: Vallarsi and Berno; columns 239–40. Pomian, Krzystof (1990), *Collectors and Curiosities: Paris and Venice 1500–1800*, Cambridge: Polity, pp. 74–8, 222–3. Findlen, Paula (1994), *Possessing Nature: Museums, Collecting and Scientific Culture in Early Modern Italy*, Berkeley: University of California, pp. 42–43, 180–3, 263–5.
7. Ray, John (1673), *Observations Made in a Journey Through Part of the Low Countries, Germany, Italy and France*, London: Martyn, pp. 218–19. Vigani, *Medulla Chymiae*, preface.
8. Peck, E. Saville (1934), 'John Francis Vigani, first professor of chemistry in the University of Cambridge', *Proc. Cam. Antiqu. Soc.* **34**, pp. 34–49, p. 35; reviews of Vigani in *Acta Eruditorum* (October 1684), pp. 394–5; *Journal des Scavans* (March, 1685), pp. 91–3.
9. Vigani, *Medulla Chymiae*, 'Letter from a friend in London to the author in the town of Newark upon Trent', sig. A4. T.R. is identified as Dr Thomas Robson in Peck, 'Vigani', p. 35; alternatively he might have been Dr Tancred Robinson, the notable physician – naturalist (see previous chapter by Knox, footnote 26). William Cavendish was in Italy with the philosopher Thomas Hobbes in 1635; Philip Stanhope was in Italy with the virtuoso John Bargrave in 1650.
10. Vigani's letters to Boyle in 1682 are mentioned in Hunter, M., Clericuzio, A., and Principe, L. (eds.) (2001), *Correspondence of Robert Boyle*, 6 vols, vol. 5, London: Pickering and Chatto, p. 284. Vigani refers his readers to Boyle's works in *Medulla Chymiae*, p. 4. Vigani's translation of Daniello Bartoli, *Del ghiaccio e della coagulatione* (Rome, 1681) is mentioned in Vigani, *Medulla Chymiae*, sig. A5; for Bartoli's attack, see Renaldo, John (1976), 'Bacon's empiricism, Boyle's science and the Jesuit response in Italy', *J. Hist. Ideas* **37**, pp. 689–95, on p. 693.
11. For Vigani and Eachard, see Jones, W. H. S. (1936), *A History of St Catharine's College Cambridge*, Cambridge: Cambridge University Press, p. 103. For Brady and Eachard's medical interests see Robb-Smith, A. H. T., 'Cambridge medicine', in Debus, Alan (ed., 1974), *Medicine in Seventeenth-Century England*, Berkeley: University of California, pp. 327–69, on pp. 350, 353. Notes on Vigani's lectures, 'Cours de chymie', University Library Cambridge, MS Dd.12.53, p. 299, refer to Brady. For the medical interests of Brady and Covel see Gascoigne, John (1985), 'The universities and the Scientific Revolution: the case of Newton and Restoration Cambridge', *Hist. Sci.* **23**, pp. 391–434, on pp. 400–1. For Vigani's advice see Vigani to Covel, 2 August 1692, British Library MS ADD 22910 fols. 410–11. For Covel's collections see Mayor, J. E. B. (1911), *Cambridge under Queen Anne*, Cambridge: Deighton, Bell, pp. 147–52.
12. Vigani to Newborough, 9 November 1696 and 8 June 1697; British Library MS 4276, fols.172, 173 mention Charles Leigh of Manchester and William Coward of Northampton.
13. Brighton, J. T. (1984), *Henry Gyles: Virtuoso and Glasspainter of York* (*York Historian*, vol. 4), pp. 8–10, 13, 42, 45; Place to Gyles, 27 January 1694, British Library MS Stowe 747, fol. 26.
14. *Ibid.*

15. Vigani to Gyles, 27 September 1698 and 13 March 1699, British Library MS Add 4276, fols. 174–5; Brighton, *Henry Gyles*, pp. 44–5; Gyles to Thoresby, 9 August 1709, in Joseph Hunter (ed.) (1832), *Letters of Eminent Men Addressed to Ralph Thoresby*, 2 vols., vol. 2, London: Colburn and Bentley, p. 63.

16. Edwards to Thoresby, 29 May 1716, in Lancaster, W. T. (ed., 1912), *Letters Addressed to Ralph Thoresby*, Leeds: Thoresby Society, pp. 242–3; Vigani to Newborough, 1696–7, British Library MSS Add 4276, fols. 171–3; Vigani to Covel, 2 August 1692, British Library MSS Add 22910, fol. 411. For Smith's operations, see Barnard, J. and McKenzie, D. F. (eds.) (2002), *Cambridge History of the Book in Britain 1557–1695*, Cambridge: Cambridge University Press, pp. 170–1.

17. de la Pryme, *Diary*, p. 34.

18. Stukeley, William (1882–7), *Family Memoirs*, ed. Lukis, W. C., 3 vols., vol. 1, Durham: Surtees Society, pp. 28, 33, 51. For student medical notes see Cambridge University Library MS Dd.12.53, pp. 295–300 on medicaments and pp. 11–12 on plant extracts. Other notes include 'Seignior Vigani's Course of Chymistry' (Queens' College Cambridge, November 1705) at Glasgow University Library MS Ferguson 62 and 'A course of chymistry under Signior Vigani' (Trinity College Cambridge, November–December 1707) at Harvard University Library MS Eng 685.

19. Gascoigne, John (1989), *Cambridge in the Age of the Enlightenment*, Cambridge: Cambridge University Press, pp. 154–5.

20. Collinson, Peter (1764), 'Memoir of Stephen Hales', *Gentleman's Magazine* **34**, pp. 273–8, on pp. 273–4; Stukeley, *Family Memoirs*, vol.1, pp. 21, 39–40, 51; Hales, Stephen (1727), *Vegetable Staticks*, republished London: Macdonald, 1961, p. 112. For Bentley's programme see Bentley to Bateman, 25 December 1712, in *Correspondence of Richard Bentley*, ed. Wordsworth, C. (1842), 2 vols., vol. 2, London: John Murray, pp. 448–51. Gascoigne, John (1984), 'Politics, patronage and Newtonianism: the Cambridge example', *Hist. J.* **27**, pp. 1–24.

21. Stukeley, *Memoirs* I, p. 50.

22. The chemistry laboratory is described in Bentley, Richard (1710), *The Present State of Trinity College in Cambridge*, London: Baldwin, cited in Willis, R. and Clark, J. W. (1886), *Architectural History of the University of Cambridge*, 4 vols., vol. 2, Cambridge: Cambridge University Press, p. 616 and in Monk, James (1833), *Life of Richard Bentley*, 2nd ed., London: Rivington, pp. 202–4. Vigani's advertisement of medicines, dated from Newark in August 1709, is in Bodleian Library, John Johnson Collection, Patent Medicines 14. For Vigani and Cotes, see Vigani to Cotes, 13 September 1707 and 14 June 1708, British Library MS Add 22911, fols. 68, 74. Stukeley visited Vigani in Newark in summer 1708; see Stukeley, *Family Memoirs*, vol. 1, p. 42. For Vigani's will and death, see Peck, 'Vigani', p. 38.

23. [Review of] 'Medulla Chymiae', *Journal des Scavans* (March 1685), pp. 91–3, on p. 92.

24. Vigani, *Medulla Chymiae*, pp. 3–4, 8; Nicolas Lefèvre, *Compleat Body of Chemistry*, first part, pp. 1, 10. For chemistry's learned status see Guerrini, Anita (1994), 'Chemistry teaching at Oxford and Cambridge circa 1700', in Rattansi, P. M. and Clericuzio, Antonio (eds.), *Alchemy and Chemistry in the 16th and 17th Centuries*, Dordrecht: Kluwer, pp. 183–99.

25. Whittet, T. D. and Newbold, M. (1978), 'Apothecaries in the diary of Samuel Newton, alderman of Cambridge', *Pharm. J.* **221**, pp. 115–18; Burnby, Juanita (1983), *A Study of the English Apothecary from 1660 to 1760*, London: Wellcome Institute, pp. 18, 32. Robb-Smith, A. H. T. (1937), 'Cambridge medicine', Berkeley: University of California Press, pp. 330, 341, 351–2. Gunther, R. T. (1937), *Early Science in Cambridge*. Oxford: Oxford University Press.

26. Rook, Arthur (1969), 'Medicine at Cambridge 1660–1760', *Medical Hist.* **13**, pp. 107–22, on p. 111.
27. Bentley, *Present State of Trinity College*, p. 60; Blomer, Thomas (1710), *A Full View of Dr Bentley's Letter*, London: Knoplock, p. 119. Compare Winstanley, D. A. (1935), *Unreformed Cambridge*, Cambridge: Cambridge University Press, pp. 143–4, 365–6 note 132.
28. Stukeley, *Family Memoirs*, vol.1, p. 39; de la Pryme to Sloane, 2 February 1702, in Pryme, *Diary*, p. 247.
29. 'Cours de chymie', University Library Cambridge MS Dd.12.53, pp. 5, 7; Vigani, *Medulla Chymiae*, pp. 6–7.
30. 'Cours de chymie', University Library Cambridge MS Dd.12.53, p. 3; 'Course of Chemistry', Harvard University Library MS Eng 685, fols. 110–12.
31. Vigani, *Medulla Chymiae*, 49; 'Course of Chymistry', Harvard University Library MS Eng 685, fol. 17. For the use of antimony, see Lefèvre, *Compleat Body of Chemistry*, second part, pp. 236–9; Coleby, 'Vigani', pp. 53–54; Thorndike, Lynn (1958), *History of Magic and Experimental Science: the Seventeenth Century*, New York: Columbia University Press, pp. 130–3.
32. Alpino, Prospero (Venice, 1591), *De Balsamo Dialogus*, preface; see Albert, Jean-Pierre (1990), *Odeurs de Sainteté*, Paris: EHESS, p. 109.
33. Vigani, *Medulla Chymiae*, pp. 28–9, 57; Coleby, 'Vigani', pp. 52–3; Thorndike, *History of Magic*, pp. 393–4.
34. Vigani, *Medulla Chymiae*, pp. 13, 53.
35. Stukeley, *Family Memoirs*, vol. 1, pp. 21, 33. See Robb-Smith, 'Cambridge medicine', pp. 354–8; Allan, D. G. C. and Schofield, R. E. (1980), *Stephen Hales: Scientist and Philanthropist*, London: Scolar, p. 12–14; Gascoigne, *Cambridge*, pp. 159–61.
36. Smith, Pamela (1994), *The Business of Alchemy*, Princeton: Princeton University Press, pp. 274–5, 277; Gunther, R. T. (1923), *Early Science in Oxford: Chemistry, Mathematics, Physics and Surveying*, vol. 1, Oxford: for the subscribers, p. 10.
37. Vigani, *Medulla Chymiae*, p. 60 and Coleby, 'Vigani', p. 59; Humphrey Newton to Conduitt, 14 February 1728, King's College Cambridge MS Keynes 135; Westfall, R. S. (1980), *Never at Rest: a Biography of Isaac Newton*, Cambridge: Cambridge University Press, pp. 283–4; Stukeley, *Family Memoirs*, vol. 1, p. 33.
38. Boyle to Lady Ranelagh, 31 August 1649, in Boyle, *Correspondence*, vol. 1, p. 83; Gunther, *Early Science in Oxford*, vol. 1, pp. 9–13, 36–42; Hannaway, Owen (1986), 'Laboratory design and the aim of science', *Isis* **77**, pp. 585–610; Shapin, Steven (1988), 'The house of experiment in seventeenth-century England', *Isis* **79**, pp. 373–404, on pp. 377–78.
39. Armytage, W. H. G. (1954–5), 'The Royal Society and the apothecaries 1660–1722', *Notes Records Roy. Soc.* **11**, pp. 22–37; Stewart, Larry (1992), *The Rise of Public Science. Rhetoric, Technology, and Natural Philosophy in Newtonian Britain, 1660–1750*, Cambridge and New York: Cambridge University Press, pp. 144–51.
40. Vigani, *Medulla Chymiae*, pp. 39, 44; 'Cours de chymie', University Library Cambridge MS Dd.12.53, p. 11. For preservation techniques see Cook, Harold J. (2002), 'Time's bodies: crafting the preparation and preservation of naturalia', in Smith, P. and Findlen, P. (eds.), *Merchants and Marvels: Commerce, Science and Art in Early Modern Europe*, London: Routledge, pp. 223–47, on pp. 223–8. For Bentley's garden see Monk, *Life of Bentley*, vol. 1, p. 204.
41. Gunther, *Early Science in Cambridge*, pp. 472–81; Peck, 'Vigani', pp. 41–8.
42. Barrera, Antonio, 'Local herbs, global medicines: commerce, knowledge and commodities in Spanish America', in Smith and Findlen, *Merchants and Marvels*,

pp. 163–81; Harris, Steven (1998), 'Long distance corporations, big sciences and the geography of knowledge', *Configurations* **6**, pp. 269–304, on pp. 287–93.

43. Peck, 'Vigani', pp. 46–8. For an exemplary north Italian mariner's cabinet with materials akin to those of Vigani, see Burnett, John (1982), 'The Giustiniani medicine chest', *Medical Hist.* **26**, pp. 325–33. For Newton's use of balsam see Iliffe, Rob, 'Isaac Newton, Lucatello professor of mathematics', in Lawrence, C. and Shapin, S. (eds.) (1998), *Science Incarnate: Historical Embodiments of Natural Knowledge*, Chicago: University of Chicago Press, pp. 121–55 on pp. 135, 152.

44. Albert, *Odeurs de Sainteté*, pp. 108, 124; Tribby, Jay (1991), 'Cooking (with) Clio and Cleo', *J. Hist. Ideas* **52**, pp. 417–39; Findlen, *Possessing Nature*, pp. 241–7, 268–87. For common viper techniques, see Lefèbvre, *Compleat Body of Chymistry*, second part, pp. 141–9.

45. Westfall, *Never at Rest*, pp. 281–301; Golinski, Jan (1988), 'The secret life of an alchemist', in Fauvel, J., Flood, R., Shortland, M. and Wilson, R. (eds.), *Let Newton be!*, Oxford: Oxford University Press, pp. 147–67, on pp. 151–2, 160. Geoghegan, D. (1957), 'Some indications of Newton's attitude towards alchemy', *Ambix* **6**, pp. 102–6, on p. 106.

46. Westfall, *Never at Rest*, pp. 191–2, 339. Notes of 'Seignior Vigani's Course of Chymistry', taken at Queens' College Cambridge from November 1705, now in Glasgow University Library MS Ferguson 62, fols. 76, 130, record Newton's invention of rectification apparatus. Newton's annotated copy of *Medulla Chymiae*, bound with Agricola, Georg (1614), *De animantibus subterraneis*, is in the Memorial Library, University of Wisconsin-Madison. For laboratory resources in Newton's mature chemistry see Dobbs, B. J. T. (1975), *The Foundations of Newton's Alchemy*, Cambridge: Cambridge University Press, pp. 213–25.

47. Thackray, Arnold (1970), *Atoms and Powers. An Essay on Newtonian Matter – Theory and the Development of Chemistry*, pp. 114–17. Cambridge, MA: Harvard University Press. For Mickleburgh on Hales, see Gonville and Caius MSS. 619/342, Day 2nd. This seems to be the second course.

48. Hales, *Vegetable Staticks*, p. xxvii; Allen and Schofield, *Stephen Hales*, pp. 39–40.

49. Underwood, E. Ashworth (1977), *Boerhaave's Men at Leyden & After*, Edinburgh: Edinburgh University Press, p. 128; Rusnock, Andrea (1996), *The Correspondence of James Jurin (1684–1750)*, Amsterdam: Rodopi, p. 13.

50. Cambridge University Archives, Original Grace Book θ, 689 (3 Aug. 1718).

51. Waller matriculated as a sizar in Bene't Hall in March 1690 [n.s.], receiving his M.A. in 1697 and his B.D. in 1705 (records of Corpus Christi College, Cambridge). He examined William Stukeley for admission to the college in 1703 (Stukeley, *Memoirs* I, pp. 20–22, 40).

52. Coleby, L. J. M. (1952), 'John Mickleburgh. Professor of Chemistry at the University of Cambridge, 1718–56', *Ann. Sci.* **8**, pp. 165–174, esp. pp. 165–167.

53. Gonville and Caius College MS. 619/342. John Mickleburgh lectures, fols. 1–2; see, for example, Burton, J. and Dack, R. in Underwood, *Boerhaave's Men*, pp. 166, 171.

54. Coleby, 'Mickleburgh', p. 171.

55. Coleby, L. J. M. (1952), 'John Hadley, fourth Professor of Chemistry in the University of Cambridge', *Ann. Sci.* **8**, pp. 293–301, esp. pp. 296–297. For Hadley on Hales, see Trinity College, Wren Library, John Hadley, MS, *An Introduction to Chemistry*, vol. I (1759), p. 15.

56. Schaffer, Simon (1994), 'Machine philosophy: demonstration devices in Georgian mechanics,' *Osiris* **9**, pp. 157–82; Golinski, Jan (1995), " 'The nicety

of experiment". Precision of measurement and precision of reasoning in late eighteenth-century chemistry', in Wise, M. Norton (ed., 1995), *The Values of Precision*, Princeton: Princeton University Press, pp. 72–91.

57. On burning lenses and degrees of heat see Mickelburgh, Gonville and Caius MSS. 619/342. Day 2nd; and Stewart, *Rise of Public Science*, p. 221. On fact production see Daston, Lorraine (1995), 'The moral economy of science', *Osiris* **10**, pp. 3–24, esp. pp. 16–17.

58. Coleby, 'John Hadley', p. 294.

59. Golinski, Jan (1983), 'Peter Shaw: chemistry and communication in Augustan England', *Ambix* **30**, pp. 19–29, esp. p. 21.

60. Coleby, 'Mickleburgh', p. 168.

61. Allen and Schofield, *Stephen Hales*, pp. 38–39; Stewart, *Rise of Public Science*, pp. 128.

62. Ihde, A. J. (1969), 'History of the pneumatic trough', *Isis* **60**, pp. 351–61.

63. Gascoigne, *Cambridge in the Age of the Enlightenment*, p. 289.

64. For Rouelle see Wren Library, Hadley MS, *An Introduction to Chemistry*, vol. I (1759), p. 9; and Beretta, Marco (1993), *The Enlightenment of Matter: the Definition of Chemistry from Agricola to Lavoisier*, Canton, MA: Science History Publications, pp. 130–2, 161–3.

65. See Bensaude-Vincent, Bernadette and Stengers, Isabelle (1996), *A History of Chemistry*, Cambridge, MA: Harvard University Press, pp. 61–2; Holmes, Frederick Lawrence (1985), *Lavoisier and the Chemistry of Life. An Exploration of Scientific Creativity*, Madison: University of Wisconsin Press, p. 24; On Nollet see Riskin, Jessica (2002), *Science in the Age of Sensibility. The Sentimental Empiricists of the French Enlightenment*, London and Chicago: University of Chicago Press, p. 74ff.

66. Hadley, *Introduction to Chemistry*, fols. 13–14; quoted in Coleby, 'John Hadley', p. 297.

67. Coleby, 'John Hadley', p. 296.

68. Millburn, John R. (1976), *Benjamin Martin: Author, Instrument-Maker and 'Country Showman'*, Leyden: Noordhoff International (Science in History series), pp. 35–8, 54–7.

69. See Golinski, Jan (1998), 'Utility and audience in eighteenth-century chemistry: case studies of William Cullen and Joseph Priestley', *Br. J. Hist. Sci.* **21**, pp. 1–31.

70. Watson, Richard (1781–1787), *Chemical Essays*, 5 vols., vol. 2, Cambridge: Cambridge University Press, pp. 39–40. Musson, A. E. and Robinson, Eric (1969), *Science and Technology in the Industrial Revolution*, Toronto: University of Toronto Press, pp. 167–170. Coleby, 'Mickleburgh', p. 115.

71. Fox, Robert (1998–99), 'Diversity and diffusion: the transfer of technologies in the Industrial Age', *Trans. Newcomen Soc.* **70**, pp. 185–96, esp. p. 186.

72. Quoted in Musson and Robinson, *Science and Technology in the Industrial Revolution*, p. 136.

73. Cf. Thackray, *Atoms and Powers*, p. 115.

74. See Levere, Trevor and Turner, Gerard l'E. (2002), *Discussing Chemistry and Steam. The Minutes of the Coffee House Philosophical Society 1780–1787*, Oxford: Oxford University Press.

75. Porter, Roy (1992), *Doctor of Society. Thomas Beddoes and the Sick Trade in Late-Enlightenment England*, London and New York: Routledge, pp. 158–60.

76. Birmingham Central Library, James Watt Papers, Beddoes to Watt, 21 April (1796?). A draft of his lectures is in Bodleian Library, Oxford, MS. Dep. C. 134/1.

77. Porter, Roy (2000), *Enlightenment. Britain and the Creation of the Modern World*, London and New York: Allen Lane, p. 427 ff.
78. Golinski, Jan (2003), 'Chemistry', and Stewart, Larry (2003), 'Global Pillage', in Porter, Roy (ed.), *The Cambridge History of Science*, vol. 4, *Science in the Eighteenth Century*, Cambridge: Cambridge University Press; on classifying see Foucault, Michel (1970), *The Order of Things. An Archaeology of the Human Sciences*, London: Tavistock; Bowker, Geoffrey C. and Star, Susan Leigh (2000), *Sorting Things Out. Classification and its Consequences*, Cambridge, MA: MIT Press.
79. Parkes, Samuel (1815), *Chemical Essays, Principally Relating to the Arts and Manufactures of the British Dominions*, 5 vols., vol. 1, London: Baldwin, Cradock and Joy, pp. 97–8.

3

Richard Watson: gaiters and gunpowder

Colin Russell

Department of History and Philosophy of Science, University of Cambridge

Richard Watson was one of the most intriguing and colourful figures in eighteenth century British science.[1] Known as the fifth Professor of Chemistry at Cambridge, from 1764 to 1771, he was also active in the spheres of politics and religion. Committed to the Whig political interest he became a rare example of a chemical bishop (or perhaps an episcopal chemist). Yet despite his elevation within the church he retained a lively concern for chemistry, and was well known to be extremely interested in gunpowder. Since he also lived in politically explosive times, the phrase 'gaiters and gunpowder' is singularly appropriate to apply to the last thirty or so years of his life.

Watson was an exceedingly controversial man, both in his lifetime and afterwards. He died in 1816, and in the following year his son (another Richard) released his father's autobiographical notes as two volumes of *Anecdotes*.[2] Whether this was wise subsequent events were to show. The immediate publication of these posthumous attempts at autobiography produced a wild variety of reactions in the press.

The *Quarterly Review* was probably the most vitriolic.[3] The *Anecdotes* were described as the results of 'more than twenty years collecting and concentrating intellectual poison', and as 'the fruits of an indolent and unlearned retreat from the duties of two important functions, the dignities and emoluments of which this prelate continued to enjoy till his death'. Watson himself was a 'squalid chemist who sallied forth from his furnace to blacken all that was elegant and ornamental in ancient literature'. Personalising its attack even further, the *Review* remarked of Watson that 'of taste he never had a tincture', and throughout his life he displayed 'a total want of delicacy'.[4]

The 1702 Chair of Chemistry at Cambridge: Transformation and Change, ed. Mary D. Archer and Christopher D. Haley. Published by Cambridge University Press. © Cambridge University Press 2005.

The *Quarterly Review* was a Tory publication, never over-kind to science in any case. But it was not untypical of some contemporary reactions to Watson. Most notable of these was a chapter in Atkinson's book, *The Worthies of Westmorland*.[5] The *Edinburgh Review* was a shade kinder, condemning 'the most extravagant vituperation from narrow-minded politicians'.[6] It wrote of Watson's 'high academical reputation . . . sustained by his valuable literary performances, extended by the firm and manly independence of his character as a politician, and his liberal and tolerant principles as a churchman'.

This journal had strong Whig sympathies.[7] So had Richard Watson and this was the reason for the venom displayed by the *Quarterly Review*. Described as 'a Whig of the straitest class', Watson accepted Locke's view that 'ultimate authority lies with the people'.[8] He supported repeal of the Test and Corporation Acts and was bitterly opposed to George III for his advocacy of war against the American colonists. He also opposed the King's growing influence on Parliament[9] and even joined the radical London Association (in 1780).[10]

Watson's birthplace was Heversham, Westmorland, a small village a few miles south of Kendal and situated on a headland overlooking the estuary of the River Kent. He was born in August 1737, and christened in the parish church on 25 September.[11] His father was the Rev. Thomas Watson, who was headmaster of the village Grammar School, remarkable for its excellence. Other pupils included another future ecclesiastic, William Preston, bishop of Ferns and Killala, Ireland, and Ephraim Chambers, of *Dictionary* and *Encyclopaedia* fame.[12]

This school, like several others in the area, had links with Trinity College, Cambridge. Following his father's death in 1753, Richard was sent there as a sizar, the poorest class of student. Having received only £300 from his father, he showed great eagerness knowing that 'my future fortune was to be wholly of my own making'. He worked at Hebrew, Greek, Latin, mathematics, natural philosophy and moral philosophy.[13] He claimed the remarkable feat of staying in the College for two years and seven months without leaving it 'for a single day'. Then, in May 1757, Watson visited his only brother (a curate in Kendal), but returned soon. By September he had decided to make Trinity 'the mother of my fortunes'.[14] He was a conspicuous member of the College:

> The narrowness of the young man's means obliged him to make his appearance in Cambridge in the then rustic dress of his native county, where his blue woollen stockings and homespun coat procured him the name of the Westmorland phenomenon.[15]

His resolve was soon rewarded with a scholarship (1757), and he became Second Wrangler in 1759 (though with a characteristic zeal for his own reputation he

Figure 3.1. Wood-engraving of Richard Watson, from a portrait by George Romney.

claimed the senior moderator had favoured one of his own private pupils for the First Wranglership).[16]

By now it was possible for Watson to supplement his income by becoming private tutor to two men of the year below him. One of these was John Luther, a young man with strong Whig sympathies who was to become a lifelong friend and Whig MP for Essex.[17]

In 1760 he was appointed Fellow of Trinity College, his M.A. following two years later. He became Moderator (examiner) and as such introduced the reform of examining candidates of similar class together instead of those from the same College.[18] A rather darkened portrait of Watson by George Romney overlooks a staircase at the College; Figure 3.1 shows an engraving of this portrait, used to illustrate the *Anecdotes*.

Shortly after his appointment as a Fellow of Trinity came Watson's first big break.

The fifth Professor of Chemistry

For some time Watson had become increasingly bored with the endless round of examinations and discussions on the subjects of the classical syllabus. An opportunity for change came with the death in 1764 of John Hadley, the fourth Professor of Chemistry at the University. None of his predecessors was particularly remarkable, though Hadley seems to have given an excellent course of lectures.

Watson decided that he would apply for the vacant chemical chair. However, there was one minor drawback: Watson wrote, 'I knew nothing at all of chymistry, had never read a syllable on the subject, nor seen a single experiment in it, but I was tired with mathematics and natural philosophy'.[19] Our own sense of impropriety at such a situation derives from the modern expectation that all professors (and lecturers too) should be experts in their subjects, and at the cutting edge of research. No such considerations applied until well into the nineteenth century, as Sedgwick discovered in 1818. When appointed to the Cambridge chair in geology he confessed that he 'knew absolutely nothing' about that subject, in contrast to a rival applicant who knew much but 'it was all wrong'![20]

Accordingly, the Chancellor of the University, the Duke of Newcastle, was informed by his chaplain that Watson 'offers himself as a candidate for the professorship of chemistry. As there is no stipend attached, I believe he will meet with no opposition. He is a man of reputation, and purposes reading lectures'.[21] The prediction was correct, and Richard Watson was unanimously elected by Senate on 19 November 1764. There was indeed no salary, but a room was made available (probably the same Queens' Lane laboratory used by Mickleburgh and Hadley, although Watson may also have experimented in the garden of his private house in St Andrew's Street, where a charcoal furnace remained into the twentieth century).[22]

However, Watson did not rest there. He would try to acquire a salary. Through his friend John Luther, now MP for Essex[18] and a man of rising influence, Watson petitioned the Chancellor that a salary be attached to the chair (1766). Newcastle regarded Watson highly and applied five times to the Marquis of Rockingham, who was then Whig Prime Minister in the coalition which lasted only a year.[23] No reply was received and Watson bluntly if tactlessly accused each of blaming the other. The Duke was deeply shocked, and sent the petitioner with his letter direct to the Prime Minister. His boldness paid off, and £100 was granted from government funds. Later this was converted into an endowment, and it remained the incumbent's salary for many years to come.[24]

The life of a professor

To overcome his complete ignorance of chemistry Watson sent to Paris for an 'operator' (named Hoffmann). From him, and from the *Manuel de Chymie* (1763) of the Parisian pharmacist Antoine Baumé, he learned quickly.

> I buried myself as it were in my laboratory, at least as much as my other avocations would permit; and in 14 months from my election I read a course of chemical lectures to a very full audience.[25]

The moment then came for the Professor to appear in public, and he began to read chemical lectures 'to very crowded audiences' each November.

> For months and years together I frequently read three public lectures at Trinity College, Cambridge, beginning at 8 o'clock in the morning; I spent 4 or 5 hours with private pupils and 5 or 6 more in my laboratory every day.[26]

Some idea of the content of those lectures may be gained from a summary Watson published a few years later. These *Lectures in Chemistry* were dedicated to Rockingham. Not many chemical syllabi can have been dedicated to a former Prime Minister, but this one was, in gratitude for his provision of a stipend for the chair.[27] The booklet (for that is all it was) included tables of salts and the famous *Table of Affinities* by E. F. Geoffroy (who had been a teacher of Baumé). The contents reveal a characteristically eighteenth century list of substances arranged in their classical 'kingdoms':

- Introduction (history of chemistry; operations etc.)
- Minerals
- Metals
- Mineral waters (including bitumen)
- Vegetables
- Vegetable juices
- Fermentation
- Animals

While occupying the chair of chemistry Watson continued with university examining. In a speech of 1766 he advocated annual examinations for all students (including the privileged classes of fellow-commoners and noblemen).[28] He also began writing what was to become his most famous and enduring work, his *Chemical Essays*.[29] Although the first volume did not appear until ten years after he had resigned his professorship and the final (and fifth) volume until 1787, he drew heavily on the work he had done as Professor of Chemistry and later. Thus the second volume was the first publication of his *Institutiones Metallurgicae*, originally written in 1768.

Watson's *Chemical Essays*, which we shall meet frequently in the next pages, was dedicated to the Duke of Rutland, a Whig peer who had once been his student and had helped him substantially in his later career. The volumes became a standard textbook of the late eighteenth century and were widely read in English and in a German translation. Thus the famous Dr Samuel Johnson, who said 'If you are to have but one book with you upon a journey, let it be a book of science', is known to have been interested; whilst en route through Bedfordshire in 1781, Johnson was 'chiefly occupied in reading Dr Watson's second volume of *Chemical Essays* which he liked very well'.[30] The *Essays* were also read by engineers such as Telford, and by budding chemical writers such as Samuel Parkes, who published five volumes of his own *Chemical Essays* in 1815, as well as his more famous *Chemical Catechism* of 1806.[31] James Parkinson, author of *The Chemical Pocket-Book* and of *Organic Remains of a Former World*, sent Watson a copy of the latter on its publication in 1804, writing that the bishop's *Chemical Essays* 'first attracted my attention earnestly to chemistry'.[32] Perhaps the most notable tribute came from Sir Humphry Davy who said he 'could scarcely imagine a time or a condition of the science, in which the Bishop's *Essays* would be superannuated'.[33]

Of course such a work was bound to receive negative comments from those who disapproved of the author or else misunderstood its intent. Examples of the former are amply provided by Atkinson, for instance in his comment that 'the man of science will hunt in vain for new or grand ideas'.[34] Much later it was remarked that the *Chemical Essays* were 'out-of-date before published' as they failed to mention Priestley's pneumatic apparatus or Scottish work on heat.[35] But other aspects of Priestley's work are mentioned, and in any case such comments entirely miss the main point of the books, as we shall see later.

One other notable event marked Watson's early years in the chemical chair. As he noted, 'My constitution was ill-fitted for celibacy and as soon, therefore, as I had any means of maintaining a family I married'. That was in 1773. His bride was Dorothy Wilson, a member of the wealthy squirearchy whose home was Dallam Tower, near Heversham. The marriage was a happy one, and produced a large family. 'She has been everything I wished her to be', wrote Watson, and the marriage lasted for over 40 years.[36] One additional reason for bliss was doubtless the annuity she brought with her of £150.[37]

A story in wide circulation relates to a time when Watson was living with his in-laws at Dallam Tower. Fond of walking in the area he encountered a hill known then as Helm Crag and today as the Helm, a few miles north and from one angle presenting a conical, almost volcanic, profile. Discovering on the top of this hill a quantity of cindery material Watson brought some home and entertained the admiring company at dinner that night with his discovery that the Helm was an extinct volcano. The following day, however, he was respectfully approached by the butler, who remarked:

> Dr Watson, Excuse me, but I thought I heard you say at the table yesterday that Helm Crag was an extinct volcano. I don't know what an extinct volcano is, as I never saw one, but I do know that when I was a lad, my father and I had a blast furnace on Helm Crag, and that's a piece of cinder from that very spot.

Money is said to have then changed hands and the 'discovery' was hopefully forgotten.[38]

Chemical researches

Watson's chemical researches were not extensive, but were sufficient to mark him out as an experimenter whose opinions carried some weight, and novel enough to contribute towards Fellowship of the Royal Society in 1769. His individually published papers fall into five classes.

First there were studies of *solutions and salts*, determining the specific gravities of saturated salt solutions. He was concerned about the possibility, mentioned by Stephen Hales, that air might exist in salts, and he disproved the theory that the volume of water was not changed when salts dissolve in it.[39] Following some exceptionally cold weather in Cambridge in February 1771, he tried to discover its effect on salt solutions. It was apparently Watson who made the first experiments on the freezing points of salt solutions, including $NaCl$, KNO_3, $MgSO_4$, $FeSO_4$, NH_4Cl, Rochelle salt and sugar. He showed that the 'resistance to congelation [lowering of the freezing point] is directly proportional to the quantity of salt dissolved'.[40] He thus anticipated Blagden's discoveries seventeen years later.[41]

A second achievement at this time was the invention of the *black bulb thermometer*. By coating a thermometer bulb with Indian ink he noted that exposure to direct sunlight led to an increase in the thermometer reading of 10 °F.[42]

Perhaps Watson's most impressive researches were in the study of metals and their ores. Some of these were reported to *Philosophical Transactions* and others were treated in detail in the *Chemical Essays*. At an early stage he made extensive studies of the lead industry in Derbyshire. One interesting discovery was that the colours in lead 'pellicles' after smelting were removable by tin or zinc (presumably by reduction of the thin films of oxide). However for Watson the formation of such colours 'may be explained from what has been advanced by Sir Isaac Newton'.[43]

Observing the large-scale operations he suggested that in a lead smelter the sulphur-rich vapours might be condensed instead of being lost into the atmosphere. He described in detail a Derbyshire furnace for preparing litharge from lead by oxidation,[44] adding that the increase in weight on calcination may be due to 'a large portion of the air' being absorbed by the lead.[45] He described the manufacture of white lead (using acetic acid and CO_2)[46] – and warned against the popular use of lead compounds as cosmetics![47]

Watson also described the smelting of zinc at Bristol, and the production of copper by the process

$$CuSO_4 + Fe \rightarrow FeSO_4 + Cu$$

Again aware of the economic and environmental disadvantages of waste he suggested recovery of the iron (II) sulphate. In order to obtain malleable iron he noted that it was better to flux cast iron with charcoal than with coke.[48] He also investigated 'onchalum', a kind of brass resembling gold.[49]

A fourth area of concern was the *sulphur wells* of Harrogate spa. Long after he vacated his chemical chair he went to the expense of having several new wells dug. Speaking of the odiferous vapours emitted he asked, 'Does this air, and the inflammable air separable from some metallic substances, consist of *oleaginous* particles in an elastic state?' He did not know then that Bergman held the same opinion, and called it 'hepatic air' (H_2S).[50]

Finally, he made important studies of the *pyrolysis of coal*. This, he found, gave rise to inflammable vapours.[51] He took 96 oz. of Newcastle coal and heated them in an earthen retort: 'during the distillation there was frequent occasion to give vent to an elastic vapour'. Measuring loss in weight as 28 oz., he concluded this to be in the same proportion as that observed by Stephen Hales in 1726. Altogether this was a much more sophisticated research than that of Clayton in 1688. Nor was gas the only product. He was aware of a patent of 1782 for extracting tar from coal in Bristol.[52] Once again he suggested a recovery process, this time for volatiles. In this respect he was proceeding along the same lines as the 9th Earl of Dundonald.

Clearly Watson did much chemical work both as the fifth incumbent of this chemical chair and afterwards. It is interesting to gain a view of his chemical philosophy. What was it that led him on to experiment and write for several decades, and which considerations gave shape to those researches?

Chemical philosophy

Watson's chemical philosophy was marked by several characteristics. First, there was a *strong empiricism*. Experimental techniques have a prominent place in his lectures and *Essays* and he was clearly a careful experimenter. One example out of many must suffice. He was concerned to know whether the old view was tenable that water may be converted into earth. On this topic he provided a detailed history, but then went on to make his own experiments. He found that water distilled in a silver retort does leave a slight tarnishing on a polished silver plate. In the light of this carefully staged experiment he concluded that Lavoisier's demonstration that the earth came only from the [glass] vessel is not completely convincing. He was wrong in his conclusions about Lavoisier, of course, but that does not deny Watson the right to be regarded as a careful and dedicated experimentalist.

In parallel with his emphasis on experiment came a deep scepticism towards theoretical schemes in general. Watson was not against theory as such and cheerfully speculated on a number of topics. Thus he devoted half a volume of his *Essays* to salt,[53] supposing that it was originally created by the seas, a view later shared by Thomas Henry.[54] He suggested (as did Black and others) that air would turn to liquid or even solid if very strongly cooled.[55]

What troubled Watson was the possibility of an over-arching scheme that threatened to dominate, even to inhibit, free experimentation. He was a self-confessed follower of Francis Bacon, who, like the alchemists, would 'call upon men to sell their books, and build furnaces'.[56] Thus he remained uncommitted to both phlogistonists and their rivals. It was as a Newtonian that he could not accept negative weight, so for him the most that could be said of phlogiston was that it is a power, not a substance.[57] He was aware of Priestley and his work, but made no reference to Lavoisier's elements.

A third feature of Watson's chemical philosophy is that the science had to have some practical use. The whole thrust of his *Chemical Essays* is to show how useful chemistry could be, particularly in an industrial context – as the quotation on p. 48 in the previous chapter illustrates. He spent much time visiting sites of industrial importance, and through his subsequent experiments was able not merely to describe what happened but also to suggest how it could be improved. Apart from the metallurgical work mentioned above he also described in detail the manufacture of copperas (ferrous sulphate), alum, sal ammoniac etc., and was especially interested in the saltpetre trade and the composition of gunpowder. On the basis of careful meteorological observations he suggested that evaporation of brine would be more efficient if it were dispersed over suspended cloths.[58] Watson is a leading figure in what may be called the 'other' chemical revolution, the one concerned with industrial practice rather than elaborate theories. With hindsight we know that both were necessary, but in the words of those pioneer historians of this chemical revolution, Archie and Nan Clow, there is 'a distinct feeling that industry was in a more receptive state for the new ideas than was purely philosophical chemistry'.[59] He was one of the first actively to promote this view in England, though Davy's 1802 discourse on galvanism at the Royal Institution was long ago described as 'one of the earliest and most remarkable examples of the constructive belief in progress',[60] and one of Watson's successors in the chemical chair, William Farish, also tended to this view of chemistry.

Finally, a fourth aspect of Watson's chemical philosophy was that, above all, *chemistry must be communicated.* It was not an esoteric science suitable only for a small elite. This is why, long after he had vacated the Cambridge chair, he continued to publish and why the *Chemical Essays* sold so well. This is also why he strongly advocated the foundation of an Agricultural College where science could be taught to farmers and landowners.

Regius Professor of Divinity

Watson's tenure of the chemical chair, though attended by 'seven years of most brilliant success', was to be brought to a dramatic conclusion.[61]

In 1771 the Regius Professor of Divinity at Cambridge, Thomas Rutherforth, suddenly and unexpectedly died. This was a most prestigious position, and one that Watson confessed 'had long been the secret object of my ambition'.[62] So he applied for the Chair, possibly because his chemical studies had raised clerical censure, but chiefly because its financial reward far exceeded the £100 received by the Professor of Chemistry. From an estimated £330 the salary was eventually raised to £1000.[63] It was not such an outrageous transition of subject, as Rutherforth himself had written books on natural philosophy. But Watson had no theological qualifications and needed a D.D. Such were academic values in late eighteenth century Cambridge that he was able to acquire the degree through the King's mandate within seven days, 'by hard travelling and some adroitness'! He was asked to answer two tricky theological questions and to submit a dissertation on a third within a fortnight. He did so and was duly elected.[64]

Watson's theology has its own fascination, but in this chapter there is need to refer only to one aspect of it. This is the relationship, if any, between the two Watsons: the chemist and the divine. Two metaphors may suffice to describe that relationship, the first from twentieth century physics and the second from nineteenth century chemistry.

Complementarity

Watson seems to have viewed his two interests, in chemistry and in theology, as complementary ways of approaching nature. There is not the slightest evidence that he saw any incongruity in his change of career, nor did the University. Admittedly at one stage he burnt his writings in chemistry lest he be lost to the church altogether, but he soon resumed his chemical studies in parallel with those in theology.[65] In one scientific paper in the *Philosophical Transactions* he claimed that chemical study 'gives to the mind of any piety the most pure and sublime satisfaction'.[66] It is sometimes said his pious remarks were mere convention, the caricature of Watson shown in Figure 3.2 illustrating how his detractors perceived tensions between his chemical, political and religious roles. However, there is no logical reason why such views should not be sincerely held. In fact the evidence points to that conclusion. For him science and divinity were complementary ways of viewing our world.

In addition to managing two careers Watson speaks of 'two books' – of nature and of Scripture and sees them as complementary accounts, not as opponents. The 'two books' metaphor came, of course, from Francis Bacon. Today some people may find it strange in the light of the 'conflict thesis' – that religion and science are always at war. However, only when that now discredited thesis is itself seen as a myth constructed in the late nineteenth century, rather than a serious historical generalisation, does it become possible to accept that in this respect at least Richard Watson did not have a problem.

Isomorphism

A further striking relation between Watson's science and religion is that ideas in one area were often reflected in the other. This is not in the least surprising on a holistic view of human nature, and is paralleled time and again in many modern studies of scientific biography. Many of Watson's chemical and religious ideas had similar 'shapes', reminiscent of isomorphous crystals. We may see this in at least four different ways.

In the first place Watson's ideas in theology as in chemistry were marked by a *strong empiricism*. Just as Bacon had advocated free enquiry based on experiment so he (and Watson) held it necessary to go to the data provided, to observed facts rather than tradition. That faith was embedded in the eighteenth century 'Latitudinarianism' to which Watson and many Whigs subscribed (he was in fact one of the best-known Latitudinarians in late eighteenth century Cambridge).[67] Latitudinarianism was a legacy of the earlier eighteenth century, and was partly a reaction to religious strife and partly a response to the

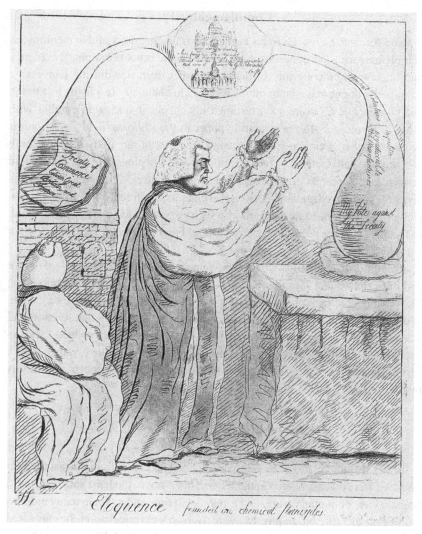

Figure 3.2. *Eloquence founded on Chemical Principles*, a caricature of Richard Watson by James Sayers.

'scientific' spirit of enquiry inspired by Newton and his successors. It steered a middle course between what were later to become 'high' and 'low' church parties in the Church of England. A passionate belief in the value of free enquiry was at its heart. In science the relevant data come from experiment; in Watson's religion they came from what he regarded as the recorded facts of history. So it was on the basis of historical evidence that he accepted the Biblical miracles and the resurrection of Christ.

The second illustration is a corollary of the first: his well-known *dislike of systems*. Watson saw that some systems of theology were as pernicious as some systems of natural philosophy (like Aristotelianism). As he put it, 'Systems in Theology have as much obstructed the progress of Revealed Truth as Systems in Philosophy [= science] have done that of Natural Truth'.[68] So he opposed 'rash expositors of points of doubtful disposition, intolerant fabricators of metaphysical creeds, and incongruous Systems of Theology'.[69] His first steps as Regius Professor were in keeping with that approach:

> I reduced the study of divinity into as narrow a compass as I could, for I determined to study nothing but the Bible, being much unconcerned about the opinions of councils, fathers, churches and bishops and other men as little inspired as myself.[70]

Though formally similar to the Bible-based theology of the rising evangelicals in England, this approach was primarily an expression of Enlightenment intellectualism and lacked both their warmth and zeal. Watson's disdain for patristic theology (and thus of the catholic faith) more closely resembled that of Samuel Clarke. Though not a Unitarian himself Watson remained indifferent to controversies about the Trinity. He sought to abolish the Athanasian Creed and proposed a thorough revision of the Thirty-nine Articles and of the liturgy 'from a genuine and intellectual desire to retain the adherence of the educated laity'.[71]

One aspect of his rejection of spurious authority raised cries of 'shame!' from the Tory press.[72] His open-mindedness presented a remarkably early example of denominational tolerance. He placed no emphasis on episcopacy as a badge of continuity and did not object to changing the present form of the Church of England 'From Episcopacy to Presbyterianism'.[73] He even refused to support the Society for the Propagation of the Gospel because it tried to convert dissenters to Anglicanism. At first suspicious of Roman Catholicism he moved towards greater toleration by the 1790s. He was in favour of Catholic Emancipation and a Catholic (as well as an Anglican) church of Ireland. He summarised his position thus:

> Now my mind was wholly unbiased: I had no prejudice against, nor predilection for, the Church of England; but a sincere regard for the *Church of Christ*, and an insuperable objection to every degree of dogmatical intolerance.[74]

His last ten words could have been equally applied to chemistry.

A third aspect of Watson's chemical and religious philosophy was that their *study should be useful*. Watson's sermon on 'The wisdom and goodness of God in having made both Rich and Poor' (1793) was a good example of the rhetorical use of theology for social purposes. He did not, however, indulge in the kind of natural theology so famously promoted by his colleague William Paley. But the form and content of his extensive theological output suggest a strong desire to influence people towards a more open-minded Christianity, and a more open society.

A final illustration of this isomorphism lies in his desire to promote *communication in theology*, as he did in science. His literary output was considerable, and some of his pamphlets went through many editions. One of his most successful theological works was his *Apology for Christianity* (1776). This was a response to the historian Edward Gibbon, who afterwards said Watson was 'the most candid of adversaries; whom I should be happy to call my friend, and should not blush to call my antagonist'.[75] Twenty years later he produced his *Apology for the Bible* in response to Thomas Paine's *Age of Reason*.

In 1781 Watson became ill, having, it is said, 'an inveterate disease, the fruit perhaps of his chemical operations'.[76] After a few years he declined to fulfil any duties associated with the Regius Professorship, but continued to draw full salary for the next 30 years. In 1787 he appointed a deputy (at a proportion of his own salary) and removed permanently from Cambridge to his native Westmorland.[77] His behaviour caused much ill feeling in the University.[78] Meanwhile, other developments were afoot.

Bishop of Llandaff

Gaiters

In 1782 the Duke of Grafton, Lord Privy Seal, wrote to the Home Secretary Lord Shelburne. He proposed that Richard Watson be appointed as Bishop of Llandaff. Through the combined influence of Grafton and the Duke of Rutland (one of Watson's former pupils) the bid was successful.

This appointment had little to do with religion but was essentially political. Bishops were expected to promote the cause of the party that had appointed them, or of the succeeding Government if different. As Dr Johnson wrote on Good Friday, 1775:

> No man can now be made a bishop for his learning or his piety; his only chance of promotion is his being connected with someone with a parliamentary interest.[79]

Thus a hope was expressed that Watson would not only speak for the Whig interest in the House of Lords (of which he now automatically became a member), but also that he 'would occasionally write a pamphlet for their administration'.[80] It is significant that 1782 was the only year for over nearly half a century in which the Whigs were solely in power. An opinion attributed to David Starkey, that 'The Church of England has nothing to do with Christianity. It's the English Shinto', gains considerable credibility when applied to the eighteenth century.

So Watson became a regular visitor to London. He had an address in Great George Street, and mixed with the great and the famous at parties given by Sir Joseph Banks and others.[81] Later he seems to have gone to London every other year.[82] Although now a bishop he continued to draw his professorial salary, and retained all except one of the preferments associated with the chair. These included two churches in Shropshire, two in Leicestershire, two in Llandaff and three in Huntingdonshire. For all these he had resident curates at about half his salary. As bishop he received income from seven more livings as 'appropriations'. Altogether he now had an income of at least £2000 p.a. from the Church. As we have seen, he had been appointing deputies since 1787.

Scandalous as these arrangements may seem to us today, there was a further aspect of Watson's life that raised howls of protest even in his own time. After his illness he resolved to move to the less stressful conditions of his native county. Aided by his considerable salaries, his wife's fortune and above all a legacy from his friend John Luther (who died in 1786), he bought an estate in Westmorland between Troutbeck and Ambleside and there built himself a splendid mansion which he called Calgarth Park (Figure 3.3). Here he spent much of the rest of his life. He lived neither in Cambridge nor in Llandaff, and thus earned himself the doubtful reputation of being the most notorious absentee bishop of the century.

Watson's absenteeism has become as famous as his chemistry and deserves some examination. With hindsight we can see that regular attendance at Llandaff could lead to difficult and even dangerous adventures. At that time travel in and to Wales presented huge difficulties, even in summer. And there were language problems everywhere except in the few big towns. Moreover he knew full well that he was expected to attend the House of Lords, which he did fairly regularly at first, and bishops could only visit their dioceses during recesses. But there was more to it than that.

Llandaff was not the attractive cathedral city it is today, for the cathedral had long been in ruins, and attempts at restoration, which had started in 1721, had extended only to the choir, presbytery and part of the nave. The rest was still ruinous in 1782. There was no bishop's house – let alone a palace – and there seemed ample reason for remaining in the Westmorland home he had by then established. Moreover, the salary for Llandaff was the lowest for all bishoprics, a mere £550 p.a. The diocese was 'the Cinderella of all brides', and during the eighteenth and early nineteenth centuries was 'at its very lowest point' and 'often used as a stepping-stone to less impoverished sees'.[83]

Alas for Watson, promotion never came. He had hoped for translation to St Asaph, Carlisle and especially to Durham (where the income was £6000 p.a.), but in vain. The reasons are simple. As we have seen, bishops were expected to spend much time in the House of Lords, upholding party line by speeches and

Figure 3.3. Calgarth Park, Westmorland.

voting. If not, further preferment was impossible. But Watson was not a party man, and ever since the coalition of 1784 he had become estranged even from the Whigs. It was said to him 'they will never make *you* an archbishop, they are afraid of you'.[84] Strangely enough, during a brief coalition in 1807 it had been agreed to make him Archbishop of York as soon as the present incumbent died.[85] Unfortunately for Watson the latter outlived the Whig coalition and his chances of preferment were lost for ever. For the rest of his life the Whigs were out of power. He blamed George III for allowing him 'to remain through life worse provided for than any bishop on the bench'.[86] However, William Pitt was his real opponent, gaining 'the merit of opposing an unyielding *No* to the many applications by Watson for higher preferment'.[87] The Bishop of Llandaff seethed with smouldering resentment, as his *Anecdotes* so sadly demonstrate.

His notoriety as an absentee bishop was widely spread in his own time and has been uncritically repeated ever since. Thus it has been alleged that he had only been once in his diocese, or that for 34 years he forgot it existed.[88] Writers have claimed that 'It may be safely asserted that, from 1782 to 1816 he was not within it [the diocese] more than *ten* times'.[89] His most uncharitable critic, Atkinson, even lambasted him as 'one of the cankerworms of the church'.[90]

Yet there is evidence that Watson has been unjustly maligned. In fact diocesan records tell a rather different story. We know of nine triennial visitations and also of ordinations in 1784, 1785, 1787, 1788, 1790, 1791, 1795, 1804, 1805, 1809, 1811, 1813, also in 1802, 1807, 1808, 1810, 1812.[91] These may not be all. Moreover he made an extensive tour of his diocese in 1809, being the first diocesan to confirm at the new industrial town of Merthyr Tydfil, for which he had a difficult journey over mountains at the age of 72.[92]

A recent historian of the church in Wales has given a more measured, revisionist view. He claims that Watson 'spent the summer months of each year in his diocese and performed all his canonical duties'.[93] Another has argued that although 'no bishop was more notorious for non-residence than Richard Watson' yet 'he performed his episcopal duties with a diligence with which he has not been credited'.[94] He was actively concerned with the education of clergy and diligent in his instructions for confirmation candidates (though these instructions were rather long and not in Welsh!). He was, in one opinion, 'the first Bishop of Llandaff to talk reform',[95] and as good a bishop as any from 1679 to 1849.

Watson's whole career illustrates a particular evil of that time: patronage. It was, of course, this system that gained him Llandaff in the first place. But without friends in the right places no man could continue to succeed, so, being a sturdy independent, he was stranded at Llandaff for life. And it must be said that he extracted further benefit from the system, being himself a patron of many clergy, including twelve associated with Llandaff Cathedral. Most of these incumbencies were filled by cronies or relatives from Westmorland.[96] In Watson's case 'It must be admitted that no such flagrant case of the abuse of patronage as his can be found to parallel it . . . Nothing can excuse his signal and wilful failure as a bishop of the church'.[97] His career also illustrates the decrepit state of the Anglican Church where some degree of absenteeism was perfectly acceptable; seven thousand of the eleven thousand clergy did not live in their parishes.[98] It is small wonder that men like Wesley and Whitefield felt it necessary to by-pass the established church in their efforts to reach the masses, while a little later in Cambridge Charles Simeon confronted the same issues while staying within the Anglican Church.

Gunpowder

Even though Watson did go to Llandaff more frequently than he has been credited with, it is clear that most of his life was centred on Calgarth. Here he belonged to a community that included the author Thomas de Quincey, the Lake poets Wordsworth and Southey, and other notables. He was, according to de Quincey, 'a joyous, jovial and cordial host'.[99] One day the bishop's son, also Richard, planned to bring his cock for a fight at Elleray (home of Christopher North). The bishop's daughter arrived with her father to announce a picnic next day by the lake, the guests to include the Wordsworths, the blind naturalist John Gough (John Dalton's former tutor) and the Sedbergh surgeon-mathematician John Dawson.[100] In 1805, Watson welcomed Sir Walter Scott to the Lakes; twenty years later Scott visited his grave.[101]

Watson's time in Calgarth was spent 'principally building farm-houses, blasting rocks, enclosing wastes, in making bad land good, in planting larches, and in planting in the hearts of my children principles of piety, of benevolence and of self-government'.[102] His rock-blasting activities attracted much attention and reflected his earlier interest in gunpowder, and it was this that brought him right back to chemistry.

Gunpowder was a huge problem for Britain in the Napoleonic Wars, and it had become painfully obvious in naval warfare that British gunpowder was inferior to that used by France. Ships of the Royal Navy were in danger of being fired upon by the French before the enemy came within their range. In 1787 the Duke of Richmond (head of the government's Ordnance Department) approached the unlikely figure of the Bishop of Llandaff for advice as to possible improvements. Watson responded positively, identifying the problem as the method of preparing the charcoal. Traditionally it had been made by slow combustion of piles of wood with air excluded as far as possible. Now, however, Watson advised distilling the wood in closed vessels. His improved gunpowder was tested at the firing ranges at Hythe and 'the improvement has exceeded my utmost expectation'. A 68 lb. ball was propelled 273 feet by powder made from the new charcoal, while under identical conditions except for the use of 'traditional' charcoal, the distance was a mere 172 feet. Further experiments confirmed that the cylinder powder effected an improvement of $100:60$.[103] Later work showed even greater improvement.

Watson's charcoal owed its improvement to several causes. One of the variables in traditional methods of manufacture was *temperature*. The higher the temperature the higher the proportion of carbon, and also the H/O ratio. But if the temperature is too high, the carbon is too dense and has a small surface area.[104] On the other hand, charcoal made at too low a temperature tends to be hygroscopic. There was also the variable of *time*: the traditional method may take up to a week; in retorts it may be as low as 8 hours.[105] Then there is *absence of air*: clearly less charcoal will be burned if oxygen has restricted access. Finally the traditional method includes *impurities*: varying quantities of earthy matter were always present in the 'pit' method. To eliminate (or at least

Figure 3.4. Charcoal cylinders as gateposts in Gatebeck, Cumbria.

reduce) these needed a high degree of control, and that can be accomplished by heating in cylinders under standard conditions.

Following Watson's advice, the government set up two cylinder charcoal works in Sussex, and others followed in some of the Cumbrian gunpowder manufactories, first at Sedgwick on the River Kent[106] and then at Lowwood.[107] At Waltham Abbey the use of *cylinders* was early recommended by Major William Congreve but not introduced till 1794.[108] Six charcoal cylinders were moved from Sedgwick when it closed in 1850 to Gatebeck where they were used till about 1865. Two of them survive as the gateposts to a modern caravan park shown in Figure 3.4.

At Faversham, one set of three cylinders could produce 2500 barrels of gunpowder per annum. This improvement was stated by an Ordnance Department official to have saved the country £100 000 a year, though Watson himself put it later (1806) at £50 000.[109] This, of course, ignores the actual effects of the improved explosive in decisive naval battles later in the war. One military historian observed: 'Watson must surely rank as a man of greater practical usefulness than any other who ever sat on the English bench of bishops'.[110]

Even that was not all. In 1794, the year after war with France was declared, the East India Company's saltpetre warehouse in London was destroyed by fire – possibly by sabotage. Since it probably contained nearly all the saltpetre in Britain, and replenishment from East India ships was slow, stringent economies were introduced, such as a ban on the firing of salutes, etc. Then, a letter from Watson to Isaac Milner in 1796 noted that the French were using ammonia to make nitre, and suggested that England should use the same process, which Milner himself had discovered.[111] Whether this advice was followed is not certain.

Gunpowder itself was also symbolic of Watson's inveterate tendency to improve on nature. This he attempted not merely by rock-blasting but also by introducing new plants and applying the latest ideas of scientific agriculture. He advocated burning local limestone with coppice wood, rather than lime expensively brought in from Kendal,[112] urged extraction of the oil formed in charcoal-making[113] and in 1792 introduced New Leicester sheep into Westmorland.[114] Above all he engaged in extensive afforestation. He particularly advocated introduction of the larch, a tree that had been introduced to Britain *c*. 1620 but not into the Lake District. In 1804–5 he made a large plantation consisting of 325 500 larches 'on two high and barren mountains' (Birch Fell and Gummer's How, at the south end of Windermere). He planted over 100 acres of high ground near Ambleside where 'firs and larches, but especially the larches, thrive as well as he could wish'.[115] It was said that his 'agricultural improvements here merit the attention of every stranger, and the imitation of every Westmorland farmer'.[116] He even wrote 'Preliminary observations' to Pringle's survey of Westmorland in 1797.[117] As ever, his commitment to utilitarian aims led him on. He argued that recovery of wasteland in Britain would ensure that 'no inhabitant of this island will be driven by distress, to seek a substitute in Africa or America'.[118] Prophetically he argued that 'The waste lands in this and other counties are a public treasure in the hands of private persons'.[119]

To all of this many objected. Wordsworth, who 'for many years had systematically abused larches and larch-planters',[120] called Calgarth 'a vegetable manufactory'.[121] A local guidebook also complained:

> Beneath us, in a marshy bottom, stood the heavy edifice of Calgarth House [*sic*], the residence of the Bishop of Llandaff; a station so unhappily selected as to exclude every interesting view of the enchanting scenery that surrounds it.[122]

Others, however, applauded, including the early feminist writer Harriet Martineau:

He built the house, he planted the woods, and he blessed the whole neighbourhood by planting the hills around, so that the Calgarth woods are the glory of the district.[123]

Elsewhere she describes the view from Miller Brow:

The Calgarth woods, for which we are indebted to Bishop Watson, rising and falling, spreading and contracting below, with green undulating meadows interposed, are a perfect treat to the eye.[124]

Another well-known guide lavished high praise:

Calgarth Park is now the property of Dr Watson, Lord Bishop of Llandaff, who has built an elegant mansion thereon, which, with other improvements in that fine situation, makes it one of the most elegant places of residence in the country.[125]

while another commentator wrote that Watson

has added greatly to the natural beauties of the estate [Calgarth] by adorning it with an elegant mansion, by a judicious management of his woods and young plantations, and by improved modes of agriculture.[126]

Not all of Watson's suggested improvements were successful. He proposed to drain Windermere to gain *c.* 100 acres of arable land from it. He thought better of the scheme when he realised that it would ruin tourism, lower average income and therefore reduce the rents obtainable.[127] The wrath of the residents was somewhat mollified by the following story which rapidly spread to their great amusement; Watson acquired extensive property in the area, including a hostelry in Ambleside known as *The Cock*.[128] The proprietor, from respect to his new landlord, renamed it *The Bishop*, with appropriate signboard. Unfortunately for him a competitor set up in business across the street and gave his new inn the discarded name and sign of *The Cock*. Customers were confused, and patronised the establishment with the familiar name. Upon this the landlord of *The Bishop*, while retaining his picture of the worthy prelate, placed under it the unfortunate – but accurate – legend 'This is *The Old Cock*.' The citizens of Ambleside were much amused.

Watson remained at Calgarth until he died, on 4 July 1816. His body was interred at Windermere churchyard, where the horizontal gravestone is sadly eroded, hardly legible and cracked throughout its length. A memorial tablet on the east wall of the south aisle in the church has fared better. In Llandaff the work of its absentee Bishop is chiefly recalled in folklore. Figure 3.5 shows a reproduction from a catalogue of a portrait of Watson by Joshua Reynolds; the original is believed to hang in the Cuban National Museum in Havana.[129]

Figure 3.5. Portrait of Richard Watson by Joshua Reynolds R. A.

More tangible memorials may also be found elsewhere, as in the iron cylinders once used to make charcoal and in the alien trees introduced into the
Lake District from his time onwards. For Watson led the way in afforestation
in Westmorland. Recognition still has to be made for the contribution of 'his'
gunpowder to the victories of British naval warfare from the 1790s. Most valuable to chemists, however, must be the permanent tradition he established of
popularising and disseminating chemistry and of stressing its role in industry.

Despite the information that has recently come to light, Watson remains an enigmatic figure. As one historian observed, 'In all he did there was a curious combination of right ends and wrong means, of real insight and great obtuseness'.[130] Truly a man of his times, he partook of their good and bad characteristics. Yet he was marked by a sturdy independence. As one of his students was told, 'Your tutor is a man of perseverance, not to say obstinacy'.[131] He was one of 'the hard progeny of the North' who were to enrich British science for a century and more. They learned the habit of independent thought, a determination to improve material conditions, a certain discipline of mind, and a love of nature from their hard upbringings in the dales, hills and towns of the North West in particular.

Acknowledgements

For much assistance I wish to thank the library staff of Cambridge University and the Open University; also the Librarians of Trinity College, Cambridge and the Department of History and Philosophy of Science at Cambridge University. In connection with the story of Watson's later life I am most grateful to staff at the Cumbria Record Office in Kendal, the Kendal Public Library, the Armitt Library at Ambleside, and the Forestry Commission at Grizedale (Cumbria) and Alice Holt (Farnham, Surrey).

Notes and References

1. There are many short biographical notes on Watson, including *D.N.B., D.S.B.*, and many books such as Atkinson's *The Worthies of Westmorland* (note 6), or, less contentiously and more recently, Musson, A. E. and Robinson, E. (1969), *Science and Technology in the Industrial Revolution*, Manchester: Manchester University Press, pp. 167–71. In the periodical literature, see Hankinson, A., 'Bishop Richard Watson', *Cumbria Life*, 1994 (Jan.), pp. 26–8; Coleby, L. J. M. (1953), 'Richard Watson, professor of chemistry in the University of Cambridge, 1764–71', *Ann. Sci.* **9**, pp. 101–23. The fullest treatment so far, chiefly from the viewpoint of social history, is Brain, T. J. (1982), 'Some aspects of the life and works of Richard Watson, Bishop of Llandaff, 1737–1816', Ph.D. thesis, University of Wales.
2. Watson, R., jr. (ed.) (1817), *Anecdotes of the Life of Richard Watson, Bishop of Llandaff; written by himself at different intervals, and revised in 1814*, London: Cadell; 2nd edn in 1818. In the following notes these are simply referred to as *Anecdotes*, the second edition being used.
3. Anonymous review of Watson's *Anecdotes*, *Quarterly Review*, 1817, pp. 229–53.
4. *Ibid.*, p. 232 and p. 249.
5. Atkinson, G. (1849–50), *The Worthies of Westmorland*, London: J. Robinson.
6. Watson, R. (Jr.) (1818), 'Anecdotes of the Life of Richard Watson, Bishop of Llandaff; written by himself at different intervals, and revised in 1814', *Edinburgh Review* **30**, pp. 206–34.
7. *The Edinburgh Review* was a Whig paper which had long 'regarded the English universities with supercilious superiority' [Ward, W. R. (1965), *Victorian Oxford*, London: Cass, p. 17].

8. See Sykes, N. (1934), *Church and State in England in the XVIIIth Century*, Cambridge: Cambridge University Press, p. 339.
9. Re Geo. III, see *Edinburgh Review*, 1818, pp. 217–19.
10. Baylen, J. O. and Gossman, N. J. (1979), *Biographical Dictionary of Modern British Radicals*, Hassocks: Harvester Press, p. 515.
11. Registers of Heversham Parish Church.
12. Bingham, R. K. (1984), *The Church at Heversham*, Milnthorpe: privately published.
13. *Anecdotes* I, p. 16.
14. *Anecdotes* I, pp. 14–16.
15. *Kendal Chronicle*, 1816.
16. *Anecdotes* I, p. 29.
17. Namier, L. and Brooke, J. (1985), *The House of Commons, 1754–1790*, London: Secker & Warburg, vol. II, pp. 63–4.
18. *Anecdotes* I, p. 30.
19. *Anecdotes* I, p. 46.
20. Clark, J. W. and Hughes, T. (1890), *The Life and Letters of the Reverend Adam Sedgwick*, Cambridge: Cambridge University Press, vol. I, pp. 160–1.
21. Talbot, W., chaplain to Duke of Newcastle, in Sykes (1934), p. 334.
22. Mann, F. G. (1957), *Proc. Chem. Soc.* **7**, p. 190. Watson's house was situated where Llandaff Chambers – named in his honour – now stands.
23. Sykes (1934), p. 336.
24. *Ibid.*, p. 335.
25. *Anecdotes* I, p. 46.
26. *Anecdotes* I, p. 53.
27. Watson, R. (1771), *A Plan for a Course of Chemical Lectures by Richard Watson, D.D., F.R.S., and Regius Professor of Divinity*, Cambridge: J. Archdeacon.
28. *Anecdotes* I, p. 48.
29. Watson, R. (1781–7), *Chemical Essays*, London: T. Evans, 5 vols (and subsequent editions).
30. Boswell, J. (1966), *Life of Johnson*, London: Oxford University Press, pp. 1155–6.
31. Musson, A. E. and Robinson, E. (1969), *Science and Technology in the Industrial Revolution*, Manchester: Manchester University Press, p. 75 and p. 136.
32. *Anecdotes* II, pp. 198–9.
33. de Quincey, T. [1834 and 1840] (1948), *Recollections of the Lake Poets*, edited with an introduction by Sackville-West, E., London: John Lehmann, p. 73.
34. Atkinson (1849–50), p. 214.
35. Golinsky, J. (1992), *Science as Public Culture: Chemistry and Enlightenment in Britain, 1760–1820*, Cambridge: Cambridge University Press, p. 53.
36. *Anecdotes* I, p. 71.
37. Wilson papers, Kendal Record Office, D11/104.
38. Atkinson (1849–50), pp. 215–6.
39. Watson, R. (1770), 'Experiments and Observations on Various Phaenomena attending the Solution of Salts', *Phil. Trans.* **60**, pp. 325–54.
40. Watson, R. (1771), 'Some Remarks on the Effects of the late Cold in February last', *Phil. Trans.* **61**, pp. 213–20.
41. Blagden, C. (1788), 'Experiments on the Effects of various Substances in Lowering the Point of Congelation in Water', *Phil. Trans.* **78**, pp. 277–312.
42. Watson, R. (1772), 'Account of an Experiment made with a Thermometer whose bulb was painted black and exposed to the direct Rays of the Sun', *Phil. Trans.* **63**, pp. 40–1.
43. Watson, R. (1778), 'Chemical Experiments and Observations on Lead Ore', *Phil. Trans.* **68**, pp. 863–83.

44. *Chemical Essays* III, p. 339.
45. *Ibid.*, p. 346.
46. *Chemical Essays* II, p. 362.
47. *Chemical Essays* III, p. 365.
48. *Chemical Essays* II, p. 344.
49. Watson, R. (1785), *Mem. Proc. Manchester Lit. & Phil. Soc.* **2**, pp. 47–67.
50. Watson, R. (1786), *Phil. Trans.* **76**, pp. 171–88.
51. *Chemical Essays* II, p. 347.
52. *Chemical Essays* III, p. 7.
53. *Chemical Essays* II, chapters 2, 3, 4 and 5; see p. 109.
54. Smith, R. A. (1883), *A Centenary of Science in Manchester*, London: Taylor & Francis, p. 113.
55. Watson, R. (1771), *Essay on the Subjects of Chemistry, and their General Division*, Cambridge, p. 3.
56. Bacon, F. [1605], 'The Advancement of Learning', in Markby, T. (ed.) (1856), *The Two Books of Francis Bacon*, London: J. W. Parker and Son, p. 64.
57. *Chemical Essays* I, p. 167.
58. *Chemical Essays* II, p. 57.
59. Clow, A. and Clow, N. L. (1952), *The Chemical Revolution*, London: Batchworth Press, pp. 381–2.
60. Woodward, L. (1962), *The Age of Reform, 1815–1870*, Oxford: Oxford University Press, 2nd edn, p. 564.
61. *Edinburgh Review*, 1818, p. 212.
62. *Anecdotes* I, p. 56.
63. *Anecdotes* I, p. 162.
64. *Anecdotes* I, pp. 60–1; however, Watson gave a 'scurrilous' speech of disapproval when Thomas Dampier was so mandated [University of Cambridge Library Add. MS. 5867, f.145, cited in Winstanley, D. A. (1935), *Unreformed Cambridge: A Study of Certain Aspects of the University in the Eighteenth Century*, Cambridge: Cambridge University Press, p. 89].
65. Clow and Clow (1952), p. 601.
66. Watson, R. (1786), *Phil Trans.* **76**, pp. 171–88 (pp. 187–8).
67. Walsh, J., Haydon, C. and Taylor, S. (1993), *The Church of England, c. 1689–c. 1833*, Cambridge: Cambridge University Press, p. 231.
68. Watson, R. [attrib.] (1790), *Considerations on the Expediency of Revising the Liturgy and the Articles of the Church of England* 'by a consistent Protestant', London: Cadell, p. 51.
69. Watson, R. (1785), *Collection of Theological Tracts*, Cambridge, vol. I, *Preface*.
70. *Anecdotes* I, pp. 62–3.
71. Sykes (1934), p. 422.
72. *Quarterly Review* (1817), p. 238.
73. Sykes (1934), p. 355.
74. *Anecdotes* I, p. 39.
75. *Anecdotes* I, p. 100.
76. *Quarterly Review* (1817), p. 230.
77. *Anecdotes* I, pp. 258–9.
78. Winstanley, D. A. (1935), *Unreformed Cambridge: A Study of Certain Aspects of the University in the Eighteenth Century*, Cambridge: Cambridge University Press, p. 108.
79. Lean, G. (1980), *God's Politician: William Wilberforce's Struggle*, London: Darton, Longman & Todd, p. 71.

80. Shelburne to Grafton, cited in *Anecdotes* I, p. 153.
81. Musson and Robinson (1969), p. 1214.
82. Sykes (1934), p. 65.
83. Newell, E. J. (1902), *Llandaff*, London: SPCK, p. 206. One of the successors of the notorious Bishop Anthony Kitchin of Llandaff commented that he was merely the Bishop of Aff since Bishop Kitchin had taken all the land.
84. *Ibid.*, p. 211.
85. de Quincey [1834 and 1840] (1948), pp. 96–7.
86. *Anecdotes* I, p. 439.
87. Overton, J. H. and Relton, F. (1906), *The English Church from the Accession of George I to the End of the Eighteenth Century (1714–1800)*, London: Macmillan, p. 283.
88. Cornish, F. W. (1910), *The English Church in the Nineteenth Century*, part I, London: Macmillan, p. 101.
89. Atkinson (1849–50), p. 222.
90. *Ibid.*, p. 203.
91. Subscription Books, Diocesan Registry, Cardiff, and R. Watson, Private List, cited in Sykes (1934).
92. *Anecdotes* I, pp. 368–71.
93. Davies, E. T. (1962), *The Story of the Church in Glamorgan, 560–1960*, London: SPCK, p. 77.
94. Walker, D. (ed.) (1976), *A History of the Church in Wales*, Penarth: Church in Wales Publications, Penarth, pp. 118–19.
95. Davies (1962), p. 77.
96. Guy, J. R. (1977), 'Bishop Richard Watson and his Lakeland friends', *Trans. Cumberland and Westmorland Antiquarian and Archaeological Society* 77, pp. 139–44.
97. Overton and Relton (1906), pp. 260–2.
98. Lean, G. (1980), *God's Politician: William Wilberforce's Struggle*, London: Darton, Longman & Todd, p. 71.
99. de Quincey [1834 and 1840] (1948), p. 76.
100. Rawnsley, H. D. (1894), *Literary Associations of the English Lakes*, Glasgow: J. MacLehose, pp. 83–4.
101. *Ibid.*, pp. 74–5.
102. *Anecdotes* I, p. 389.
103. *Anecdotes* I, p. 241–3.
104. Lewes, V. B. and Brame, J. S. S. (1906), *Service Chemistry*, 3rd edn, London and Greenwich: Glaisher, p. 273.
105. At Faversham, 580 lb of wood in each of three cast-iron cylinders was heated for 8 hours.
106. See Patterson, E. M. (1995), *Blackpowder Manufacture in Cumbria*, The Faversham Society, 2nd impression, p. 9. Manufacture of gunpowder in Cumbria greatly increased in the eighteenth century, partly because of the abundance of coppice wood for the charcoal, and partly because of the considerable water-power to drive the mills. See also Tyler, I. (2002), *The Gunpowder Mills of Cumbria*, Keswick: Blue Rock Publications.
107. Cocker, G. (1988), *The Lowwood Gunpowder Works*, Cartmel: R. E. Harvey, p. 6.
108. Buchanan, B. J. (1998–9), 'Waltham Abbey Royal Gunpowder Mills: The Old Establishment', *Trans. Newcomen Soc.* 70, pp. 221–50; Gray, E., Marsh, H. and McLaren, M. (1982), 'A short history of gunpowder and the role of charcoal in its manufacture', *J. Mat. Sci.* 17, pp. 3385–400.
109. *Anecdotes* I, pp. 240–2, and II, p. 282.

110. Glover, R. (1963), *Peninsular Preparations: The Reform of the British Army, 1795–1809*, Cambridge: Cambridge University Press, pp. 68–9.
111. Partington, J. R. (1962), *A History of Chemistry*, vol. iii, London: Macmillan, p. 344; Isaac Milner was President of Queens' College and had been the first Jacksonian Professor of Chemistry at Cambridge, as Chapter 4 details.
112. Garnett, F. W. (1912), *Westmorland Agriculture 1800–1900*, Kendal: Titus Wilson, p. 54.
113. *Ibid.*, p. 258.
114. *Ibid.*, p. 145.
115. Pringle, A. (1797), *A General View of the Agriculture of the County of Westmoreland*, Newcastle, p. 278.
116. Housman, J. (1816), *A Descriptive Guide to the Lakes, Caves and Mountains and Other Natural Curiosities in Cumberland, Westmorland, etc.*, 7th edn, Carlisle, p. 204.
117. Pringle (1797), pp. 245–6. Watson estimated the area of the county by cutting a sheet of paper around a trace obtained from a map, and then comparing its weight with that of a rectangular sheet representing a known area. His figure of 540 100 acres was incorporated into government statistics.
118. *Ibid.*, p. 248.
119. *Ibid.*, p. 252.
120. de Quincey [1834 and 1840] (1948), p. 304.
121. *Dictionary of National Biography*, article 'Watson, Richard'.
122. Travis, B. (1806), *A Description of the Lakes of Cumberland and Westmoreland in the Autumn of 1804*, London, p. 61.
123. Martineau, H. (2002), 'Diary for June 21, 1845', in Todd, B., *Harriet Martineau at Ambleside*, Carlisle: Bookcase, p. 93.
124. Martineau, H., *Guide to Windermere*, Windermere: Garnett, 2nd edn [c. 1884], p. 15.
125. West, T. (1807), *A Guide to the Lakes of Cumberland, Westmorland and Lancashire*, Kendal: W. Pennington, p. 62.
126. Britton, J. and Brayley, E. W. (1814), *The Beauties of England and Wales*, vol. 15, *Westmorland*, London: J. Harris, p. 216.
127. Letter from Miss Weeton to Mrs. Chorley, September 5, 1810, in Hall, E. (ed.), *Miss Weeton: Journal of a Governess, 1807–1811*, London: Oxford University Press, 1936, p. 288.
128. Carnie, J. M. (2002), *At Lakeland's Heart – Eighteen Journeys into the Past of Ambleside and its Locality from Rydal to Clappersgate*, Windermere: Parrock Press.
129. From Mannings, D. (2000), *Sir Joshua Reynolds: A Complete Catalogue of his Paintings*, New Haven: Yale University Press.
130. Cragg, G. R. (1960), *The Church and the Age of Reason, 1648–1789*, Harmondsworth, Middlesex: Pelican Books, p. 170.
131. *Edinburgh Review*, 1818, p. 209.

4

Lavoisier's chemistry comes to Cambridge

Christopher Haley and Peter Wothers

Department of Chemistry, University of Cambridge

At the end of the eighteenth century, chemistry underwent a revolution. In 1787, Lavoisier, Fourcroy, Berthollet and de Morveau published their *Méthode de Nomenclature Chimique*, followed two years later by Lavoisier's *Traité élémentaire de Chimie*. The consequences of these two publications are still apparent today. Such familiar terms as oxygen, hydrogen, sulphate, sulphide, and so on, are now ingrained in many languages. Lavoisier's chemistry not only gave the world a new nomenclature, but one which focused attention on the elemental composition of substances and which helped forge chemistry into a coherent international discipline. From this time, chemistry took a new direction.

Lavoisier's central doctrine concerning the combustion and the calcinations of metals released chemists from Becher and Stahl's phlogiston theory, the problems of which had become increasingly evident throughout the century.[1] In retrospect, the adoption of Lavoisier's chemistry by other chemists both in France and abroad might seem inevitable, but it was not necessarily immediate, nor a simple model of either unresisted diffusion or irrational resistance. Britain in particular seemed rather slow to accept the French ideas.[2] To accept even the new nomenclature was to accept the new French theories, and for some British chemists, such as the staunch phlogistonist Priestley, this was never acceptable. British textbooks of the time reflect this reluctance. For instance, in the first edition of his *Dictionary of Chemistry* (1795), Nicholson states that:

> In the controversy respecting Phlogiston, or the nature of Combustion, not a little remains to be done. I have asserted under the latter article, that the doctrine which rejects phlogiston, or a common inflammable matter, appears to me to be much more simple, and consequently probable. But I have not adopted the nomenclature

The 1702 Chair of Chemistry at Cambridge: Transformation and Change, ed. Mary D. Archer and Christopher D. Haley. Published by Cambridge University Press. © Cambridge University Press 2005.

of the Antiphlogistians. We are so continually misled by words that it would, no doubt, be of great advantage if a consistent and uniform nomenclature were generally adopted. The French nomenclature, though not without its faults, appears to be more perfect than any other which has been offered: but I did not think myself at liberty to anticipate the public choice, by using it in an elementary work.[3]

It was, as may be expected, the French books that opened the eyes of British chemists to Lavoisier's work. By 1802, Lavoisier's textbook had been through five English editions, but the ideas contained within the book were put to immediate use in the popular texts of the Frenchmen Fourcroy and Chaptal. It is through such books that the new chemistry came to Cambridge.

Isaac Pennington

Interestingly, the New Chemistry of Lavoisier was first taught in Cambridge not by the Professor of Chemistry, but by another. However, before coming to this, we should first describe the events following Richard Watson's resignation of the Chair of Chemistry in 1771, following his election to the chair of divinity. This resignation prompted a struggle for control. The way in which Watson had used the Chair of Chemistry as a step towards fame and fortune had not gone unnoticed, and the stipend that he had procured from the Crown made the position now far more attractive in its own right. In all previous elections to the Chair, candidates had effectively stood unopposed, either through a genuine lack of competition, or because informal soundings and the desire to save face had led to a gentlemanly agreement between potential candidates. Following Watson's resignation, however, five candidates expressed an interest, leading to a direct contest.[4] Moreover, the two favourites were the junior dean of Trinity, William Hodson, and a rising star from the rival college of St John's, Isaac Pennington (1745–1817), shown in Figure 4.1.

Since neither the Trinity nor the Johnian candidate was prepared to withdraw, the Master of Trinity, John Hinchliffe, argued that the new professor should be elected by popular vote.[5] St John's resisted this move, hoping perhaps that they might prevail upon the influence of the Professor of Physic, Russell Plumptre, who had a significant hand in the election of John Hadley, as the Grace of Hadley's election indicates.[6] Pennington was well known to Plumptre through Addenbrooke's Hospital: as a prospective physician, Pennington was a regular attendant of the weekly hospital governors' meetings from 1771 onwards, which brought him into close contact with Plumptre in his capacity as a governor and physician to the hospital.[7] Both Trinity and St John's may thus have believed that Pennington would be Plumptre's preferred choice for the chair, and that Plumptre would use his influence to secure Pennington's election.

Figure 4.1. Isaac Pennington (1745–1817).

Despite St John's protestations, in November 1773 Trinity eventually succeeded in their bid to change the procedure.[8] The resulting ballot, held that
December, 'was as much contested an election, and brought as many people
together and from as great a distance, as had been known'.[9] Nevertheless,
Pennington ultimately defeated Hodson by 148 votes to 129 and was thus
declared the sixth Professor of Chemistry at Cambridge.[10]

Regrettably, Pennington's life is largely uncharted, due in part to a paucity of surviving documents. It is known that he was born in 1745 near Furness Fell in Lancashire.[11] The son of a merchantman captain, the young Isaac was educated at Sedburgh School before being admitted to St John's as a sizar in August 1762.[12] As an undergraduate he evidently displayed some interest in natural philosophy, for in 1766 the college 'agreed to allow Pennington fifteen pounds a year for the care of the Observatory and for making observations to be delivered to the Master & Seniors'.[13] One may imagine that, as a sizar, Pennington was compelled by financial reasons to volunteer for such duties, but there was nevertheless some prestige attached to these tasks: the eighteenth century was a vibrant time for astronomy within the colleges, and at St John's there was particular interest in the 1760s. Built atop the Shrewsbury tower in the college, the St John's observatory had been founded in 1765 under the direction of the astronomer (and butler of Pembroke College) Richard Dunthorne.[14] In 1766, therefore, the observatory generated much interest within college.

The following year Pennington graduated as thirteenth wrangler, with Richard Watson, then still Professor of Chemistry, being one of the examiners.[15] Shortly afterwards, Pennington was elected a Fellow of St John's, whereupon he turned his attention towards medicine. From whom he learned the subject is not clear, although he would almost certainly have attended Plumptre's lectures. What is clear is that, from 1771 onwards, Pennington was a regular attendant at Addenbrooke's Hospital, which had opened on Trumpington Street a few years earlier in 1766.[16] There being a limited number of positions for physicians and surgeons within the hospital, Pennington was not actually appointed a physician there until March 1773, when a position became vacant through the retirement of Robert Glynn (1716–1800).[17]

The extent to which Pennington's work at Addenbrooke's Hospital was connected with his appointment to the 1702 Chair a few months later is still uncertain, but whatever the influence of the Regius Professor of Physic, it seems highly probable that Pennington's medical interests swayed the election in his favour; whilst Watson – and, to an extent, Hadley – had promoted the more industrial applications of chemistry, the medical associations of the subject nevertheless remained strong at this time. Unfortunately, Pennington's own views about the purpose and applications of chemistry may remain a mystery, since despite the fierce contest for the Chair, there is no evidence to suggest that Pennington ever lectured.[18] After all, chemistry was not yet examinable, nor was there any formal requirement for the Professor to teach. Indeed, treating a stipendiary professorship simply as a sinecure was not uncommon throughout the eighteenth and nineteenth centuries.[19] In this manner, Pennington appears to have considered the post simply as a means to support his medical studies at Addenbrooke's and his various duties within St John's.

Isaac Milner and the Jacksonian chair

However, we should not imagine that under Pennington's twenty-year reign, chemistry found no outlet within the university. The year after Pennington's election to the Chair of Chemistry, the Senior Wrangler was a student of Queens' College, Isaac Milner (1750–1820), shown in Figure 4.2. Born near Leeds, and educated in part by his elder brother Joseph (who had earlier obtained a sizarship at Catharine Hall, now St Catharine's College), Milner was admitted to Queens' as a sizar in 1770. He so impressed the 1774 examiners that he was declared '*incomparabilis*' among his year.[20] After taking holy orders and being ordained as a deacon, Milner was elected a Fellow of Queens' in 1776, and from 1778 he served as Rector of St Botolph's Church, near Queens' College.

Milner's interest in chemistry dates from around this time and was probably aroused through his membership of the 'Hyson Club', an elite society established by the senior wranglers of 1757.[21] The Club counted among their number the mathematician Edward Waring (1734–98) and the theologian William Paley (1743–1805), as well as Richard Watson. Milner certainly received some instruction in the subject from Watson, and the two 'maintained a correspondence of chymical topics' long after Watson's transition to the chair of divinity.[22]

Sometime around 1779, Milner started to conduct his own experiments, although these early trials were not without their hazards. As Milner's niece recalled, 'it was about this time, that, by incautiously inhaling some noxious gas, he laid the foundation of a serious pulmonary complaint, from which he never entirely recovered'.[23] Nevertheless, Milner grew more proficient, and by 1782 was reading public lectures in chemistry, apparently as Pennington's deputy.[24] These lectures were very well received, with one contemporary recalling that:

> Dr Milner was always considered as a very capital lecturer. The chemical lectures were always well attended; and what with *him*, and what with his German assistant, Hoffman, the audience was always in a high state of interest and entertainment.[25]

Milner rapidly became well known throughout the university. As a result of several mathematical papers communicated to the Royal Society, he was elected a Fellow of that Society in 1780, being sponsored by Waring among others; he acted as Junior Proctor in 1781 and, like Watson, served both as Moderator and

Figure 4.2. Isaac Milner (1750–1820).

Sir Robert Rede's Lecturer in Philosophy.[26] Milner also made himself known in other ways. Whilst his interest in science owed a debt to Watson and Paley, he diverged from them on matters of religion. From his pulpit in St Botolph's, he embarked upon an increasingly evangelical battle against the 'profane and licentious spirit of infidelity and irreligion', attracting the attention of a number of low Anglican Fellows in Cambridge, and subsequently earning the approval of the 'Clapham Sect' of London evangelicals.[27] This was furthered by the formation within the university of a Bible Society with the help of Milner's friend and fellow evangelical Charles Simeon (1759–1836), later vice-provost of King's College and a founder of the Church Missionary Society.[28]

An opportunity arose in 1783, when a bequest from Richard Jackson, a Fellow of Trinity, permitted the establishment of a new chair. Jackson had left the majority of his estate to provide for 'a Lecturer, Professor, or Demonstrator of Natural Philosophy . . . qualified by his knowledge in Natural Experimental Philosophy and the practical part thereof, and of Chemistry, to instruct the Students'.[29] The terms of the bequest permitted a liberal interpretation of natural philosophy as including 'Anatomy, Animal Economy, Chemistry, Botany, Agriculture or the Materia Medica'.[30] There was one stipulation: evidently troubled by a common disease possibly not unconnected with the indulgent reputation of Trinity's Fellows, Jackson specified that any holder of the chair should 'have an eye more particularly to that *opprobrium medicorum* called the gout'![31]

With the emphasis that the Jacksonian bequest laid on lecturing, Milner was an obvious candidate for the new position. In December 1783, he stood unopposed for the chair and was thus appointed the first Jacksonian professor. He dedicated himself to experimental chemistry, although by the terms of the bequest, each lecture course had to be different from the preceding year's, and so Milner alternated his course of chemistry with one in 'experimental philosophy', incorporating lectures on air pumps and steam engines.[32]

There was, however, no dedicated room for Milner's use. From 1782, he performed chemical experiments in the stable yard of Queens' College, but as this would not suffice as a permanent arrangement, a syndicate was formed to find a location in which a new lecture room could be built specifically for the Jacksonian professor.[33] In 1784, it was reported that: 'the Syndics appointed to erect a Botanical and Chemical Lecture Room [are agreed] that the piece of Ground at the South East corner of the Botanical Garden is a proper Spot'.[34] In fact, the syndicate had actually been formed for the purpose of finding accommodation for the Jacksonian and Botanical professors, but as the above quotation demonstrates, chemistry was now firmly associated with Milner rather than Pennington.

The Botanical Garden or Physick Garden had been established two decades earlier on the former site of the Austin Friars, following the bequest of Richard Walker (1679–1764), an avid botanist and Vice-Master of Trinity.[35] One of the purposes of the garden was to cultivate plants for medicinal use, and so there were obvious links with chemistry. Indeed, the first professor of botany, Richard Bradley (*c.* 1688–1732) lectured on many chemical subjects, including topics as seemingly unrelated to botany as metal smelting.[36] It was not too

Figure 4.3. The Schools in the Botanic Garden, later the New Museums Site.

incongruous, then, that the new professor of experimental philosophy should share quarters with the professor of botany. Moreover, one fifth of Jackson's bequest also provided an income for the head gardener of the Botanic Garden itself, and so it was deemed fitting that some of the funds be used to provide joint accommodation.[37] Thus in 1784, at a cost of some £1600, a building was constructed in the south-east corner of the Botanic Garden – the first seeds of what would later become known as the New Museums Site, the first central location for science in Cambridge. Figure 4.3 (and also the top drawing in Figure 6.4, shown later) illustrates this 1784 building, which provided a room for Milner and another for the professor of botany, Thomas Martyn, together with a central lecture theatre and small laboratory behind this for preparation and experiment.[38]

Milner's resources increased in other ways, too. The founder's bequest attached a stipend to the Jacksonian chair of £72 per annum, which Milner was able to raise by £100 p.a. by appealing to the Crown with the experienced help of Richard Watson.[39] However, with the grant of an additional stipend came a requirement that Milner attract at least a dozen students to his lectures.[40] Given that students were expected to pay a fee of three guineas to attend the course, Milner was under pressure to maintain the popularity of his lectures. In this he seems to have succeeded, for he wrote in 1791 that the 'number of pupils has of late years . . . increased to about 150'.[41] Perhaps inevitably, however, this need for continual 'interest and entertainment' was seen by some as flippancy. The diarist Henry Gunning recalls that in Milner's lectures on natural philosophy:

> He did not treat the subjects under discussion very profoundly, but he contrived to amuse us, and we generally returned laughing heartily at something . . . His experiments in Optics were very little more than exhibitions of the Magic Lanthorn on a gigantic scale . . . I cannot say that I benefited much by my attendance on these lectures.[42]

Milner's chemical lectures were better received. 'These, I understood from persons much better qualified than myself to judge of their merits, were very excellent', recorded Gunning.[43] Milner himself was at pains to point out that his course 'was *truly* an experimental one & required many hours to be spent every day in preparation; that is, in the adjusting of instruments & the management of chemical processes'.[44]

As for his theoretical views, Milner was initially a phlogistonist; whether he completely embraced Lavoisier's chemistry in later years is not clear. He used books by Macquer and Boerhaave, both phlogistonists, but he is also known to have been influenced by Black, who in turn 'subscribe[d] to almost all Mr Lavoisier's doctrines'.[45] A greater proportion of Milner's manuscripts are devoted to the doctrine of phlogiston, but he did also teach Lavoisier's chemistry. However, it is not clear whether Milner was just putting phlogiston into historical perspective with later lectures (now lost) devoting more space to Lavoisier's ideas, or whether he simply still had more faith in the old ideas. In the copy of his Jacksonian Lectures deposited at Trinity College, Milner writes:

> We now proceed to consider the Opinions of philosophical Chemists concerning Phlogiston – an intricate Subject of great importance & extensive application! What part of Chemistry is there that does not in some way or other involve the doctrine of Phlogiston? It is true, the very existence of it has lately been doubted, but that does not render a knowledge of the arguments by which its existence is supported less necessary.[46]

Milner goes on to state what exactly was meant by phlogiston and how this concept evolved over time:

> From observing that the residuum of an inflammable body was no longer capable of Inflammation, Chemists were led to conclude that during this process something was lost. This something, they denominated Phlogiston or the principle of Inflammability. The Ancient Chemists indeed do not pronounce clearly concerning the Identity of the principle of Inflammability; that is they do not appear to have formed an opinion, that Phlogiston was always the same thing in different Inflammable bodies. They rather speak of it as a different thing; yet with a want of

precision, which forbids us to conclude that they had any decided Sentiments on this head. As oils & sulphurs are remarkably inflammable, they sometimes call it oil & sometimes sulphur and seem to use these terms indiscriminately.

But modern Chemists, and among these particularly the illustrious Stahl, describe the Phlogiston of all inflammable bodies to be one and the same thing. Neither oils, nor Sulphur are the phlogiston, though they contain it in abundance: They contain it but not in its simple state; it is united in each body to a peculiar basis . . . This beautiful Theory has for many years been embraced by the most intelligent & philosophical Chemists and is supposed to rest on the most solid foundation.

Milner's 'modern chemists' were not Lavoisier and his followers, but rather Stahl, suggesting that he still clung to these earlier theories. After giving some of the experimental evidence for phlogiston, Milner described in his lectures the then-classic examples of the combustion of iron and zinc to give the calces (oxides), and the revivication of the metals by heating with charcoal. Of the reduction of calces using inflammable air (hydrogen), he comments:

These experiments proved most acceptable to the advocates for phlogiston, who now began to defend their doctrines by arguments that seemed direct and irresistible. They had been repeatedly ridiculed for believing the existence of a substance, which they could neither exhibit to the senses, nor demonstrate to be possessed of any one known property of matter. But this reproach seemed now entirely done away. The inflammable principle was capable of being weighed and measured like other bodies; it was confined in bladders and other vessels; and though it could not be represented as an object of sight, its existence seemed as certain as that of the Atmosphere we breath.

However, some new and unexpected discoveries concerning the nature of inflammable air enabled the Antiphlogistians to content this important conclusion with great plausibility and consistent arguments.

But before we proceed to give the history of these discoveries and the uses made of them, it will be proper to give a more particular account of the opinions of the Antiphlogistians to describe their mode of explaining the facts adduced in favour of Phlogiston, and to state the principal arguments upon which their opinions are founded.

Unfortunately, no copy of Milner's final Jacksonian lectures, detailing the opinions of the 'Antiphlogistians', seems to have survived, if indeed they were ever given.

Milner was credited by his peers with at least one novel development: a new means of producing 'nitrous acid', by forcing 'volatile alkali' through a red-hot gun-barrel filled with 'calx of manganese'.[47] This was published in the language of the phlogistonists (whilst admitting that for 'those who chuse to reject the doctrine of phlogiston . . . the reasoning will be much the same'), and the process would now be interpreted as the oxidation of ammonia to nitric acid by manganese dioxide. This reaction was potentially of great industrial and political significance. The importance of gunpowder at this time has been discussed in the previous chapter and, like many European countries, Britain depended heavily on colonial imports of potassium nitrate for its manufacture. This was an obvious weakness, leaving the country vulnerable to blockades or colonial disturbances. It was possible to produce potassium nitrate from nitric acid and potash, as Richard Watson had pointed out a few years earlier, but given that most nitric acid was itself produced from potassium nitrate, this was not much help.[48] However, with an independent method of producing nitric acid, the required nitrates could be produced domestically. For Milner to have discovered a method of doing this, wrote Watson, was 'a very high honour' indeed.[49]

In 1788, Milner was elected President of Queens' College, a position which earned him a further £200 per annum and enabled him to further his evangelical views and pursue his campaign against 'Jacobins and infidels'.[50] Soon afterwards, however, he 'suffered two distinct attacks of violent fever', apparently connected with his earlier lung complaint, which he claimed reduced him to an invalid and left him 'with almost constant pains in the head and intestines'.[51] The symptoms persisted, confining Milner to bed and forcing him to turn to opium to relieve the pain.[52] Milner, by then the leader of a significant group of Tory Evangelicals within the university, begged assistance from his political patrons.[53] In late 1791 he wrote to William Pitt, the Tory Prime Minister whom he knew through his close friend William Wilberforce, informing him that:

> These several strokes of distemper have now so shattered my constitution that there appears no probability left that I shall ever be able to undergo the fatigue of reading public lectures again.[54]

Pitt – who from 1784 was parliamentary representative for Cambridge University, and who, as Prime Minister, dispensed most ecclesiastical preferments – came to Milner's aid.[55] He appointed him Dean of Carlisle, although Milner requested to be excused the duties of the office on grounds of ill health.[56] Preferments such as this had long been used to provide income for distant or illegitimate offspring of royalty and it is likely that all concerned anticipated its being a sinecure for the ailing Milner, whose installation as Dean was conducted by proxy. Certainly, as soon as his income was provided for, Milner resigned the Jacksonian chair (1792), effectively marking the end of his chemical career. However, in keeping with his personal faith, he did not neglect his clerical duties entirely and, aided by Paley (who was archdeacon of the diocese from 1782 until 1804), he appears to have initiated something of an Evangelical revival within Carlisle.[57]

Shortly after his appointment to Carlisle, Milner was also elected Vice-Chancellor of the university, and so was presented with an opportunity to repay some of his debt to Pitt. As with his appointment as Dean, on accepting the position he pleaded ill health 'to attempt . . . to get me totally excused' from the more routine administrative obligations.[58] He did however use the position to promote his religious and political interests, as exemplified by his expulsion from the university of William Frend, a Fellow of Jesus College accused of prop-agating 'Socinian principles', whose trial has been described by Gascoigne as 'the great cause célèbre of late eighteenth century Cambridge'.[59] The Socinians were a rationalist religious sect which denied the virgin birth and the Holy Trin-ity and advocated the separation of church and state. As such, they threatened the orthodoxy of Anglican Cambridge. More generally, however, such radicals caused alarm to allies of the Pitt government, who were increasingly concerned by the course of the French Revolution. Fearing that 'English Jacobins' would incite similar turmoil at home, they saw religious reform as dangerously linked with social revolution. In suppressing radicals such as Frend, therefore, Milner was not only acting in accordance with his own faith, but also furthering the interests of his political friends and patrons.

Francis Wollaston and the Jacksonian chair

On Milner's resignation of the Jacksonian chair, two men put their name forward for consideration as his successor: Francis John Hyde Wollaston (1762–1823), the Senior Wrangler of 1783, and William Farish (1759–1837), the Senior Wrangler of 1778.

The junior of the two, Francis Wollaston, had entered Sidney Sussex College in 1779 from Charterhouse.[60] He came from a distinguished dynastic family which had strong connections both to science and to Sidney Sussex: his great-grandfather William (1659–1724) was well-known in his time as a religious philosopher; his grandfather Francis (1694–1774) was a Fellow of the Royal Society, as was his father, also called Francis Wollaston (1731–1815), his uncles George (1738–1826) and Charlton (1733–64), and several other relatives.[61] The career of his younger brother, William Hyde Wollaston (1766–1828), is discussed further in the next chapter. There were intricate connections between the family and other men of science: Francis' uncle George, for instance, was on intimate terms with William Paley and Richard Watson, whilst Henry Cavendish was a close family acquaintance.[62]

After graduating as Senior Wrangler, Francis Wollaston had been appointed Taylor Lecturer within Sidney Sussex, a position previously held by his uncle

George.[63] This was a position akin to a Fellowship in mathematics, but which – crucially, as it would later transpire – did not confer full status as a Fellow. Thus when in 1783 a Fellowship became vacant in Trinity Hall, Wollaston applied, was elected, and migrated to that college, where he was also employed as a tutor.[64]

Wollaston's activities between his election to a Trinity Hall Fellowship in 1783 and his election to the Jacksonian chair in 1792 are not clear, although it is interesting to note that this coincides with the period in which his brother William was a medical student at Caius. Francis was an examination moderator in 1788 and 1789, implying a measure of academic recognition, but he published nothing during this period to indicate a concern with natural philosophy. Nevertheless, he was elected F.R.S. in 1786 and given his family's activities would have been sufficiently well connected to find strong support for a bid as Milner's successor.

The other candidate, William Farish, was the son of a Carlisle clergyman. He attended Carlisle Grammar School, but a large part of his education was conducted by his father, then vicar of Stanwix (a post later occupied by William Paley).[65] The natural philosopher Benjamin Franklin and the chemist and physician William Brownrigg (1711–1800) both mention an 'ingenious' Reverend Farish of Carlisle who impressed them with his 'great learning and extensive knowledge in most sciences'.[66] It was no doubt through this route that the young Farish acquired the seeds of an evangelical faith and also his first taste for natural philosophy.[67]

Farish had entered Magdalene as a sizar in 1774. Having 'carefully avoided unprofitable and contaminating society', he was not only declared the senior wrangler of 1778, but was also awarded the first Smith's Prize – a mathematical honour which was competed for in a series of additional examinations.[68] After being elected a Fellow of Magdalene, Farish was appointed a tutor of that college, and a variety of college and university appointments soon followed: Praelector (1781), Barnaby Lecturer in Mathematics (1783), Taxer (1783), member of the Caput Senatus (1784), Sir Robert Rede's Lecturer (1785), Moderator (from 1790) and Senior Proctor (1792–3).[69]

By 1792, then, both candidates for the Jacksonian chair were respectable university figures, noted for their skills in mathematics. Neither had an outstanding reputation for natural philosophy, although both clearly exhibited some interest in the subject: both were members of the Society for the Promotion of Philosophy and General Literature, which was formed within the university in 1784 – seemingly an early version of the Cambridge Philosophical Society. This society drew together men such as Milner, Busick Harwood, Smithson Tennant,

Thomas Martyn and Samuel Vince, but dissolved in 1786 through lack of support.[70] Both Farish and Wollaston were also appointed to the Observatory Syndicate by a grace of 28 May 1790, although their astronomical activities are not clear.[71] Come the election for the Jacksonian chair, then, there was arguably little to mark the men two apart, and indeed, this is reflected in the closeness of the result, with Francis Wollaston winning by the narrow margin of 35 votes to 30.[72]

More will be said about Farish later. As for Wollaston, on assuming the Jacksonian chair, he took off where Milner had stopped, with Scheele's theory of heat and light.[73] As with Milner, Wollaston alternated between chemical lectures and those on natural philosophy (mainly electricity), making use of the substantial collection of philosophical instruments that Milner had accumulated. Valued by Milner at 'some thousands of pounds', this collection included such items as the delicate torsion balance constructed by the Woodwardian professor John Michell;[74] Wollaston passed this to his acquaintance Henry Cavendish, who subsequently used it in determining the density of the Earth.[75]

Fortunately for historians, the notes of Wollaston's 'Course of Demonstration, Exhibitions and Lectures' for the period 1792–1799 are preserved in Trinity College library.[76] Notes taken by his students during these lectures have also survived. The lectures trace the adoption of Lavoisier's chemistry in Cambridge. The first of these, delivered in 1792, begins:

> The Master of Queen's [Milner] having already treated of the subject of Heat in four Essays which he delivered to the University in conformity with the Will of the Founder of the Professorship to which the Senate has done me the Honour to appoint me his successor, I should not attempt to add anything to what he has so fully and so ably discussed had not a treatise of the celebrated Mr Scheele expressly on the subject of fire been unnoticed by him or rather been mentioned in such a manner only as might induce a person to imagine that had his health permitted him to continue his chemical researches we should in a future essay have been favoured with an examination of it.

The treatise mentioned here is presumably Scheele's *On Air and Fire*, the English version of which was published in 1780.[77] In trying to understand heat, this treatise focused on the role of air during combustion, and therefore provided a link between Milner's work on heat and Wollaston's own interests in chemistry. In his work, Wollaston outlined Scheele's theories, and then gave an interpretation using Lavoisier's system, noting that the 'practice of giving both interpretations' was one which he 'regularly observed', since at this stage he was 'unwilling either immediately to discard the old opinions and language of Chemistry or much more to reject a theory which by its simplicity and accordance with facts is daily gaining ground even among its most strenuous opponents'.

Wollaston quoted Scheele's experiments for the absorption of the 'vital air' (oxygen) from the atmosphere by compounds such as 'alkaline liver of sulphur' (a substance prepared by fusing potassium carbonate and sulphur), and calcareous hepatic sulphuris (calcium sulphide), noting in each case that the remaining air no longer supported combustion. The explanation for these observations was first given in terms of the phlogiston theory:

> Hence according to the Phlogistic theory, the sulphur losing its phlogiston, which is attracted by the pure part of the atmospheric air to which the hepar [sulphide] is exposed, is converted into vitriolic [sulphuric] acid, uniting in that state with the alkaline or earthy base as vitriolated tartar [potassium sulphate] or gypsum [calcium sulphate] . . . But what becomes of the phlogiston then disengaged and the vital air with which it is combined? Do they exist in the remaining air thus united and deprived of their elasticity? If so, it should be specifically heavier: which is contrary to experience. Or is the air lost united and fixed in the hepar sulphuris and other substances exposed? Mr Scheele concludes that it is not because it is not reproducible as fixed air; no precipitate being occasioned by lime water. – Before we follow him in the pursuit of this lost air, let us see what interpretation may be given of the same phaenomena by the antiphlogistic theory. Here the Vital air (I avoid calling it oxygen gas as the use of a term would intimate a predilection for the theory) which Mr Lavoisier considers as the principle of acidity and which Mr Scheele himself suspects to be so likewise, as being a compound of an extremely subtle acid and phlogiston. This vital air uniting with the sulphur, when assisted by the affinity of the alkali or calcareous earth for the compound, forms vitriolic acid, the result being vitriolated tartar or gypsum.

Clearly, at this point in time Wollaston was on the cusp between the two theories. His transition between them came whilst taking a break from this field. Since the Jacksonian regulations demanded alternation between chemistry and natural philosophy, Wollaston's lectures for the crucial years 1793–6 were concerned not with this area of chemistry, but with Leiden jars ('93 and '95), fulminating silver ('94) and the freezing of mercury ('96). These lectures on natural philosophy ended in 1796, when Samuel Vince succeeded the notoriously reticent Anthony Shepherd as Plumian Professor of Astronomy and Experimental Philosophy. Since Vince expressed a determination to lecture, Wollaston passed much of his collection of physical apparatus to him, declaring that: 'In consequence of the election of Mr Vince to the Plumian Professorship, Mr Wollaston will discontinue his lectures in Experimental Philosophy, and intends to read *Chemistry* annually'.[78] When Wollaston's chemical lectures resumed in 1797, they were once again concerned with the problem of combustion; however, whilst he still gave explanations using both systems, Wollaston now used Lavoisier's nomenclature.

In 1794, the year in which Lavoisier was guillotined, Wollaston published *A Plan of a Chemical Course of Lectures*, with a second edition identical to the first appearing in 1805. As was common for students' course notes of this period, the book was bound with blank pages interleaved throughout. Four copies of this work with copious manuscript annotations by his students have been located; these are held by Cambridge University Library (UL)[79], the Whipple Library (WL)[80], the library of Trinity College, Cambridge (TL)[81] and the British Library (BL).[82] The notes in the TL copy were taken by Charles Christopher Pepys (1st Earl of Cottenham) of Trinity College between 1799 and 1801. The notes in the BL copy were recorded by Walter Blackett Trevelyan of St John's College.[83] Unfortunately, the TL copy is the only one to bear a date, but it is clear from the references contained within the notes that the UL and BL annotations were made after 1800. In the BL and UL copies, Wollaston gives numerous references to Chaptal's *Elements of Chemistry* (either the first English edition of 1790 or the second one of 1795) and Gren's *Principles of Modern Chemistry* (1800). Occasional references are also made to Fourcroy, Lavoisier's *Elements of Chemistry* (English editions published in 1790, 1793, 1796, 1799 and 1802) and Macquer's *Dictionary*. In the TL copy, Wollaston defines chemistry thus:

> Chemistry as a Science is defined; that which estimates and accounts for the Change produced in Bodies by Motions in their Parts too minute to affect the Senses individually. As an Art it consists in the application of Bodies to each other in such Situations as are best calculated to produce those Changes.

The lecture course began with a discussion on the different opinions of heat. Wollaston reduced the different opinions to two alternative hypotheses: first, that 'heat is a material fluid' and, second, that it is 'only an affection of the particles of the body'. Under the first theory, he distinguished Stahl's opinion 'that the material fluid of heat is loosely united to bodies [in] which state it is sensible and also that it becomes a constituent part of ye body in w^h state he calls it phlogiston'[84] from Lavoisier's theory:

> Lavoisier, who has destroyed Stahl's Doctrine of phlogiston, supposes heat to be a material fluid to w^h he gives the name of calorique but that it is not a constituent part of the Body.[84]

Wollaston, however, seemed to prefer the anti-caloric theory, as propounded by Rumford in 1798, that heat is best explained as a 'rapid motion of the minute particles of Bodies' and cites evidence from the production of heat by agitation, friction, and so on.[85] Wollaston may have illustrated this idea with a practical demonstration:

Experiment: A small bar of Iron may be struck on an anvil till it be red hot, hence a good workman will keep up for some time the heat of a piece of Iron by the rapidity of his blows as is particularly the case in the making of Anchors.[86]

Wollaston clearly explains why the doctrine of phlogiston is no longer considered to be satisfactory and why Lavoisier's theory is superior. The following is taken from the annotations in the BL copy:

According to ye old Theory Phlogiston was considered as a something necessary to the existence of metals in their metallic state but of w^h they were deprived by Heat during their Calcination & according to this Theory the Calx of Lead was considered as a more simple substance than Lead in its metallic form in w^h state it was considered as the calx of Lead united with Phlogiston.

The New Theory, on much more substantial proof, considers metals in their metallic state as more simple than in their calcined state in w^h they are much heavier than they were before Calcination & this increase of weight is found to be in addition of pure air (*oxygen*) so that according to the old Theory, a principle called Phlogiston was taken away from Metals during their calcination & according to ye new Theory, there was an addition made to them in that state of pure air or oxygen.[87]

Wollaston gave a number of 'experimental facts' to support the new theory, chiefly concerned with the combustion of charcoal and the reduction of red lead with charcoal, concluding that:

These facts are a strong proof in favour of the New Theory that the calcination of metals is the addition, by means of Heat, of oxygen to the original metals instead of a Subtraction from them of an unknown principle, Phlogiston, according to ye old Theory: in favour of w^h Dr Priestly [sic] made the following experiment; to show that by uniting the calx of a metal with phlogiston the metal would be produced in its metallic state. – He put some red lead in an earthen support, surrounded the whole by a jar containing Hydrogen (w^h was conceived to be ye most pure Phlogiston) & exposed the red lead to ye focus of a burning glass by w^h the red lead was reduced to its metallic state and the Q[uanti]^{ty} of Hydrogen diminished; but this is satisfactorily explained in the New Theory as it may be proved that the union of Oxygen with Hydrogen produces water. Since then the Old Theory went on the supposition of something being lost when there was a manifest addition and of something being added when there was a manifest diminution, the circumstance of the weight alone seems to give a just preference to the New Theory.[88]

There is a similar section in the UL copy, which concludes with the comment:

In nothing does the absurdity of the old system appear so plain as in the metal acquiring weight by driving off phlogiston because this phlogiston must then have negative weight.

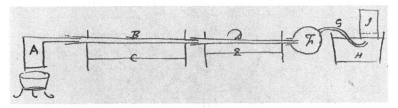

Figure 4.4. Wollaston's apparatus for the decomposition of water. Steam generated in A passes through heated gun barrels B and D which are filled with iron wire, producing hydrogen which is collected at I.

A central part to Lavoisier's theory is the understanding of the composition of water. Wollaston clearly recognised the importance of this discovery. His copy of the lecture given in 1799 is devoted to this topic. In it he states:

> Till within a very few years water has always been considered as an elementary substance, incapable of resolution into more simple constituents and with chemists also it has been ranked among the elements, as it has evaded every analytical [unintelligible] for this decomposition, and as every attempt to form it synthetically has also failed . . . Within a short time however it has been established beyond a doubt, that water does consist of two distinct substances, chemically combined; and the proof is confirmed both by disuniting them and collecting them in their separated states, and by uniting them again and reproducing the same quantity of water which was first employed in the experiment. For in this important case both the analysis and the synthesis are complete, and when taken together form as full a demonstration as can be given of any chemical fact whatever.[89]

Wollaston considered the composition of water 'a Discovery so highly valuable in itself and upon which so very much depends', that he described his apparatus for both the quantitative synthesis and decomposition of water 'in a full and satisfying manner' in his 1799 lecture.[89] A more concise summary appears in the student version with minor alterations, no doubt made by Wollaston over time. The sketches of the apparatus shown in Figures 4.4 and 4.5 are taken from Wollaston's own notes of 1799, where they are followed by details of the experiments performed using the apparatus. During the demonstration of the decomposition of water, Wollaston found that 320 grains of water produced 46 grains of hydrogen gas and caused an increase of 255 grains in the mass of the iron wire due to the formation of iron oxides. Without quoting his source, he states that theoretically 46 grains of hydrogen should be produced from 306 grains of water and should cause the mass of the iron to increase by 260 grains. (For comparison, a modern calculation predicts that 320 grains of water should produce 36 grains of hydrogen gas and cause the iron to increase in mass by 284 grains.) In later parts of the course, Wollaston also examines the synthesis of ammonia by driving nitrous acid vapour through a red-hot gun barrel filled

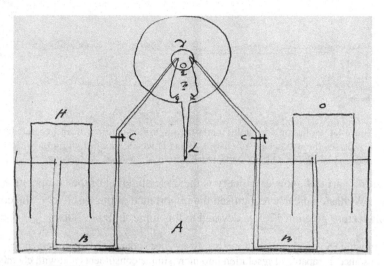

Figure 4.5. Wollaston's apparatus for synthesis of water. Hydrogen and oxygen from H and O are fed through the neck of round-bottomed flask D and burn at point F. Condensed water collects in quill L.

with charcoal (effectively the reverse of Milner's experiment for the production of nitrous acid, mentioned above).

Later in the course, during a discussion of gunpowder, Wollaston notes that 'soft woods make the best charcoal for gunpowder; hard woods for chemical purposes. By the new method of making Charcoal in close vessels 1/3 of the charge is saved. 2 parts going as far as 3' – the 'new method' presumably being Watson's method as described in the previous chapter.

One particularly intriguing annotation is found in the WL copy, where a student records Wollaston's preparation of nitrous oxide (laughing gas) from ammonium nitrate. The notes appear to demonstrate that Wollaston was among the first to teach the peculiar physiological effects of this substance:

> Nitric ammonia put in a retort over slow heat air [nitrous oxide] comes over. [It] is a very curious preparation . . . no residuum left; all this air which we know because nitre is oxygen and azot ammonia oxygen [sic] and hydrogen. All these combined [illegible] air which produces exhilaration etc when inhaled. Wollaston not so much affected as others [illegible] have been.[90]

It is unfortunate that the WL copy is not dated; presumably it must date after 1800 when Davy published his pioneering work on nitrous oxide (*Researches Chemical and Philosophical Chiefly concerning Nitrous Oxide or Dephlogisticated Nitrous Air*, 1800), but no reference to this work is given.

Wollaston continued in the Jacksonian chair until his resignation in 1813. However, he might have retired earlier had it not been for an unfortunate episode. In 1807, following the death of William Elliston (1733–1807), Wollaston applied for the Mastership of Sidney Sussex, the college where he had studied as an undergraduate. He was elected Master, but less than a year later, the election was overturned by one of the Fellows (with the help of Trinity College) on a technicality.[91]

William Farish and the Chair of Chemistry

What, though, of Wollaston's unsuccessful competitor for the Jacksonian chair, William Farish? As luck would have it, Farish (shown in Figure 4.6) did not have to wait long after the 1792 election for a second chance. In 1793, the Regius Professor of Physic, Russell Plumptre, died. This was of no direct use or interest to Farish, but was precisely the event for which Isaac Pennington – then still occupying the Chair of Chemistry – had been waiting. Pennington lost no time in registering his interest in succeeding Plumptre and, being successful in his bid, thus became the second Professor of Chemistry to resign the Chair in favour of another. After two decades in which the Chair of Chemistry had been treated as a sinecure, it was once again vacant.

Despite having no particular proficiency in the subject beyond a general interest in natural philosophy, Farish seems to have had no qualms about standing for the Chair of Chemistry, which is perhaps unsurprising given the behaviour and qualifications of the two previous occupants of the position. Francis Wollaston attempted to persuade his brother William to stand against Farish. As we shall see in the next chapter, William Hyde Wollaston was by now a practised chemist who had just become both a doctor of medicine and a Fellow of the Royal Society. William occasionally experimented in his brother's laboratory, and the thought of the brothers simultaneously holding both the Jacksonian and the Chemistry professorships was no doubt attractive to them.[92] However, William was trying to establish his own medical practice and declined to enter the contest, leaving Farish to stand unopposed.

Farish was duly elected to the Chair of Chemistry in 1794. On assuming the position, he reported that since 'the province of reading Lectures on the principles of Chemistry [was] already ably occupied by the Jacksonian professor', he was 'obliged to strike out a new line'.[93] Unashamedly ignoring the medical side of chemistry in favour of a course on industrial chemistry and mechanical technology, the lectures which Farish delivered were thus focused upon 'Arts and Manufactures . . . such as relate to Chemistry'.[94] Divided into four sections,

Figure 4.6. Engraving of William Farish (1759–1837) in 1815 by H. Dawe.

the lectures started with *Metals and Minerals*, before proceeding to *Agriculture*, *Construction of Machines* and finally *Waterworks and Navigation*. The first and largest section included some items and processes relating to chemistry, such as mining and smelting, whilst the remainder consisted in the main of a discussion of various engines, pumps and machines, together with such diverse devices as whaling harpoons, windmills and salmon traps! This was not to suggest that the gentlemen who attended these lectures would ever be directly involved in the manual work of such activities as smelting or whale-harpooning, but rather indicates, as discussed in Chapter 1, an awareness of the dramatic industrial and technological changes which were transforming all levels of British society at this time, and which Cambridge graduates were expected to take a role in directing.

Yet whilst it is true that Francis Wollaston was delivering chemical lectures, so forcing Farish to find other subjects on which to lecture, Farish would probably have delivered such a course of lectures in any event. In truth, he does not appear to have been very interested in chemistry, publishing nothing of relevance to the subject besides his lecture list, and preferring instead to spend his time experimenting with 'cogs and wheels and all mechanical contrivances'.[95] In fact, it seems that he only accepted the Chair of Chemistry because he had competed unsuccessfully for the Jacksonian chair the previous year. Certainly, as soon as the Jacksonian professorship became vacant following Francis Wollaston's resignation in 1813, Farish lost no time in applying to change chairs. Although William Wollaston is said to have expressed an interest in the chair, Farish again stood unopposed, and so was elected the new Jacksonian professor.[96] It seems likely, however, that the change in chairs made little difference to the topic of his lectures, since his *Plan of a Course of Lectures* from 1796 was reissued virtually unchanged in 1813 and 1821.[97] He enjoyed several years lecturing on engineering and mechanics, illustrating his talks on the construction and use of machines with working models which he built using an assembly kit of reusable parts – an early version of Meccano, which he thought sufficiently novel and interesting to describe to the fledgling Cambridge Philosophical Society.[98] Indeed, Farish's Jacksonian lectures are considered by some to mark the beginning of regular engineering teaching at Cambridge.[99]

Farish was known by his contemporaries as a conscientious and dedicated educator, driven by his evangelical faith.[100] His lectures were always well attended, despite his notoriously weak voice. This handicap was even more problematic for Farish when, from 1800 until his death, he served as vicar of St Giles' Church near Magdalene College. He overcame his impediment with ingenuity, constructing a parabolic sounding-board 'likened to a tin coal-scuttle bonnet' above the pulpit, so as to help his parishioners hear his sermons. What his audience thought of this is not recorded, although Farish himself found that the reflector also worked in reverse, and 'conducted not a few whispers to his ear'![101]

A further case of Farish's engineering ingenuity is illustrated by an additional anecdote. At his home in Merton Cottage, at the back of St John's College, he installed a moveable partition which could be used to divide the ground floor rooms, or raised through the ceiling to partition the room above it. A nineteenth-century commentator records how this caused some consternation when:

> One evening having almost sat-out his dining room fire in some dynamical calculation, being suddenly seized with a desire to make himself more snug, he suddenly let down the division from above, forgetful of his guests on the upper floor, who awoke from their first sleep to find themselves bewitched into a double bedded room![102]

A further development with which Farish has been credited is the introduction of quantitative marking in the Senate House examination.[103] Whilst he was Moderator, Farish suggested that by assigning specific marks to individual questions, the wranglers and senior optimes could be better evaluated and differentiated, and the subjectivity and biases of individual examiners – of which Richard Watson had complained bitterly in the 1750s – more easily avoided.[104] This may be seen as one development in a series of eighteenth-century reforms in which Senate House examinations were sat by greater numbers of students, with a shift from oral to written assessment and increased emphasis on the relative ranking of candidates.

Before concluding, we should briefly trace the remainder of the lives of the four principal characters of this chapter. Farish retained the Jacksonian chair until 1837, when he died at the rectory of Little Stonham in Suffolk.[105] Isaac Milner's health recovered sufficiently for him to assume some further duties, and in 1798 he succeeded Waring as Lucasian Professor of Mathematics.[106] Although he never resumed lecturing, Milner held this chair until his death at the home of his friend William Wilberforce, in the spring of 1820; one contemporary remarked of him that: 'The University, perhaps, never produced a man of more eminent abilities'.[107] Francis Wollaston served as rector of Cold Norton in Essex from 1813, being declared archdeacon of Essex later that year; he died a decade later, on 12 October 1823.[108] As for Isaac Pennington, having been elected President of St John's in 1787 – a position akin to Vice-Master – Pennington was knighted in December 1795, and held the Regius chair of physic until his death in February 1817.[109]

The brief period in which the teaching of chemistry became closely associated with the newly established Jacksonian chair meant that, somewhat ironically, Lavoisier's chemistry was first introduced to Cambridge by someone other than the Professor of Chemistry – although, being relatively unencumbered with the medical associations of chemistry, the holders of the new position were arguably more at liberty to adopt the new terminology. Whilst Milner appears to have been on the verge of accepting the new theory, there is no doubt that his Jacksonian successor, Francis Hyde Wollaston, fully embraced the ideas and incorporated them into his lectures. Indeed, his adoption of Lavoisier's nomenclature makes his lecture notes strikingly intelligible to the chemist of today. On Farish's transfer to the Jacksonian professorship, the teaching of chemistry returned to its expected place when the Chair of Chemistry was filled by Smithson Tennant. As Melvyn Usselman notes in the following chapter, Tennant – who had initially considered standing for the Jacksonian chair at the same time as Farish and Wollaston – had been one of the very earliest converts to Lavoisier's theory; by the time of his appointment in 1813, such ideas had gained almost universal acceptance.

The four decades between Richard Watson's and Smithson Tennant's occupancy of the 1702 Chair were an interesting if rather confusing time for chemistry in Cambridge. The period covered not only a theoretical transformation, with the introduction of Lavoisier's theories and nomenclature, but also saw the development of the Chair itself, with the teaching of chemistry emerging as a more organised, and in some ways more professional, enterprise after a period of unsettlement. Indeed, one might remark that Tennant was the first holder of the Chair since Vigani to have been appointed specifically on the basis of his chemical accomplishments, rather than general scholarship alone. From this point on, as the sciences became increasingly professionalised, the position of Professor of Chemistry at Cambridge demanded that those appointed were among the most notable chemists in the country.

Notes and References

1. See, e.g., Crossland, M. (1962), *Historical Studies in the Language of Chemistry*, London: Heinemann.
2. See, e.g., Bensaude-Vincent, B. and Abbri, F. (eds.) (1995), *Lavoisier in European Context: Negotiating a New Language for Chemistry*, Canton, MA: Science History Publications.
3. Nicholson, W. (1795), *Dictionary of Chemistry*, London: Printed for C. G. and J. Robinson, pp. vi–vii.
4. Coleby, L. J. M. (1954), 'Isaac Milner', *Annals of Science* 10, pp. 234–57.
5. St John's College Cambridge (1903), *Admissions to the College of St John the Evangelist*, Cambridge: Cambridge University Press, Part III, p. 683.

6. *Admissions to the College of St John the Evangelist*, Part III, p. 683; Cambridge University Archives *Original Grace Book* (κ) number 257.
7. Minutes of Addenbrooke's Hospital in Cambridge, ref. AHM.1, 28 Oct. 1771 onwards.
8. Cambridge University Archives *Original Grace Book* (Δ) number 12; *Admissions to the College of St John the Evangelist*, Part III, p. 683.
9. MSS. Cole, xxxiii, British Museum Addl. MSS 5832 p. 158, cited in *Admissions to the College of St John the Evangelist*, Part III, p. 683.
10. Cambridge University Archives O.XIV.15; *Admissions to the College of St John the Evangelist*, Part III, p. 683. Hodson was later Vice-Master of Trinity.
11. Moore, N. (1895), 'Pennington, Sir Isaac, M.D. 1745–1817', in Leslie, S. and Lee, S. (eds.), *Dictionary of National Biography*, London: Smith, Elder & Co., p. 776.
12. Moore (1895), p. 776.
13. St John's College Archives C5.2 [f. 179].
14. St John's College Archives C5.2 [f. 162].
15. Moore (1895), p. 776.
16. Rook, A., Carlton, M. and Cannon, W. G. (1991), *The History of Addenbrooke's Hospital, Cambridge*, Cambridge: Cambridge University Press, p. 1.
17. Minutes of Addenbrooke's Hospital in Cambridge, ref. AHM.2, 29 March 1773; Rook, Carlton and Cannon (1991), p. 59.
18. Coleby (1954), p. 234.
19. *Ibid.*
20. Milner, M. (1842), *The Life of Isaac Milner, D.D. F.R.S., Dean of Carlisle, President of Queens' College and Professor of Mathematics in the University of Cambridge*, Cambridge: J. & J. J. Deighton, p. 8.
21. 'L. S.' (1895), 'William Paley', in *Dictionary of National Biography*; Milner (1842), p. 9. The club was presumably named after the tea, which was enjoying a particular fashion at this time.
22. Milner (1842), pp. 33 and 109.
23. Milner (1842), p. 14.
24. Milner (1842), p. 15.
25. William Smyth (Regius Professor of Modern History at Cambridge, 1807–49) quoted in Milner (1842), p. 32. The Hoffman mentioned here may well be the Parisian 'operator' first employed by Richard Watson, since the diarist Henry Gunning records that 'Hoffman . . . though a German, was known to the villagers by the name of the *French Doctor*' [Gunning, H. (1855), *Reminiscences of the University, Town, and County of Cambridge from the Year 1780*, 2 vols., 2nd edn, London: George Bell, vol. 1, pp. 236–7]. However, by this time he appears to have been much more than a basic 'operator': Gunning relates how Hoffman himself had an assistant and prescribed medicine, and if so, he is almost certainly the same Hoffman found in the Addenbrooke's minutes of this period [AHM.1, 25 Nov. 1771].
26. Proctor in 1781–2, Moderator in 1780 and 1783, Sir Robert Rede's Lecturer in 1783, Milner (1842), p. 14; Venn, J. (1954), *Alumni Cantabrigienses*, Cambridge: Cambridge University Press, vol. 4, p. 422.
27. Milner (1842), p. 97.
28. Cunningham, G. G. (1837), *Lives of Eminent and Illustrious Englishmen*, Glasgow: A. Fullarton & Co., pp. 36–7.
29. Cambridge University Archives CUR 39.20[1].
30. *Ibid.*
31. *Ibid.*

32. Milner (1842), p. 18.
33. Twigg, J. (1987), *A History of Queens' College, Cambridge 1448–1986*, Woodbridge: Boydell, p. 213.
34. University of Cambridge Archives, Minutes of Syndicates 1778–1803, reproduced in Willis, R. and Clark, J. W. (1886), *The Architectural History of the University of Cambridge, and of the Colleges of Cambridge and Eton*, 4 vols., Cambridge: Cambridge University Press, vol. 3, p. 153.
35. Edwards, G. M. (1899), *University of Cambridge College Histories: Sidney Sussex College*, London: F. E. Robinson & Co., p. 193.
36. Bradley, R. (1730), *A Course of Lectures upon the Materia Medica, Antient and Modern. Read in the Physick Schools at Cambridge, upon the Collections of Doctor Attenbrook and Signor Vigani, deposited in Catherine-Hall and Queens' College*, London: Charles Davies.
37. University Archives CUR 39.20[1]; Coleby (1954), p. 236.
38. Smith, G. (1936), *Plans and Notes on Development of the New Museums and Downing Street Sites 1574–1936*, Cambridge University Archives ref. UA PVIII 1–10, plan 3; Willis & Clark (1886), vol. 3, p. 153.
39. Letter from Milner to Pitt, 7 Nov. 1791, Cambridge University Library Add. MS. 6958(c) [1016]; Milner (1842), p. 33; Milner (1842), pp. 33–4.
40. Jackson's will originally specified that the lecturer produce 'a certificate, signed by eight scholars at least, who have attended his Lectures for twenty days out of sixty' [Cambridge University Archives CUR 39.20(1)]. However, as one of the conditions for the additional stipend, this was raised to twelve [Letter from Milner to Pitt, 7 Nov. 1791, Cambridge University Library Add. MS. 6958(c) (1016)].
41. Letter from Milner to Pitt, 7 Nov. 1791, Cambridge University Library Add. MS. 6958(c) [1016].
42. Gunning, H. (1854), *Reminiscences of the University, Town, and County of Cambridge from the Year 1780*, 2 vols., London: Bell, vol. 1, p. 236. [Parts cited in Wordsworth (1877), *Scholae Academicae: Some Account of the Studies at the English Universities in the 18th Century*, Cambridge: Cambridge University Press, p. 193; and Searby, P. (1997), *A History of the University of Cambridge*, vol. 3, Cambridge: Cambridge University Press, p. 219.]
43. *Ibid.*
44. Letter from Milner to Pitt, 7 Nov. 1791, Cambridge University Library Add. MS. 6958(c) [1016].
45. Black, J. (1803), *Lectures on the Elements of Chemistry*, Edinburgh: John Robison, vol. I, pp. 548–9. That said, Black's view of Lavoisier's theory changed over time, and he noted that 'the person who learns chemistry by Lavoisier's scheme may remain ignorant of all that was done by former chemists, and unable to read their excellent writings' [*ibid.*]
46. Cited in Coleby (1954), p. 252.
47. Milner, I. (1789), 'On the production of nitrous acid and nitrous air', *Phil. Trans. Roy. Soc.* **79**, pp. 300–13. The term 'nitrous acid' was applied both to what would now be considered nitric acid (HNO_3) and nitrous acid (HNO_2).
48. Watson, R. (1781–7), *Chemical Essays*, 5 vols., Cambridge: Cambridge University Press, vol. 1, p. 254, cited in Coleby (1954), p. 241.
49. Milner (1842), p. 109.
50. Milner (1842), p. 243, cited in Gascoigne, J. (1989), *Cambridge in the Age of Enlightenment*, Cambridge: Cambridge University Press, p. 230.
51. Letter from Milner to Pitt, 7 Nov. 1791, Cambridge University Library Add. MS. 6958(c) [1016].

52. Milner (1842), p. 148; Letter from Milner to Pitt, 7 Nov. 1791, Cambridge University Library Add. MS. 6958(c) [1016].
53. Hilken, T. J. N. (1967), *Engineering at Cambridge 1783–1965*, Cambridge: Cambridge University Press, p. 37.
54. Letter from Milner to Pitt, 7 Nov. 1791, Cambridge University Library Add. MS. 6958(c) [1016]. For Milner's relationship with Wilberforce, see Milner (1842), chapters 2 and 3.
55. Duffy, M. (2000), *The Younger Pitt*, London: Longman, p. 41.
56. Letter from Milner to Pitt, 3 Dec. 1791, Cambridge University Library Add. MS. 6958(c) [1026]; Milner (1842), pp. 70–4.
57. Bouch, C. M. L. (1948), *Prelates and People of the Lake Counties: A History of the Diocese of Carlisle 1133–1933*, Kendal: Titus Wilson and Son, p. 374.
58. Milner (1842), pp. 79–80.
59. Milner (1842), pp. 84–100; Gascoigne (1989), p. 228.
60. Clark, J. W. (1900), 'Wollaston, Francis John Hyde' in *Dictionary of National Biography* vol. 21, pp. 779–80.
61. Royal Society Sackler Archives. Francis Wollaston (1694–1774) was also father-in-law to the physician William Heberden (1710–1801).
62. 'E.I.C.' (1900), 'Wollaston, Francis' in *Dictionary of National Biography* vol. 21, pp. 778–9; Jungnickel, C. and McCormach, R. (1998), *Cavendish: The Experimental Life*, Lewisburg, Pennsylvania: Bucknell, p. 442.
63. 'E.I.C.' (1900), p. 779.
64. Clark, J. W. (1900), 'Wollaston, Francis John Hyde' in *Dictionary of National Biography*, vol. 21 pp. 779–80; Royal Society Sackler Archives.
65. Bouch, C. M. L. (1948), *Prelates and People of the Lake Counties: A History of the Diocese of Carlisle 1133–1933*, Kendal: Titus Wilson and Son, p. 364.
66. Brownrigg, W., Franklin, B. and Farish, W. (1774), 'Of the stilling of waves by means of oil. Extracted from sundry letters between Benjamin Franklin, LL. D. F. R. S., William Brownrigg, M. D. F. R. S. and the Reverend Mr. Farish', *Phil. Trans. Roy. Soc.* **64**, pp. 445–60. For a biography of Brownrigg, see Russell-Wood, J. (1950–1), 'The scientific work of William Brownrigg MD FRS (1711–1800)', *Annals of Science*, vol. 6, pp. 436–47; vol. 7, pp. 77–94 and vol. 7, pp. 199–206.
67. 'Obituary of Rev. William Farish' [anonymous] (1837), *The Christian Observer*, vol. 429, pp. 611–13; vol. 430, pp. 674–7; vol. 431, pp. 737–74, p. 612; Cunich, P., Hoyle, D., Duffy, E. and Hyam, R. (1994), *A History of Magdalene College Cambridge 1428–1988*, Cambridge: Magdalene College, pp. 186–7; Hilken, T. J. N. (1967), *Engineering at Cambridge 1783–1965*, Cambridge: Cambridge University Press, pp. 38–44.
68. 'Obituary of Rev. William Farish', (1837), p. 612.
69. *Ibid.*; Cunich *et al.* (1994), pp. 186–7; Hilken (1967), pp. 38–44.
70. Milner (1842), p. 19; Edwards, G. M. (1899), *University of Cambridge College Histories: Sidney Sussex College*, London: F. E. Robinson & Co., pp. 196–7.
71. Cambridge University Archives, Index of Syndics.
72. Clark, J. W. (1900); Gunther (1937), p. 81.
73. Trinity College library R.8.42[6].
74. Letter from Milner to Pitt, 7 Nov. 1791, Cambridge University Library Add. MS. 6958(c) [1016]; Royal Society Sackler Archives ref. GB 117/EC/1760/06.
75. Cavendish, H. (1798), 'Experiments to determine the density of the Earth', *Phil. Trans.* **88**, pp. 469–526; Jungnickel and McCormach (1998), p. 442.
76. Trinity College library MS R.8.42. Each lecture is accompanied by a certificate, signed by at least twelve people who attended the lectures. James Cumming, who

was later to hold the 1702 Chair as its ninth incumbent, was among those who signed the certificate for 1798–1799.

77. Scheele, C. W. (1780), *Chemical Observations and Experiments on Air and Fire* [translated from the German by J. R. Forster], London: J. Johnson.

78. Edwards, G. M. (1899), *University of Cambridge College Histories: Sidney Sussex College*, London: F. E. Robinson & Co., pp. 196–7. Also Wordsworth (1877), p. 193.

79. UL: 7360.d.11. Another copy, without marginalia, is referenced Cam.c.794.10.

80. WL: Store 65:7

81. TL: Add MS.c.204(21c)

82. BL: 1142.h.21

83. Admitted 29 Sept. 1792, died 1818 [Venn (1954) Part II, vol. VI, p. 229].

84. MS annotations in British library copy, leaf after title page.

85. For Rumford's cannon-boring experiment, see Rumford, Count (1798), 'An inquiry concerning the source of the heat which is excited by friction', *Phil. Trans.* **88**, pp. 80–102.

86. MS annotations in British library copy, leaf after title page.

87. BL copy, MS annotations 2 leaves after p. 4, recto. Original parenthesis.

88. BL copy, MS annotations 1 leaf after p. 4, recto.

89. Trinity College library, R.8.42.

90. The student presumably means that nitrates are a combination of oxygen and nitrogen (azot), whilst mistakenly writing that ammonia is a combination of oxygen and hydrogen.

91. Thomas Hoskins (who had voted for himself as Master), objected that Wollaston's earlier Mathematical Lectureship did not count as a Fellowship. If this was the case, then for historical reasons, the college was obliged to consult Trinity College. See uncatalogued MS in Sidney Sussex College Archives [MR.113], particularly 'Statement and Memorial of the Revd John Green'; 'Draft petition from Hoskins, Chafy and Davie to John Shelley Sidney'; 'In the Matter of Sidney Sussex College Cambridge, Expleo The Master Fellows & Scholars of Trinity College'; 'Letter from the Master of Trinity College to the Visitor of Sidney Sussex, 27 Feb. 1807'. Also Edwards, G. M. (1899), *University of Cambridge College Histories: Sidney Sussex College*, London: F. E. Robinson & Co., p. 210.

92. Gilbert, L. F. (1952), 'W. H. Wollaston MSS at Cambridge', *Notes and Records of the Royal Society* **9**, pp. 311–22.

93. *University Calendar* (1794), cited in Mills, W. H. (1953), 'Schools of chemistry in Great Britain and Ireland', *J. Roy. Inst. Chem.* **77**, pp. 423–31 and 467–73, p. 429.

94. Farish, W. (1796), *A Plan of a Course of Lectures on Arts and Manufactures, more particularly such as relate to Chemistry*, Cambridge: J. Burges, Printer to the University; Farish, W. (1803), *A Plan of a Course of Lectures on Arts and Manufactures, more particularly such as relate to Chemistry*, Cambridge: Cambridge University Press.

95. Cunich *et al.* (1994), p. 186.

96. Gascoigne (1989), p. 292.

97. Indeed, the title page of the 1821 edition still reads 'Professor of Chemistry in the University of Cambridge'.

98. Farish, W. (1822), 'On isometrical perspective', *Trans. Cam. Phil. Soc.* **1**, pp. 1–20, Cambridge: Cambridge University Press.

99. Hilken (1967), pp. 38–44.

100. 'Obituary of Rev. William Farish' (1837), p. 612; Cunich *et al.* (1994), p. 186.

101. Wordsworth (1877), pp. 40–1; Smyth, C. (1937), 'William Farish, 1759–1837', *Magdalene College Magazine*, vol. 76, pp. 281–6.

102. Wordsworth (1877), pp. 40–1.
103. Hilken (1967), p. 40; Hoskin, K. (1979), 'The examination, disciplinary power and rational schooling', *History of Education* **viii** no. 2, pp. 135–46; Postman, N. (1993), *Technopoly*, New York: Alfred A. Knopf, p. 13. For an eighteenth–century account of the Senate House examination, see Wordsworth (1877), chapter v. For some of the proposed reforms, see Gascoigne (1989), pp. 203–6.
104. Hoskin (1979), p. 143; Watson, R. (1817), *Anecdotes of the Life of Richard Watson, Bishop of Llandaff*, London: Cadell and Davies, p. 10.
105. Cooper, T. C. (1888), 'Farish, William, 1759–1837' in *Dictionary of National Biography*, p. 1072.
106. Tanner, J. R. (ed.) (1917), *The Historical Register of the University of Cambridge*, Cambridge: Cambridge University Press, p. 83.
107. Wordsworth (1877), p. 78; Cunningham (1837), p. 37; Gunning (1855), vol. 1, p. 234.
108. Clark, J. W. (1900), 'Wollaston, Francis John Hyde', *Dictionary of National Biography*, vol. 21, pp. 779–80.
109. Moore (1895), p. 777; *Admissions to the College of St John the Evangelist*, Part III, p. 683.

5

Smithson Tennant: the innovative and eccentric eighth Professor of Chemistry

Melvyn Usselman

Department of Chemistry, University of Western Ontario

The opening years of the nineteenth century were exciting ones in the emerging discipline of chemistry, and most of this excitement was generated in England. In early 1800, Alessandro Volta sent an account of his researches on devices for the production of electricity in two letters to Joseph Banks, the President of the Royal Society, with the request that the results be published in the *Philosophical Transactions*.[1] By the time Volta's letters were read to the Royal Society on 26 June, William Nicholson and Anthony Carlisle had verified his remarkable results, created 'voltaic piles' of their own and announced the electrical decomposition of water. These experiments initiated a rush of investigations into the chemical effects of electrical current and were, as Davy later described them, 'the true origin of all that has been done in electrochemical science'.[2] The majority of chemists had accepted Lavoisier's operational definition of a chemical element as the last point of analysis, and qualitative tests for the known elements came into general use. Qualitative tests on unusual minerals often led to puzzling results, which were suggestive of unknown chemical constituents. Techniques for concentrating the unknown constituents, and purifying and characterising them, soon followed, with the consequent unexpected (and for some unsettling) discoveries of new chemical elements. These developments in practical chemistry and new technology were contemporaneous with the bold thinking of John Dalton, who began to reveal the central components of his atomic hypothesis in lectures and publications from 1803 onwards. Smithson Tennant, the future Professor of Chemistry at Cambridge, was a significant figure in this dynamic chemical community, although neglect by historians has done much to dim the contemporary view of Thomas Thomson that:

The 1702 Chair of Chemistry at Cambridge: Transformation and Change, ed. Mary D. Archer and Christopher D. Haley. Published by Cambridge University Press. © Cambridge University Press 2005.

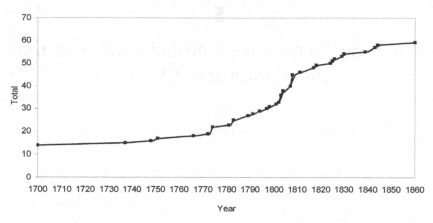

Figure 5.1. Discovery of the elements: graph showing the number of known elements by date.

No man that Great Britain has produced was better fitted to have figured as an analytical chemist, both by his uncommon chemical skill, and the powers of his mind, which were of the highest order, than Mr Smithson Tennant.[3]

The first decade of the nineteenth century is clearly revealed as a momentous one in chemistry by the plot shown in Figure 5.1 of the number of chemical elements against the year of their discovery. Fifteen new elements were characterised in the period 1800–1810, and eleven of them were the result of work carried out in London.[4] Using conventional chemical techniques, Charles Hatchett discovered niobium (1801, then named columbium), William Hyde Wollaston palladium (1802) and rhodium (1804), and Smithson Tennant osmium and iridium (1804). Using a powerful voltaic pile, Humphry Davy isolated sodium and potassium (1807) and barium, strontium, calcium and magnesium (1808). The contributions of these men were acknowledged by the Royal Society, which bestowed on each the Copley Medal – then the most prestigious award for scientific achievement in Britain. By 1800, the quality of science being done and the number of its practitioners had increased to the point where medal winners 'had to have made a significant discovery or at least have contributed several worthy papers to *Philosophical Transactions*'.[5] Discovery and characterisation of a new element clearly merited consideration, as did other discoveries in the fashionable discipline of chemistry, for chemists received seven of the twelve awards made between 1798 and 1810. Hatchett was awarded his in 1798, Wollaston in 1802, Tennant in 1804 and Davy in 1805.[6]

It is clear, then, that Smithson Tennant had established himself as a chemist of note in the early nineteenth century, and Banks' Presidential discourse on

the occasion of the award summarises Tennant's contributions to the burgeon-
ing science, but ends with words of encouragement that draw attention to his
lack of resolve, a characteristic that was to define and erode his subsequent
contributions to science:

> Into your hands then, Mr Hatchett, as the friend of Mr Tennant, I deposit this
> unequivocal testimony of the gratitude and esteem of his applauding brethren.
> Exhort him, Sir, I entreat you to continue his scientific labours and to increase if
> possible his diligence and assiduity. Talents like his deserve to be cultivated with
> increasing industry and uninterrupted patience; his chemical rivals admire him, the
> Royal Society esteem him, and the public looks up to him for further improvements
> in his most useful pursuit.
>
> Assure him, Sir, that confiding in the gratitude with which the sight [of] this
> medal will inspire him, I offer myself a willing pledge to the chemist, to the Royal
> Society, and to the public, that their expectations shall not be disappointed, but that
> his diligence in unravelling the mysteries of nature shall be unabated in future and
> that he will deserve at least as eminent success hereafter in disclaiming her eternal
> laws, as he has hitherto engaged.[7]

The effusive praise in Banks' oration was characteristic of his many award
eulogies, but the exhortation to increased 'diligence and assiduity', desired
by 'the chemist, . . . the Royal Society, and . . . the public' is quite unusual.
Clearly, Banks was putting into words, at an opportune time, the sentiments
of his scientific confidants – that Tennant had exceptional talents, but was not
using them to the full. For reasons unknown, Smithson Tennant, F. R. S., was
not at the anniversary meeting to hear the approbation, or the exhortations, and
the historian has to look elsewhere to learn if the compliments and the cavils
were appropriate, and whether or not their delivery would have had an impact.
Tennant's only contemporary biographer, his close friend John Whishaw, intro-
duces his life of Tennant by quoting from Samuel Johnson's *Life of Edmund
Smith*: "Mr Tennant may be considered as one of those 'who, without much
labour, have attained a high reputation, and are mentioned with reverence rather
for the possession than the exertion of uncommon abilities'".[8] This overly harsh
assessment may well correlate lack of focus with lack of exertion, a correlation
implicit also in Banks' comments, and one that may be more illusory than real.
Let us now look to Tennant's life for answers.

The making of the chemist

Smithson Tennant was born on 30 November 1761 in Selby, Yorkshire, the
only child of the Reverend Calvert Tennant and his wife Mary (née Daunt).[9]

His father had graduated from St John's, Cambridge, with a B.A. (1739), M.A. (1743) and B.D. (1751). He had been ordained as a priest in 1741 and was made a Fellow of St John's in 1743, a fellowship he retained until his marriage in 1759. He was granted the living of the Rectory of Great Warley, Essex, in 1758, but moved upon his marriage to his wife's home in Selby. Mary, the only surviving child of William Daunt, apothecary of Selby, inherited her father's estate after his death in 1758 and her mother's in 1770. Smithson's early education was provided by his father, who began to teach him Greek at the age of five, but upon the father's death in 1772, his mother sent him for schooling at Yorkshire Grammar schools at Scorton (Swaledale), Tadcaster (Wharfedale) and Beverley (near Hull). At Tadcaster, Smithson attended the presentations of the itinerant lecturer Adam Walker, and evidence of his precocity first appears:

> During the time he was at school at Tadcaster, he happened to be present at a public lecture given by Mr Walker, formerly well known as a popular teacher of experimental philosophy. Although then very young, he put several pertinent questions to the lecturer respecting some of the experiments, and displayed so much intelligent curiosity as to attract the attention of the audience, and give great additional interest to the lecture. Mr Walker, sensible of the effect which the boy's presence had produced, requested that he would continue to attend his lectures during the remainder of the course.[10]

Although distant from the population centres of England, or perhaps because of this, the schools of the north were generally of high calibre, and its graduates were at no intellectual disadvantage when competing for honours at the great universities. Tennant left Beverley with the desire, encouraged by his mother, to study natural philosophy under the tutelage of Joseph Priestley, but these plans fell through about the same time that tragedy struck. While riding with him in July 1781, his mother was thrown from her horse and died on the spot, leaving Smithson at 20 years of age without parents or siblings, but sole heir to the Tennant and Daunt estates.

After his mother's death, Tennant set off for Edinburgh, where he enrolled in classes in anatomy and surgery, chemistry and *materia medica*, perhaps to begin the study of medicine in the footsteps of his grandfather. A few notebooks in Tennant's handwriting from the Edinburgh period, including one on Joseph Black's famed chemical lectures, have been preserved and are in the archives of Cambridge University Library. Whishaw relates that

> Mr Tennant being at breakfast with him [Black] one morning, the conversation turned upon the new doctrine concerning heat; and some experiment was made, in which the temperature of a liquid was to be tried by the thermometer. Mr Tennant, immersing the instrument in the liquid, instead of waiting during a long interval, until the mercury became stationary, noted the point to which it immediately rose;

and then heating it above the temperature of the liquid, noted the point to which it fell: upon which Dr Black observed, "I see, young man, that you know how to make an experiment".[11]

The fact that Tennant had breakfast with Black, and conducted experiments with him, suggests that the Yorkshire student had quickly earned Black's respect and quite probably learned a great deal through conversations with the celebrated chemistry teacher. Nonetheless, Tennant's stay at Edinburgh was brief, for shortly after his twenty-first birthday and attainment of his inheritance, he moved in October 1782 to Cambridge and entered Christ's College as pensioner with the desire of studying chemistry and botany, but 'disliking the ordinary discipline and routine of an academical life, he obtained an exemption from those restraints by becoming shortly afterwards a Fellow Commoner'.[12]

It is well known that Cambridge in the eighteenth century has been accused of 'violating its statutes, misusing its endowments, and neglecting its obligations',[13] but it was also true that a Cambridge degree, deservedly or not, was nevertheless a mark of cultural attainment and served as passport into the higher echelons of English society. For someone like Tennant, who had financial independence and connections remaining from his father's time there thirty years previously, Cambridge offered social and cultural opportunities as well as intellectual ones. The evidence that we have suggests that Tennant took advantage of all that Cambridge had to offer. Henry Warburton, a nineteenth century philosophical radical and parliamentarian who collected a great deal of Tennant material in support of a planned Wollaston biography, wrote of him:

> [Tennant] had a wonderfully comprehensive mind, and in all matters speculative, singular latitude of judgement. The more vast the data, and bewildering [to] ordinary minds, the more certainly would he arrive at a just conclusion. To this was added a playful fancy, a flowing diction, readiness of wit, an easy, negligent, fascinating manner, and at this point of his life a buoyancy of spirits, which made him the life of the society in which he moved.[14]

Whishaw adds:

> Yet, although he was incessantly employed, there was a singular air of carelessness and indifference in his habits and mode of life; and his manners, appearance, and conversation, were the most remote from those of a professed student. His College rooms exhibited a strange disorderly appearance of books, papers, and implements of chemistry, piled up in heaps, or thrown in confusion together. He had no fixed hours or established habits of private study; but his time seemed to be at the disposal of his friends; and he was always ready either for books or for philosophical experiments, or for the pleasures of literary society, as inclination or accident might determine. But the disadvantages arising from these irregular habits were much more than counterbalanced by extraordinary powers of memory and

understanding; and especially by a faculty, for which he was remarkable, of reading with great rapidity, and of collecting from books, by a slight and cursory inspection, whatever was most interesting and valuable in their contents.[15]

One other account by a friend substantiates these recollections that portray Tennant as intellectually gifted and socially active, but utterly unrestrained by prevailing norms of behaviour.

> During his residence at Emmanuel [begun in 1786], he agreed to accompany a friend to France, who, knowing his want of punctuality, thought it better they should travel to town in a post-chaise, as Tennant would be sure to miss the coach. The time was fixed for starting; the request that he would be ready when his friend called for him, was faithfully promised but not adhered to, for when the post-chaise stopped to take him up, he had not finished breakfast. His friend complained bitterly at the detention, and his annoyance was much increased when Tennant said, "I have only to drink my cup of tea, and I shall then have nothing to do but to pack up."
>
> This unlooked-for information was scarcely to be endured; but when witnessing the process of packing up, his anger was converted into a hearty laugh. Tennant first removed the breakfast things, and then spread the table-cloth on the floor; upon this he emptied, with the utmost composure, the contents of a drawer which contained his linen; then getting a second table-cloth of larger dimensions, he emptied into that the contents of another drawer, consisting of coats, waistcoats, etc.; to these he added shoes, boots and brushes; and tying up the corners in the same manner that college laundresses carry away the dirty linen, he announced he was ready. These two bundles were crammed into the chaise, and the two friends started.[16]

Tennant had decided to take up the study of medicine at Cambridge, a course of study that imposed the fewest demands on its students of all Cambridge degrees. Attendance at lectures was not enforced and, consequently, 'all that was asked in practice of a candidate for a first medical degree was to keep his name on the books of a college for five years, to reside nine terms, to witness two dissections, and keep one act'.[17] That students needed to do no more than this is corroborated by Warburton, who wrote of Wollaston's medical studies at Cambridge in the 1780s:

> Small is the number of students who profess either Law or Physic; little emulation prevails amongst them; little motive to exertion on the part of the Professor. If this is the case with the Law students, still more the case with those in Physic; who, in general, if they reside the requisite number of terms, conduct themselves without gross impropriety, attend a very limited course of Lectures given by the Professor of Anatomy and Physic, pay the Professor fees and maintain their theses without betraying gross ignorance, are considered entitled to their degree, and to practise the healing art.[18]

The requirements for an advanced M.D. degree, the basic requirement for membership in the Royal College of Physicians and a potentially lucrative practice in London, were monetary and institutional ones only – a candidate had to keep his name on the books of his college for five years and an M.D. degree, together with a copy of the Aphorisms of Hippocrates, was his.

Tennant became close friends with a fellow Christ's student, Busick Harwood (later Sir), who had entered Cambridge in his mid-thirties after time spent in India as a surgeon. Harwood had a great interest in blood transfusion and read a thesis on that controversial topic for his M.B. degree in 1785; he became a Fellow of the Royal Society in 1784 and was elected Professor of Anatomy in 1785. He had a considerable reputation in the colleges as a *bon vivant*, 'profligate and licentious in the extreme',[19] with 'a vast fund of obscene stories, of which one was so particularly filthy that no one, as he boasted, could hear it without being sick'.[20] So close was the bond between the two that they toured the Low Countries together in 1783, and Tennant later transferred with Harwood from Christ's College to Emmanuel in December 1786. Their frequent presence thereafter in the after-dinner Combination Room is noted in the Parlour Books of Emmanuel College.[21]

In the summer of 1784, Tennant travelled to Sweden and Denmark and there he met Scheele, the most industrious chemist of his time, whose 1777 book *Chemical Observations and Experiments On Air and Fire* had appeared in English translation in 1780. In Stockholm, Tennant also met Johann Gahn, from whom he learned the techniques of mineral analysis by use of the blowpipe. On his return to England, he proudly showed his Cambridge friends a variety of minerals given him by Scheele and demonstrated several experiments he had learned from him. His scientific credentials and Cambridge supporters were now sufficient for him to be successfully elected F.R.S. in January 1785. That year, Tennant set off for the continent once again, this time to visit France (where he met Berthollet and Delamethérie), Holland and the Netherlands. At this point in his career, near the endpoint of the five-year registration period required for the initial medical degree, Tennant had made himself into a competent, well-informed chemist and his first contributions to the subject began. He mentioned to his friends an apparatus he had designed to demonstrate the amount of latent heat released by the condensation of steam, an idea clearly emanating from Black's studies on heat. By using the heat released by the condensation of steam from a first boiler to evaporate more water from a second boiler held at reduced pressure, he found that he could increase the quantity of distilled water by 75%. The details of this apparatus, shown in Figure 5.2, were not published until 1814, when its use as a lecture demonstration device prompted Tennant to write up the research.[22] More significantly, Tennant had become an adherent in 1784 of the antiphlogistic theory and 'entirely satisfied himself as to the truth of this doctrine',[23] probably because of his exposure to the

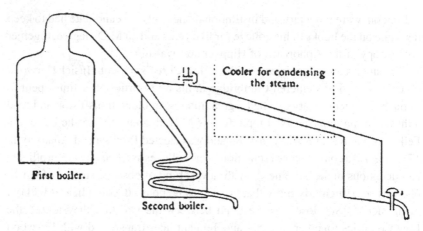

Figure 5.2. Double distillation by the same heat.

arguments in its favour that he would have read in the literature, coupled with his vaunted ability readily to grasp the salient points of scientific claims and counterclaims. Carleton Perrin has suggested that the antiphlogistic doctrine found its first favourable British reception in Edinburgh and Glasgow in 1784, but Cambridge must now be added to the locales where the French theory of combustion had at least one noted proponent.[24] Since Tennant would almost definitely have attended Milner's chemistry lectures, Milner's own exploration of antiphlogistic theory – as discussed in the previous chapter – would certainly have been no discouragement.

About this time Tennant made the most consequential of all his Cambridge acquaintances, that of William Hyde Wollaston, a younger medical student at Caius who differed from Tennant in almost every way. Warburton comments on the meeting of Tennant and Wollaston, and their differing personalities:

> A meeting at the room of a common friend was the commencement of the acquaintance of these two men. Medical studies which they pursued in common and the discovery which a superior understanding makes of another, improved this acquaintance. Tennant had enthusiasm for chemistry. "Though the chemist cannot create elements" he would say, "yet he can combine them in ways which nature never thought of; and that is the next thing to creation." He strengthened in Wollaston that passion for the science which Milner had kindled. Wollaston held Tennant's knowledge as a chemist in profound admiration, and a year or two after this period, expressed his despair of ever becoming Tennant's equal.[25]

In these early years of their acquaintance, the older, more broadly educated and effusive Tennant had a decisive impact on Wollaston, and perhaps on the future course of chemistry, for Warburton also recounts that

> With desires of emulating his distinguished friend [Tennant] he [Wollaston] applied
> sedulously to chemistry, not making many experiments in his own rooms, but
> availing himself of a Laboratory which his brother Francis had fitted up. Platina
> even at that time engaged his attention; he made some persevering attempts to fuse
> it in a Blacksmith's forge, aided by Dr Pemberton, then of the same College.[26]

The Cambridge interaction between the two did not last long. In 1788, after keeping the requisite Act, which an associate claimed to be 'of the highest order though marked by great eccentricity',[27] Tennant obtained his M.B. degree and left Cambridge for London, where he needed to occupy himself for five years before becoming eligible for an M.D. degree. Little is known about the first few years of his life in the city, except for the publication of his first scientific paper in 1791, which cleverly exploits the combined attractions of phosphorus and calcareous earth for vital air to decompose fixed air (carbon dioxide, released by calcareous earth) into elemental carbon, with the co-production of a compound of lime and phosphoric acid:

$$\text{phosphorus} + \text{calcareous earth} + \text{lime} \rightarrow \text{phosphate of lime} + \text{charcoal}$$
$$4\,P \quad + \quad 5\,CaCO_3 \quad + \quad CaO \rightarrow \quad 2\,Ca_3(PO_4)_2 \quad + \quad 5\,C$$

Although Lavoisier had earlier synthesised fixed air by the combination of vital air and charcoal, Tennant was the first to devise a method of overcoming the strong affinity between the two constituents to liberate uncombined charcoal.[28]

A year after the publication of this paper, Milner vacated his Jacksonian Chair at Cambridge and Tennant, at the urging of his friends, offered himself as a candidate. However, learning that William Farish and Francis John Hyde Wollaston were also candidates, Tennant withdrew his name and F. J. H. Wollaston (the eldest brother of Smithson's friend William) was elected to the Chair in 1792.[29] Undeterred, Tennant went abroad again that year on an extended tour of France, Switzerland, Italy and Germany. At Lausanne, he met Edward Gibbon, the famed author of *Decline and Fall of the Roman Empire*, and at Rome and Florence he became enthralled by the beauty of its ancient and modern art. He was surprised that many German chemists remained adherents of alchemical doctrines and in Paris he was taken aback by 'the gloom and desolation arising from the system of terror then beginning to prevail in that capital'.[30] Shocked by the changes that the Revolution was wreaking in France, Tennant returned to London, where he took up residence at 4 Garden Court, Temple, just east of Somerset House, then home to the Royal Society. This remained his principal residence until his death.

Tennant received his M.D. from Cambridge in 1796, and gave some thought to the practice of medicine. He read widely in the discipline, made extensive notes and attended regularly at London's great teaching hospitals, but the experience convinced Tennant that the life of a physician was not for him, for Whishaw reports that

> he had suffered very greatly, during his attendance at the hospitals, in consequence
> of the acute and painful emotions he had constantly experienced from those
> sights of hopeless misery which he had so often occasion to witness. He justly
> apprehended that the frequent recurrence of such scenes, unavoidable in medical
> practice, would be destructive of his comfort and happiness.[31]

Although the career for which his education had best prepared him seemed now to be undesirable, chemistry remained a passionate interest, and Tennant published two papers in 1797. The first, 'On the Nature of the Diamond', established that heating powdered diamond with nitre (potassium nitrate) in a small gold tube gave fixed air as the only product, and in amounts consistent with Lavoisier's published values for the carbon content of fixed air. Thus, Tennant was able to conclude that diamond 'consists entirely of charcoal, differing from the usual state of that substance only by its crystallised form'.[32] The second paper, 'On the Action of Nitre upon Gold and Platina', revealed that both gold and platina reacted with nitre on strong heating to form soluble calxes.[33] These papers established Tennant as an imaginative and skilled chemist, and gave promise of a productive career in the science, but Whishaw, then a neighbour at Garden Court, provides a cautionary note:

> It is worthy of remark, that Mr Tennant had ascertained the true nature of the
> diamond some years before he made the above communication to the Royal
> Society. In conversing about this time with a particular friend, . . . he happened to
> mention the fact of this discovery. His friend, who had often lamented Mr T.'s
> habits of procrastination, urged him to lose no time in making his experiment
> public; and it was in consequence of these entreaties that the paper on the diamond
> was produced.[34]

Whishaw's comments illuminate those of Banks cited earlier in this paper, and portray a man who was not only an innate and gifted seeker of knowledge, but also utterly unaffected by social and scientific custom and undesirous of fame, certainly not the type of man one would expect to become a chemical entrepreneur. But such was what Tennant was soon to become.

The Wollaston/Tennant partnership

In late 1797, William Hyde Wollaston left his medical practice in Bury St Edmunds and moved into a house at 18 Cecil Street, London just off the

Strand to the west of Somerset House. There he hoped to establish a larger medical practice and to engage more fully in the scientific life of the capital, as he had become a Fellow of the Royal Society in 1793. Once settled in London, Wollaston renewed his acquaintance with Tennant and joined with him in the execution of chemical experiments. In fact, he had assisted Tennant even before his move to London, probably on one or more of his trips to London for meetings of the Royal Society, for Thomson recounts that

> A characteristic trait of Mr Tennant occurred during the course of [the diamond experiment], which I relate on the authority of Dr Wollaston, who was present as an assistant, and who related the fact to me. Mr Tennant was in the habit of taking a ride on horseback every day at a certain hour. The tube containing the diamond and saltpetre were actually heating, and the experiment considerably advanced, when, suddenly recollecting that his hour for riding was come, he left the completion of the process to Dr Wollaston, and went out as usual to take his ride.[35]

One suspects that Wollaston's high regard for Tennant's chemical talents would have been shaken by events such as this, but the two continued their close association and perhaps discussed plans to pursue chemistry for profitable ends, as events were unfolding which would give Wollaston the same degree of financial independence as Tennant.

In February 1797, Wollaston's wealthy uncle West Hyde had died, and his entire estate was willed to George Hyde Wollaston, the second son of Francis and William's older brother. George came into possession of the estate in early 1799, and on 1 March 1799 transferred £8000 in 3% annuities to the account of William.[36] In December 1800, Wollaston announced to a friend that he had made the decision to leave medicine

> ... I cannot help thinking that I have at various times given you reason to think with me that the practice of physic is not calculated to make me happy. I am now so fully convinced of it & have so well satisfied those who are most interested for my welfare & whom I thought it most prudent to consult that I have fully determined & now declare that I have done with it. What I shall do instead I do not yet know. I feel no doubt of finding employment & turning my time to account in some way or other less irksome to me, for even if I turn waiter at a tavern ready to say "Yes Sir" to everyone that calls at any hour of the day or night, I cannot be a greater slave.[37]

Wollaston was not completely candid in this letter, for he made no mention of the fact that he and Tennant had decided to enter into a chemical partnership in which expenses incurred in the preparation of chemical commodities, and profits that might accrue from their sale, would be shared equally. The partners decided to keep their collaboration secret, and were so successful that knowledge of the partnership has only been uncovered by twentieth century historians.[38] Account books of the chemical business are extant and they reveal that Wollaston and

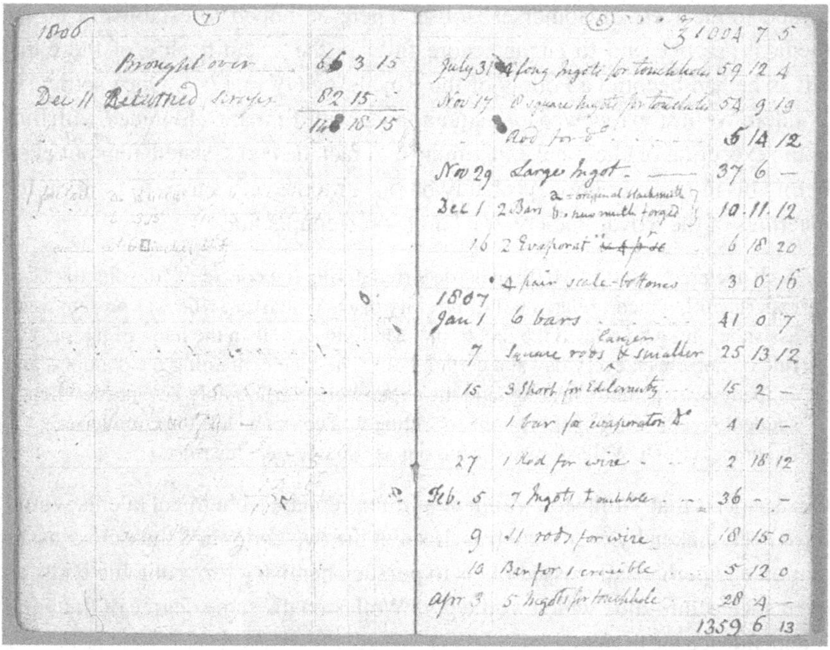

1806		
	Brought over	66 3 15
Dec 11	Returned scraps	82 15 –
		148 18 15

		1004 7 5
July 31	4 long ingots for touchholes	59 12 4
Nov 17	8 square ingots for touchholes	54 9 19
	Rod for do.	5 14 12
Nov 29	Large ingot	37 6 –
Dec 1	2 Bars (a = original blacksmith, b = new smith forged)	10 11 12
16	2 Evaporat[illegible]	6 18 20
	4 pair scale-bottoms	3 0 16
1807 Jan 1	6 bars	41 0 7
7	Square rods larger & smaller	25 13 12
15	3 short [illegible]	15 2 –
	1 bar for evaporator &c	4 1 –
27	1 Rod for wire	2 18 12
Feb. 5	7 Ingots touchhole	36 – –
9	11 rods for wire	18 15 0
16	Bar for 1 crucible	5 12 0
Apr 3	5 ingots for touchhole	28 4 –
		1359 6 13

Figure 5.3. Ledger of Wollaston and Tennant's sales of platinum during 1806–7, showing the varied uses of the metal.

Tennant planned chemical ventures in two promising areas, the production of malleable platinum from the crude alluvial ore and the marketing of organic substances isolated from natural sources.

The business began on 24 December 1800, with the purchase of 5959 ounces of platina ore for £795, of which Wollaston contributed £475 and Tennant £320.[39] This was a significant financial investment, and not one without risks, for it was not certain that the two men could develop a process that would yield pure, malleable platinum from the intractable alluvial ore. Both men were aware that platinum artefacts had been produced in Europe and basic chemical purification processes had been published, but no one had been able to bring the process to the point where a consistent product could be reliably produced.[40] Wollaston and Tennant knew that malleable platinum would find use in applications which required a heat-resistant and chemically inert metal. After marketing of the metal began in 1805, Wollaston's account book for platinum sales records some of the predictable applications, such as evaporating pans, scale bottoms and crucibles, together with innovative, high-volume applications such as touch-holes for guns (Figure 5.3).[41]

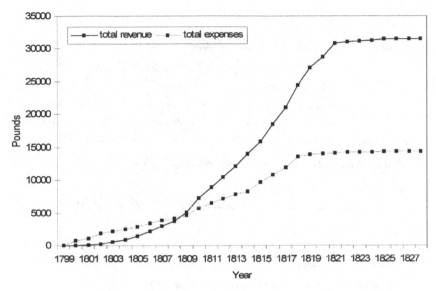

Figure 5.4. Cumulative revenue and expenses from the Wollaston/Tennant platinum business.

The crude ore was a by-product of Spanish gold-mining operations in the Viceroyalty of New Granada (now Colombia), and its availability was tightly controlled by the Spanish government.[42] The ore that found its way to England had been smuggled out of New Granada to Kingston, Jamaica, where it became available for legitimate purchase at a few pence per ounce. The evidence suggests that Wollaston and Tennant hoped to achieve a monopolistic position in the platinum business by purchasing all the platina ore that became available in England, and they ultimately purchased and processed over 47 000 ounces of the crude ore. The platinum business turned out to be an immensely profitable one, for the pure, malleable platinum was sold at an average price of 16 shillings an ounce, and its sale over the early years of the nineteenth century netted the partners a profit of nearly £17 000 (Figure 5.4).

It is apparent that Wollaston took on the responsibility for the chemical and metallurgical processing of platinum, for in 1801 he moved to a more spacious home at No.14, Buckingham Street, Fitzroy Square, London, where he established a laboratory at the rear of the house. There, with the help of a paid assistant, he processed all the platinum in 16–30 ounce batches over a period of 20 years. There is no evidence that Tennant did anything other than pay for a few initial expenses; up to the time of Tennant's death in 1815, Wollaston had paid out £3998 in expenses and Tennant a mere £74. There can be little doubt

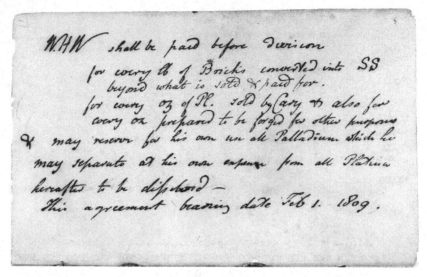

Figure 5.5. Wollaston/Tennant financial agreement, 1809.

that the inequity of the situation troubled Wollaston, and in April 1809 the new agreement on the division of profits shown in Figure 5.5 was reached. Entries in the account books reveal that, from that date forward, the profits were no longer shared equally and Wollaston's share increased to 55%.

The platinum venture paid dividends in other ways as well, for during the search for an economical and successful purification procedure, Wollaston discovered that the crude platina contained some new chemical elements. His preliminary treatment of the crude ore with *aqua regia* yielded a large soluble component and a small insoluble component. Wollaston focused on the soluble fraction because it contained the platinum, and gave the insoluble portion to Tennant for analysis. After precipitation of platinum from the soluble component, Wollaston was able to isolate palladium (1802) and rhodium (1804) from the supernatant solution.[43] In an effort to keep details of the platinum business secret, he offered palladium for sale anonymously before publishing the discovery in 1805; he had published the discovery of rhodium earlier, in 1804.[44] Wollaston's discoveries were hailed as a triumph of his analytical skills as the elements were present in only small amounts in the crude ore. However, what Wollaston did not reveal in his publications, nor did anyone suspect, was that he was dealing with relatively large amounts of metallic mixtures much enriched in the new elements. In fact, palladium comprised about 13% and rhodium 11% of the nearly 300 ounces of 'second metallic precipitate' Wollaston had accumulated by the end of 1803, and from which they were isolated.

The insoluble residue from Wollaston's chemical treatment of platina was, as noted above, given to Tennant for analysis, and it is likely that Tennant's successful analysis (published in 1802) of the intractable and quite different substance, emery, was part of the investigation.[45] As a result of Wollaston's labours, Tennant would have had nearly 100 ounces of the insoluble platina residue to work on and a strong suspicion that the residue might contain new elements, since its chemical behaviour differed so greatly from that of platinum and the metals Wollaston was working to isolate. In 1803, Tennant learned that French researchers were also investigating the properties of insoluble platina residues, and had suggested that new metals might be present.[46] Perhaps to establish his priority, Tennant wrote to Joseph Banks that the insoluble residue 'did not, as was generally believed, consist of plumbago, but contained some unknown metallic ingredients'.[47] He then set to work to isolate and characterise them. After fusing the platina residue with soda, and repeatedly treating the resulting mixture alternately with caustic alkali and marine acid (HCl), Tennant obtained a basic solution containing 'the oxide of a volatile metal', and an acid solution which contained a metal that gave 'a red colour to the triple salt of platina with sal-ammoniac'.[48] Treatment of the alkaline solution with sulphuric acid followed by distillation gave an aqueous solution of the volatile metal oxide [OsO_4, a quite toxic substance] which had a 'sweetish taste' and a 'pungent and peculiar smell'. Addition of a base metal such as copper to the aqueous solution gave a black metallic powder, which Tennant named Osmium because of the distinctive odour of its gaseous oxide (Greek ὀσμή smell, odour). Slow evaporation of the acidic solution gave a solid, which on subsequent solution in water gave a 'deep red coloured solution' and which on evaporation gave 'distinct octahedral crystals'. Heating the crystals gave a new metallic element '... in a pure state ... [which] appeared of a white colour, and was not capable of being melted, by any degree of heat I could apply'. Tennant named this metal Iridium (Greek ἴριδος, rainbow) because of 'the striking variety of colours which it gives, while dissolving in marine acid'.

The discovery of these two metals, made possible only by the prior undisclosed labours of Wollaston, deservedly placed Tennant's name in the top rank of the world's chemists, and earned for him the praise meted out by Banks in his Copley address. One measure of the status Tennant had achieved as a chemist is that his name was submitted, along with those of Davy, Wollaston, Chenevix and Dalton, as a candidate for election as a foreign member of the Académie de Sciences in 1805.[49] But the discovery of two new metals also marked the apogee of Tennant's chemical career, for his notorious lack of resolve prevented him from publishing any further significant research. In the paper announcing his discovery of palladium, Wollaston mentioned some iridescent metallic grains

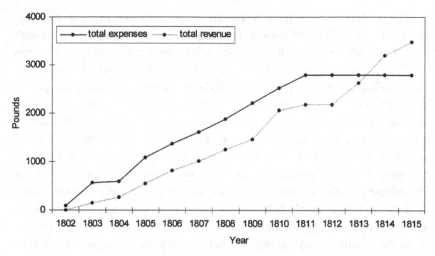

Figure 5.6. Cumulative revenue and expenses from the organic chemicals business.

of anomalously high specific gravity that he had separated from the insoluble platinum residues, and wrote:

> I have selected a portion of them, and have requested Mr TENNANT to undertake a comparative examination, from whose well-known skill in chemical enquiries, as well as peculiar knowledge of the subject, we have every reason [^] to expect a complete analysis of this ore.[50]

In his offprint of this publication Wollaston has pencilled in, at the marked insertion point, the exasperated comment 'barring indolence';[51] and indeed, Tennant never did publish an analysis of the ore.

The organic chemicals part of the chemical business was a dismal failure by comparison with the platinum business, largely because the partners could not find profitable markets for their products. There is evidence that Tennant was active in the marketing of some of the products but, once again, he did none of the chemical work. The core of the business was the isolation of tartaric acid from argol, the inexpensive and abundant solid residue of wine making. Some tartaric acid was sold to textile producers, who employed it as a mordant in wool dying, and some was oxidised to oxalic acid and then converted to salt of sorrel (potassium hydrogen oxalate) and/or ammonium oxalate.[52] Over the period 1802–15, nearly 17 000 lbs of argol were processed to yield 9700 lbs of tartaric acid. Most of the tartaric acid was converted first to 8400 lbs of oxalic acid, and then to 6300 lbs of salt of sorrel. Expenses exceeded revenue until 1813, when purchases of argol ceased. When the production of products ceased in 1815 on Tennant's death, the business had shown a net profit of £685 (Figure 5.6); of this 45% went to Tennant.

The business enterprise of Wollaston and Tennant was of enormous benefit to both men, for it returned a healthy profit and led each into novel chemical researches. Wollaston, however, well earned his return by the many hours
of labour he invested in the batch processes he devised for both the platinum
and the organic processes. Tennant was the fortunate recipient of his partner's
labours, for all evidence suggests that his principal contribution to the success
of the business was provision of capital, although his part in initiating the business was probably crucial and should not be disregarded. But what, one might
wonder, was Tennant doing over the many years while Wollaston was working
assiduously in the back of his house? He had become a gentleman farmer.

Tennant and agriculture

In the mid 1790s, Tennant's health began to decline and he started to take
daily rides on horseback as a restorative, for the medical profession recommended riding for improved blood circulation and consequent physical and
mental well-being. His rides became a regular feature of his daily schedule
and soon took precedence over his scientific work, as we noted earlier with
the diamond experiments. The rides appeared to rejuvenate Tennant, and many
excursions on horseback were lengthy trips to places as far afield as Liverpool
and Selby. On one trip north to Lincolnshire, Tennant learned that farmers were
experimenting on newly enclosed land with novel crops, such as rapeseed and
root vegetables, introduced from Holland. Intrigued by this, Tennant purchased
seven acres near Epworth in Lincolnshire in the summer of 1799 and became a
frequent visitor to the area thereafter.[53]

Acting on claims from resident farmers that there were considerable differences in the two types of limestone used to increase the fertility of the soil,
Tennant initiated a chemical investigation. The research, published in 1799,
revealed that calcareous limestone was beneficial to emerging seedlings, but
magnesian limestone was injurious to them.[54] He identified several quarries
which had high levels of the magnesian limestone and others which contained
only the beneficial limestone, thus instigating changes in liming practices in
English agriculture. Determined now to try his own hand at improving crop
yields by applying scientific principles, but unable to expand his holdings in
Lincolnshire, in late 1799 Tennant purchased 500 acres of newly enclosed
land near Shipham in Somerset. He soon after sold off the greater part, but
retained 165 acres, on which he built a summer home. For the remainder of
his life, Tennant divided his time between the farm, where he spent most of
the warm months, and London, where he resided during the cool parts of the
year.

A near neighbour to the Shipham property was a fellow Emmanuel student, Thomas Smith, who owned the Manor of Easton Grey in Wiltshire. Tennant was a frequent guest at Easton Gray, and there befriended many of the leading figures of English political and economic life: David Ricardo, Thomas Malthus, Leonard and Francis Horner, Samuel Romilly, Lord Brougham and others who moved among the Whigs of Holland House, the Kensington mansion of Lord and Lady Holland. He soon gained access to their London gatherings as well, and in 1799 he became a member of a dining and conversation club known as the 'King of Clubs'. The club members met on the first Saturday of each month at the Crown and Anchor in the Strand and a typical meeting 'consisted chiefly of literary reminiscences, anecdotes of authors, criticisms of books, etc.',[55] made palatable by bottles of sherry, madeira, port, bucillas and claret, as the bill for one dinner indicates. Such close interaction with the leading intellectuals of the day made Tennant a keen student of political economy, and he even contemplated founding a chair of political economy at Cambridge and writing a book on the subject. These agricultural and socio-political initiatives in the early years of the nineteenth century must have given Tennant ample reason to neglect the science that 'the chemist, the Royal Society, and the public' (as Banks had phrased it), and Wollaston especially, anticipated from him. One member of Tennant's new social circle, John Whishaw, observed:

> The change in his habits, occasioned by his agricultural engagements, was not equally favourable to his scientific pursuits. His spirits were often exhausted, and his mind fatigued and oppressed, by the attention which he thought it necessary to bestow upon the correspondence with his agents, the examination of his farming accounts, and other details equally tedious and minute; and it is impossible to reflect upon the time thus consumed, without lamenting that it was not employed for purposes more beneficial to mankind, and more worthy of his genius and understanding.[56]

The years following his discovery of osmium and iridium were scientifically fallow ones for Tennant, and an interesting insight into his scientific decline is given by Berzelius, who visited England in 1812 and met most of England's best chemists. He and Tennant rode on horseback to inspect the oat field which Tennant had limed in controlled amounts, and Tennant pointed out the well-developed oats at the more heavily limed end and the weaker plants at the unlimed end. Despite these pleasant times, Berzelius recognised that Tennant's abilities were eroding:

> Tennant is of about the same age as Wollaston, but is gray-haired and looks like an old man. He is a charming man, gets off a lot of droll ideas which entertain any sort of society, scientific or otherwise. He is a rather good, reliable chemist, but doesn't have either Wollaston's or Davy's head; and now he has lost much of his memory,

so that one can tell him the same thing on two successive days with full assurance that it will be new to him. He is badly dressed, is careless of his appearance, and makes a poor showing. His chemicals are so helter skelter that he gets permission to pull out all the table drawers in the parlour to convince himself of the absence of what one would never expect to find except in a laboratory.[57]

Despite the decline in his mental acuity, and his persistent lack of resolve, Tennant retained the ability to grasp and clarify the subtleties of complex issues, and to express his thoughts with wit and intelligence. In an effort to capitalise on his many personal talents, his friends steered him at this point in his life to a new career as Cambridge's eighth Professor of Chemistry.

The eighth Professor of Chemistry

Tennant's fondness for intellectual debate and social intercourse made him an effective communicator, and he made his first foray into informal scientific lecturing in 1812 as a consequence of his interests in mineralogy. In 1807, several members of the Askesian Society and the British Mineralogical Society with a shared interest in geology banded together to form the Geological Society of London.[58] In 1808, Tennant's close friend Francis Horner became a member and in 1809 Tennant too was elected. In 1809 the Society engaged rooms in Horner's name at 4 Garden Court, Temple (Tennant's address) to hold and display its extensive mineral collection, and this proximity to such mineralogical riches rekindled Tennant's interest in the subject. One result was a pedestrian publication on the analysis of volcanic boracic acid, to which Horner had drawn his attention.[59] A more important result was the inclusion of mineralogical topics into the morning parties that he frequently hosted in his rooms at Garden Court. A promise to present a mineralogical discourse to a group of his friends, male and female, was expanded at their request into a short lecture series, which he delivered in 1812. Whishaw, who was in attendance, states that:

> The great clearness and facility of his statements, the variety and happiness of his illustrations, and the comprehensive philosophical views which he displayed, were alike gratifying to every part of his audience. He delivered about four lectures, each of which was of great length; yet the interest of his hearers was never in the least suspended. Though his style and manner of speaking were raised only in a slight degree above the tone of familiar conversation, their attention was perpetually kept alive by the spirit and variety with which every topic was discussed, by anecdotes and quotations happily introduced, by the ornaments of a powerful, but chastised, imagination; and, above all, by a peculiar vein of pleasantry, at once original and delicate, with which he could animate and embellish the most unpromising subjects.[60]

This account of Tennant's lecturing technique, although embellished by the euphuistic style of his biographer, rings true; his noted lack of resolve at scientific investigations (termed 'indolence' by Wollaston) in no way impeded his talents at scientific communication. The lecture attendees noted his impressive performance, and in early 1813 he was invited to give a lecture on the principles of mineralogy to the Geological Society. This lecture established his reputation as a geologist sufficiently for him to be elected as a Vice-President of the Society in the same year, and to bring his name to the attention of those who sought to fill a more prestigious position.

In 1813, Francis John Hyde Wollaston resigned his Jacksonian professorship at Cambridge to become rector of Cold Norton in Essex. William Farish was elected as his successor, and thereon vacated his chair as Professor of Chemistry. Whishaw states that Tennant was urged to submit his name as a candidate by his many Cambridge friends, and agreed to stand for election in the knowledge that lecturing responsibilities would force him to focus once again on chemistry. As Brock notes in the next chapter, James Cumming, who was ultimately to succeed Tennant in the Chair, also declared for the position, but

> the exertions which were made by Mr Tennant's friends, and the assurances of support which he received, greatly exceeded what had ever been known on any similar occasion. The opposition being withdrawn, he was elected professor in May, 1813.[61]

The election of Tennant was significant, for he was the first Professor of Chemistry to have gained the position with an established reputation in research, the first non-resident to hold the position, the first to have strong personal connections to the leading European chemists and the first with an avowed desire to lecture and promote the subject.

The timing of Tennant's appointment was too late to mount a lecture course in 1813, so he began preparations for a course of 20 lectures to be given in April and May of 1814. He continued, however, to divide his time between London and Shipham while he was Professor, residing at Cambridge only during lecture time. From letters to Berzelius, we know that Tennant was in Cambridge for a few days in July 1813 to inspect the condition of the laboratory, but he left soon after for Shipham. On his return to London, Tennant worked diligently to assemble chemical apparatus and to prepare his lectures. He oversaw the construction of a galvanic machine powerful enough to melt iridium and to serve both as a demonstration device for the power of voltaic electricity and to showcase one of the elements he had discovered; he built a working version of his apparatus for the double distillation of water (originally conceived when he was an undergraduate, but not published until 1814) to accompany

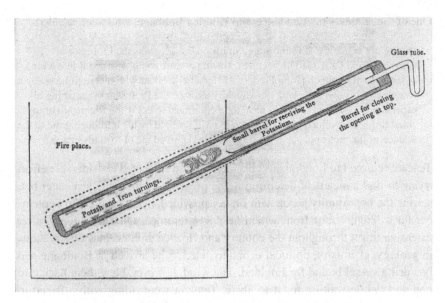

Figure 5.7. Tennant's improved apparatus for potassium production.

his discussion of latent heat. Also in 1814, he published the improvement on Gay Lussac and Thenard's method for producing elemental potassium shown in Figure 5.7.[62]

Tennant wrote to Berzelius for clarification on the Swede's preference for the traditional oxymuriatic acid interpretation of chlorine and, just days before the first lecture, he wrote, 'I am eager to demonstrate to my hearers some new fact which has come from you, or some ingenious method of making clear the old truths and of being able to boast that I owe the knowledge of it to your friendship.'[63] Clearly, Tennant was re-energised as the eighth Professor of Chemistry and his lectures must have been interesting, entertaining, factual and current, as the few known descriptions suggest. Whishaw says:

> During the months of April and May [in 1814] he delivered his first and only course of academical lectures, which was attended by a very numerous class of students. The greater part of these lectures were spoken from notes containing the order of the subjects, and the principal heads of discussion. But the introductory lecture was written at length . . . [and] presents a rapid and masterly outline of the history of chemistry, interspersed with many original and striking remarks on the nature of the science itself, on its extensive application, and prodigious effects in promoting the civilisation of mankind, and on the merits and discoveries of some of its most distinguished professors in different ages and countries . . . The impression made by these lectures will not soon be forgotten in the University.[64]

One of the students in attendance was Charles Babbage, who later recalled:

> During my residence at Cambridge, Smithson Tennant was the Professor of
> Chemistry and I attended his lectures. Having a spare room, I turned it into a kind
> of laboratory, in which [John Frederick] Herschel worked with me, until he set up a
> rival one of his own. We both occasionally assisted the Professor in preparing his
> experiments . . . I had hoped to have long continued to enjoy the friendship of my
> entertaining and valued instructor, and to have profited by his introducing me to the
> science of the metropolis, but his tragical fate deprived me of that advantage.[65]

Tennant returned to London after his lectures and expended considerable effort
trying to find a means of determining the presence of iodine in sea water but,
seizing the opportunity peace with France provided, he sailed there in September
for a lengthy tour from which he never returned. On 22 February, after
extensive travel throughout the country and 'loaded with curious observations
in geology, chemistry, political economy, etc.',[66] he arrived at Boulogne and
boarded a vessel bound for England. The wind, however, blew them back into
port and, while waiting for it to abate, Tennant went riding with a friend to
view Buonaparte's Pillar, which was nearby. On their return, Tennant rode his
horse across the drawbridge of an old fort which gave way under him and he
plummeted into the gully below, his horse landing on top of him. He died within
the hour and was buried in the Boulogne public cemetery. Cambridge had lost
its innovative and eccentric eighth Professor of Chemistry.

There is no known likeness of Tennant, which is almost certainly due to his
refusal to allow one to be made. His old Emmanuel colleague Busick Harwood
had covered his college walls with paintings of his friends, but even he could
not induce Tennant to sit for a portrait and, perhaps as a result, Tennant is not
included in the 51 men depicted in the popular engraving 'The Distinguished
Men of Science of Great Britain living in 1807–08, assembled at the Royal
Institution'. Again one must rely on Whishaw for a description:

> Mr. Tennant was tall and slender in his person, with a thin face and light
> complexion. His appearance, notwithstanding some singularity of manners, and
> great negligence of dress, was on the whole striking and agreeable . . . The general
> cast of his features was expressive, and bore strong marks of intelligence; and
> several persons have been struck with a general resemblance in his countenance to
> the well-known portraits of Locke.[67]

Tennant lived in a period of fundamental chemical discovery and he recognised
that he had been fortunate to have been a participant in it. He had progressed
from a humble and unfortunate childhood to become one of England's most
notable chemists, and his comment to a friend after hearing Davy's lecture on
the decomposition of the alkalis is appropriate for his epitaph, 'I need not say

how prodigious these discoveries are. It is something to have lived to know them.'[68]

Notes and References

1. See Partington, J. R. (1972), *A History of Chemistry*, London: Macmillan, vol. 4, pp. 6–22, for a brief account of Volta's work and its early development in England. The research was published (in French) in *Phil. Trans.* **90** (1800), pp. 403–31.
2. Davy, H. (1826), 'The Bakerian Lecture: on the relations of electrical and chemical changes', *Phil. Trans.* **116**, pp. 383–422.
3. Thomson, Thomas (1830), *The History of Chemistry*, 2 vols, London: H. Colburn and R. Bentley, vol. 2, p. 232.
4. The chronology of element discovery is taken from Ihde, Aaron J. (1964), *The Development of Modern Chemistry*, New York: Harper & Row, pp. 747–9.
5. For a comprehensive discussion of the Copley Medal, see Yakup Betkas, M. and Crosland, Maurice (1992), 'The Copley Medal: the establishment of a reward system in the Royal Society, 1731–1839,' *Notes Rec. Roy. Soc. Lond.* **46**, pp. 43–76.
6. A list of all Copley Medal awardees to 1847 is given in Weld, Charles Richard (1848), *A History of the Royal Society*, London: Cambridge University Press, vol. II, pp. 566–72.
7. Royal Society Journal Book, Minutes of Weekly Meetings (1793–1829), 30 Nov. 1804, Royal Society archives.
8. Anon. [J. Whishaw] (1815), 'Some account of the late Smithson Tennant, Esq.', *Ann. Phil.* **6**, pp. 1–11, 81–100; the quotation is on p. 1. The author, assisted by Henry Warburton, was the lawyer John Whishaw. He had a small number of offprints published under his name for distribution to Tennant's friends. This biography is the primary source of information on Tennant's life and is invaluable for its insight into Tennant's personality, his evolving interests and social environment. Whishaw is unusually candid in his assessment of Tennant's character but, as a non-scientist, tends to underestimate the quality of his subject's scientific work. This weakness is counterbalanced in the unpublished thesis of A. E. Wales; see Wales, A. E. (1940), 'The life and work of Smithson Tennant', Dip. Ed. Thesis, University of Leeds. An abbreviated version of Wales' biography is Wales, A. E. (1961), 'Smithson Tennant, 1761–1815', *Nature* **192**, pp. 1224–6. Other useful works are McDonald, Donald (1962), 'Smithson Tennant, F.R.S. (1761–1815)', *Notes. Rec. Roy. Soc. Lond.* **17**, pp. 77–94; and Webb, K. R. (1961), 'Smithson Tennant, 1761–1815', *J. Roy. Inst. Chem.* **85**, pp. 432–4. Brief, derivative biographies are given by Harden, A. in the *DNB* and Goodman, D. C. in the *DSB*.
9. *Ibid.*, Wales (1940). Unattributed biographical details are taken from this reliable source.
10. Whishaw (1815), pp. 2–3.
11. Quoted in McDonald (1962), p. 79.
12. Whishaw (1815), p. 4.
13. Winstanley, D. A. (1935), *Unreformed Cambridge*, London: Cambridge University Press, p. 3.
14. Warburton, Henry (undated), preparatory notes for Wollaston biography, Cambridge University Library, Add. MSS 7736.
15. Whishaw (1815), p. 5.
16. Gunning, Henry (1854), *Reminiscences of the University, Town, and County of Cambridge*, London: George Bell, vol. 2, pp. 60–1.
17. Winstanley (1935), p. 61.

18. Warburton (undated).
19. Gunning (1854), vol. 1, p. 53.
20. Winstanley (1935), p. 261.
21. Wales (1940), vol. 1, pp. 72–3.
22. Tennant, Smithson (1814), 'On the means of producing a double distillation by the same heat', *Phil. Trans.* **104**, pp. 587–9.
23. Whishaw (1815), p. 5.
24. Perrin, C. E. (1982), 'A reluctant catalyst: Joseph Black and the Edinburgh reception of Lavoisier's chemistry', *Ambix* **29**, pp. 141–76.
25. Warburton (undated).
26. Warburton (undated).
27. Gunning (1854), p. 61.
28. Tennant, Smithson (1791), 'On the decomposition of fixed air', *Phil. Trans.* **81**, pp. 182–4.
29. Whishaw (1815), p. 7; Wales (1940), Part II, p. 80.
30. Whishaw (1815), p. 8.
31. Whishaw (1815), p. 9.
32. Tennant, Smithson (1797), 'On the nature of the diamond', *Phil. Trans.* **87**, pp. 123–7, especially p. 124.
33. Tennant, Smithson (1797), 'On the action of nitre upon gold and platina', *Phil. Trans.* **87**, pp. 219–21.
34. Whishaw (1815), pp. 10–11.
35. Thomson (1830), vol. 2, p. 236.
36. Bank of England archives, folio 11292.
37. Letter from W. H. Wollaston to Henry Hasted, 29 December 1800; copy in Gilbert papers, University College Library, London.
38. See Gilbert, L. F. (1952), 'W. H. Wollaston MSS. at Cambridge', *Notes Rec. Roy. Soc. Lond.* **9**, pp. 310–32, and Usselman, Melvyn C. (1978), 'The platinum notebooks of William Hyde Wollaston', *Plat. Met. Rev.* **22**, pp. 100–6.
39. Details of the platinum business are available in two papers by Usselman, M. C.: (1980), 'William Wollaston, John Johnson and Colombian alluvial platina: a study in restricted industrial enterprise', *Ann. Sci.* **37**, pp. 253–68, and (1989), 'Merchandising malleable platinum: the scientific and financial partnership of Smithson Tennant and William Hyde Wollaston', *Plat. Met. Rev.* **33**, pp. 129–36. Unreferenced details of the platinum business are taken from these sources.
40. An excellent source of information on platinum is McDonald, Donald and Hunt, Leslie B. (1982), *A History of Platinum and its Allied Metals*, London: Johnson Matthey.
41. Wollaston Mss., Cambridge University Library Add MSS 7736, notebook I, pp. 7–8.
42. See the two papers by Vallvey, Luis Fermin Capitan: (1994), 'The Spanish monopoly of platina: stages in the development and implementation of a policy', *Plat. Met. Rev.* **38**, pp. 22–31; and (1994), 'The Spanish monopoly of platina: Part II: first attempts at organizing the collection of platina in the viceroyalty of New Granada', *Plat. Met. Rev.* **38**, pp. 126–33.
43. Kronberg, B. I., Coatsworth, L. L. and Usselman, M. C. (1981), 'The artefact as historical document Part 2: the palladium and rhodium of W. H. Wollaston', *Ambix* **28**, pp. 20–35.
44. Wollaston, W. H. (1804), 'On a new metal, found in crude platina', *Phil. Trans.* **94**, pp. 419–30, and 'On the discovery of palladium with observations on other substances found with platina', *Phil. Trans.* **95** (1805), pp. 316–30. For an analysis of the controversy surrounding Wollaston's discovery of palladium, see Usselman,

Melvyn C. (1978), 'The Wollaston/Chenevix controversy over the elemental nature of palladium: a curious episode in the history of chemistry', *Ann. Sci.* **35**, pp. 551–79.

45. Tennant, Smithson (1802), 'On the composition of emery', *Phil. Trans.* **92**, pp. 398–402.
46. Collet-Descotils, H. V. (1803), *Ann. Chim.* **48**, pp. 153–76; Fourcroy, A. F. and Vauquelin, N. L. (1803), *Ann. Chim.* **48**, pp. 177–83.
47. McDonald and Hunt (1982), p. 150.
48. Tennant, Smithson (1804), 'On two metals, found in the black powder remaining after the solution of platina', *Phil. Trans.* **94**, pp. 411–18.
49. Wales (1940), Part II, p. 42.
50. Wollaston (1805), pp. 317–18.
51. Wollaston Mss., Cambridge University Library Add. MSS 7736, Wollaston's published papers.
52. Ball, Matthew and Usselman, Melvyn C., unpublished research.
53. Wales, A. E., unpublished notes. I am grateful to the late Mr. Wales for a copy of his extensive notes on Tennant; most of the details of Tennant's land purchases are taken from these notes.
54. Tennant, Smithson (1799), 'On different sorts of lime used in agriculture', *Phil. Trans.* **89**, pp. 305–14.
55. Wales (1940), Part II, p. 99.
56. Whishaw (1815), p. 84.
57. Letter from Berzelius to Gahn, 25 January 1813; Soderbaum, H. G. (1912–14), *Jac. Berzelius Bref*, Upsala: Almqvist and Wilsells, vol. 9, p. 73.
58. Woodward, Horace B. (1907), *The History of the Geological Society of London*, London: Geological Society of London, pp. 6–7.
59. Tennant, Smithson (1811), 'Notice respecting native concrete boracic acid', *Geol. Soc. Trans.* **1**, pp. 389–90.
60. Whishaw (1815), pp. 90–1.
61. Whishaw (1815), p. 92.
62. Tennant, Smithson (1814), 'On an easier mode of procuring potassium than that which is now adopted', *Phil. Trans.* **104**, pp. 578–82.
63. Letter from Tennant to Berzelius, 17 April 1814; copy in Wales (1940), Part III.
64. Whishaw (1815), p. 92.
65. Babbage, Charles (1864), *Passages in the Life of a Professor*, London: Longman, p. 38.
66. Letter from Marcet to Berzelius, 29 March 1815; Soderbaum (1912–14), vol. 1, part 3, pp. 117–19.
67. Whishaw (1815), p. 97.
68. Whishaw (1815), p. 98.

6

Coming and going: the fitful career
of James Cumming

William Brock

Centre for Cultural and Historical Studies of Science,
University of Kent at Canterbury

In that well-loved Victorian classic of the 1890s, *The Diary of a Nobody*, the Savoyard brothers George and Weedon Grossmith record their mundane anti-hero Mr Pooter puzzling over why his city friends Cumming was 'always *going*' while Gowing was 'always *coming*'.[1] Pooter's pun did not go down well, and we may well imagine that James Cumming (Figure 6.1), who must have had his leg pulled at school in a similar manner, never found it amusing. Nevertheless, the punning metaphor of 'coming and going' seems apt when applied to a peripatetic career that involved over 40 years of travelling between Cambridge and King's Lynn and continually changing hats between that of a university professor of chemistry and a clergyman of the Church of England.

James Cumming was born in Westminster, London on 26 October 1777.[2] His forbears were Scots and his grandfather, John Cumming, is reputed to have left Altyre for East Anglia after the Jacobite defeat at Culloden in 1746. His father, also named James Cumming (1745–1804), was an hotelier. In 1791 he became the proprietor of the Buxton Hall Hotel (currently the Old Hall Hotel), leasing the property from the fifth Duke of Devonshire. The hall had been erected by the Earl of Shrewsbury in 1572 as a secure house for Mary Queen of Scots whenever she came to take the Buxton waters, and passed to the Cavendish family in the eighteenth century. Under James Cumming's proprietorship the Hall hotel continued to cater for visiting aristocracy, gentry and bishops. Clearly, Cumming, Sr was socially rather more than an ordinary village innkeeper. Figure 6.2 shows his family tree: he was twice married and a son (John, 1770–1858) by his first marriage was a watchmaker in Oxford Street. James Cumming, Jr was the eldest of four surviving sons by his father's second marriage to Alice Atherton (1749–1811) in January 1777, and the only one to

The 1702 Chair of Chemistry at Cambridge: Transformation and Change, ed. Mary D. Archer and Christopher D. Haley. Published by Cambridge University Press. © Cambridge University Press 2005.

Figure 6.1. The only known photograph of 'the venerable' Reverend Professor James Cumming (1777–1861), probably taken in the 1850s when Cumming was in his seventies.

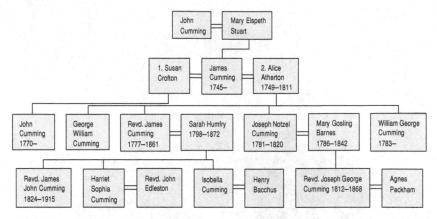

Figure 6.2. James Cumming's truncated family tree, based upon information supplied by David Bates and the author's own research.

follow an academic bent. His brother Joseph Notzel (1781–1820) ran the Old Bath Hotel in Matlock, and had a son, the Reverend Joseph George Cumming (1812–68), who was to enjoy an even more distinguished career than his uncle, as an educationist, Manx antiquarian and geologist.[3] James's two other brothers were George William Cumming, who became a captain in the Royal Navy; and William George Cumming (1784–1813), who fought and died in the Peninsular War. These facts suggest that the Cummings were from the rentier class.

Cumming was sent to school at Marlborough in Wiltshire,[4] from where (like Newton 135 years before him) he was admitted to Trinity College, Cambridge as a sizar in September 1796.[5] A sizarship, it should be recalled, was effectively a scholarship by which the college gave a student free board, lodgings and tuition in exchange for some domestic duties. In almost all cases a sizarship was a mark of intellectual promise, and in this Cumming was following the same route by which Waller, Mickleburgh, Pennington and Farish came to Cambridge. At the time Cumming matriculated, Trinity, like the other Cambridge colleges, was undergoing a scholarly renaissance after decades in what the poet Wordsworth described as the 'unscoured' doldrums.[6] Admissions were slowly rising to about 50 students each year and the quality of the tuition was improving. However, there were only two tutors and students who aspired to doing well in examinations commonly paid for extra-mural tuition. Whether Cumming sought extra coaching is not known. His mathematics tutor was Henry Porter (*c*. 1757–1822), who had been fourth wrangler in 1780, elected a Fellow in 1782, and served as vicar of Great St Mary's church in Cambridge between 1792 and 1795.[7] Cumming evidently showed great academic promise since he was elected a Scholar in 1800, the year before he graduated as tenth wrangler in the Mathematical Tripos examination. There had been a college scandal in the mid-1780s when it was discovered that Fellows were being elected by favouritism without formal examination;[8] but by the time Cumming was made a Fellow of Trinity in 1803 the honour was conferred strictly on academic merit. It was obligatory among Fellows at Cambridge (except for those intending to follow a medical career) to take holy orders, but Cumming had already taken this step in 1802 under the instruction of the Bishop of Lincoln, George Tomline.

Cumming's reticence in putting pen to paper means that we do not know how he was affected by the development of evangelicalism at Cambridge. Was he inspired by the preaching of Charles Simeon at Holy Trinity Church, as his fellow wrangler Henry Martyn was? Did he take part in the campaign to establish a branch of the British & Foreign Bible Society in Cambridge?[9] Evangelical commitment is certainly implied by his decision to take holy orders a year before he obtained the Trinity Fellowship. Small clues also suggest that he was of low rather than broad church disposition in later life. For example, in June 1860 Cumming told Joseph Romilly, the University Registrary, how angry he was that the Revd. Charles Kingsley had been appointed Regius Professor of History. Presumably his anger was due to the fact that the latitudinarian Kingsley was not a serious historian, but a writer of historical novels, and that Kingsley had been sympathetic to the Chartists.[10] An anonymous Cambridge obituarist (Revd. James Edleston), speaking from inside knowledge, was also explicit that Cumming (like his friends Henslow and Sidgwick) was dismissive of Darwin's theory of evolution.[11] It may also be significant that in 1818 Cumming briefly served as a University Proctor during the campaign by Christopher Wordsworth (master of Trinity) to improve the morals of students accommodated in lodgings in the town.[12]

What did Cumming do between graduation and his election to the chemistry chair in 1815? The records do not say. It was not uncommon in unreformed Cambridge for Fellows to lose their academic interests and to hang on in college until a benefice came their way, allowing them to leave and marry.[13] Despite his invisibility to the historian, Cumming does not seem to have fallen victim to this academic malaise. Yet his silence is surprising given the exciting developments in chemistry during this period – the complete acceptance and consolidation of Lavoisier's new system of chemistry, the debates over the nature of galvanism (electrolysis), the discovery of new elements by Davy, and the announcement by Dalton of the atomic theory. Although there were two monthly science periodicals, *Nicholson's Journal* and the *Philosophical Magazine* (and from 1813, *Thomson's Annals of Philosophy*), Cumming was not moved to publication.

This is curious, given that his fellowship involved no teaching duties. He was Junior Dean from 1807 to 1814, which gave him administrative responsibilities for maintaining order and discipline among Trinity's undergraduates, but he would have had abundant leisure for scientific pursuits. The only clue that we have of his experimental interests comes from an obituary statement that, as an undergraduate, and possibly therefore, as a graduate, he had become a close friend of Francis John Hyde Wollaston (1762–1823), the Jacksonian professor of natural philosophy, assisting him at undergraduate lecture demonstrations.[14] Wollaston, it should be recalled, had elected to lecture on chemistry rather than natural philosophy, while William Farish, the holder of the 1702 Chair from 1793, followed Richard Watson's style in offering lectures on the 'application of chemistry to the arts and manufactures of Britain'.

There is no evidence that Cumming sought out the evangelical Farish's company, and his closeness to Wollaston must have been the key factor in Cummings's application for the chemistry chair in 1813 and 1815. These elections (decided since 1793 by an open poll of the Senate)[15] came about through a series of 'comings and goings'. On Wollaston's resignation in 1813 to become an archdeacon in Essex, a game of musical chairs developed. William Farish decided to move to the Jacksonian chair, leaving the 1702 Chair vacant. The 36-year-old Cumming was initially a candidate, but as Usselman noted in the previous chapter, he withdrew his name on learning that Smithson Tennant was the preferred choice.[16] However, the latter's accidental death in France in 1815, after delivering only one course of lectures, led to Cumming's re-application. Several Cambridge M.A.s (including Francis Wollaston himself, as well as the Professor of Mineralogy, Edward Daniel Clarke) hoped that Wollaston's brilliant brother, William Hyde Wollaston (1766–1828), would apply. There is evidence in Wollaston's correspondence that had he done so, Cumming would have withdrawn his name again.[17] In the event, Wollaston declined to be a candidate, largely on the grounds that, unlike Cumming, he had had no teaching or demonstrating experience. In any case, he clearly preferred the scientific excitements of London to Cambridge. Although neither man knew the other at this time, it is interesting to read that Wollaston recommended to Henry Warburton and Edward Clarke that Cumming's candidature should be supported. That suggests that, despite not having published anything, Cumming's experimental and lecturing prowess had come to the attention of Cambridge's external natural philosophy network.[18] Because he was unopposed, only seven Cambridge electors needed to cast their votes.[19] In fact there had been a rival candidate, the 23-year-old John Herschel, but he graciously retired from the competition before it went to the vote. As Herschel told Charles Babbage:

I made my retreat with as good a grace as might well be and have nothing for my pains but knowing that if Tennant had lived a twelvemonth or two longer I should in all probability . . . have been his successor. I do not care much about it and on the whole I believe it is better as it is. I have been attending Clarke's lectures and am become half a mineralogist – and have been analysing some of his specimens for him.[20]

Would Cambridge chemistry have developed differently had either Wollaston or Herschel been elected? While Cambridge might well have become the centre where photography was developed, one may doubt whether either chemist could have done anything more than Cumming did to expand the teaching of chemistry. Medical teaching and the Cambridge examination system had to be first reformed before chemistry could become a distinctive discipline and department in the university. Like Cumming, too, they would in all probability have been forced by financial necessity to being peripatetic professors.

The fact that Cumming had published nothing in chemistry obviously did not count against him; in any case, some of his predecessors had known nothing of the subject when appointed. What counted in his favour was undoubtedly the fact that he had proved a sound college man. His abilities as an experimentalist and demonstrator, which Francis Wollaston would have noted, must also have counted strongly in his favour. He was the ninth chairholder. There had been eight before him in 113 years; he was destined to hold the 1702 Chair for 46 years.

The syllabus of Cumming's first course of 28 lectures in 1815–16 has survived.[21] The course structure was Lavoisian – that is, indebted to the format of exposition laid down by Lavoisier in the *Traité élémentaire de chimie* in 1789 and popularised in Britain in successive editions of Thomas Thomson's *System of Chemistry*. Chemical materials were divided into imponderable agents of heat, light, electricity and galvanism and ponderable bodies, the chief of which is the combustible agent, oxygen. The latter was then used to examine each of the oxidation products of ponderable bodies, beginning with water, and followed by the mineral acids and alkalis. The course then examined the properties of metals, classified according to whether they were malleable, brittle (but easy to fuse) or brittle and difficult to fuse. A very brief coda (probably just two lectures) referred to vegetable and animal compounds, the analysis of minerals and mineral waters and 'electro-chemistry'. A mention of iodic acid (first prepared by Gay-Lussac in December 1813) shows that the materials were up to date, as a reference to the atomic theory in the context of the combination of acids and bases to form neutral salts confirms. Altogether Cumming defined 350 subtopics that his auditors were expected to note. He ended the course with a review of the history and present state of chemistry.

Another printed syllabus for 1825 – interleaved with blank sheets for the convenience of students' notetaking – shows some subtle changes.[22] The combustion supporters were now extended to include the halides. Animal and vegetable chemistry shows more sign of theoretical interest, with vegetables generalised as:

$$\text{Ternary} \quad \{C\}\text{ in} \qquad \{O_2 + H_2\} \rightarrow \text{gums}$$
$$\text{compounds} \quad \{H\}\text{ excess} \quad \{O_2 + C\} \quad \rightarrow \text{oils}$$
$$\text{of} \quad \{O\}\text{ with} \qquad \{H_2 + C\} \quad \rightarrow \text{acids}$$

and animals as:

$$\text{Quaternary compounds of }\{O_2\,H_2\,C\,N\}$$

The inclusion of Brodie's experiments on blood and animal heat, Prout's discovery of hydrochloric acid in gastric juice, and information on human bladder calculi demonstrate that Cumming was trying to make the syllabus appeal to medical students. The only 'chemical' paper he published was a purely qualitative analysis of a huge bladder stone that had been exhibited in the library of Trinity College.[23] Nevertheless, Cumming also included his own current research on electrochemistry, and the 'construction of the galvanoscope'. A reference to ultimate analysis by copper oxide implies that he explained Berzelius's method of organic analysis. There were lectures devoted to the theory of atoms, volumes and equivalents, and Wollaston's slide rule was demonstrated. Equivalent weights were bracketed after each element. Further indicators that Cumming was up to date in his reading are shown by the inclusion of hydrogen peroxide and Berzelius's new element, selenium. Metals were, however, still classified qualitatively as before in 1816.

In 1831, as a result of changes in the medical curriculum, Cumming increased the number of his lectures to 50, dividing the course into two parts, inorganic and organic. A third surviving interleaved syllabus dates from 1834 and provides only the inorganic part of this course.[24] The inorganic syllabus was again right up to date with chemical symbols now being continually used, and considerable attention being reserved for electrochemistry and the 'recent researches of Faraday'. Although there are no surviving syllabuses from the 1840s and 1850s, all the evidence points to Cumming as a conscientious teacher who was well aware of contemporary developments.

Recognition from the wider scientific community followed Cumming's election to the 1702 Chair. He was admitted to the Fellowship of the Royal Society on 11 January 1816 (along with Lord Byron) on the recommendation of Francis Wollaston, Henry Warburton (then a Fellow of Trinity and a close friend

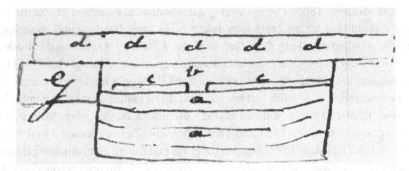

Figure 6.3. Sketch of James Cumming's laboratory by Erasmus Darwin.

of William Hyde Wollaston), John Herschel (then still a Fellow of St John's College), George D'Oyly (the theologian), Thomas Harrison (a mathematician and barrister), John Whishaw (a geologist-barrister and former close friend of Tennant's), and William Dealtry (the professor of mathematics at the East India Company's college at Haileybury).[25] Significantly, these were all Cambridge men. Cumming also joined the Geological Society in 1816, but never participated in its lively meetings. His involvement in plans for the new Cambridge observatory in 1820 were probably purely administrative, rather than reflecting a genuine interest in astronomy – though he was possibly chosen because of his prowess with scientific instrumentation.[26]

Who attended Cumming's lectures? We do not know how many students attended Cumming's early classes, though the 1822 sketch of the lecture room by Erasmus Darwin (Charles's elder brother) shown in Figure 6.3 implies that there were no more than three rows of seats – allowing possibly 18–20 auditors at the most. His sketch showed two large tables arranged in front of an array of furnaces with a blacksmith's chimney, with the rows of auditors viewing the tables. Erasmus Darwin found Cumming's lectures 'entertaining', writing to his brother Charles:

> He has 3 men to assist him so that we get over a good deal of ground in an hour. He lectures every day, all this [autumn] term.[27]

One of these 'assistants' was the young John Stevens Henslow (1796–1861), who was controversially elected to the chair of mineralogy in 1822 and to the chair of botany in 1825.[28] A much later assistant was the brilliant Scottish fifth wrangler Duncan Farquharson Gregory (1813–44) who, despite his mathematical bent, founded a short-lived Cambridge chemical society at which Cumming lectured.[29] The close relationship between the Mathematical Tripos and Cumming's interests is also shown in the encouragement he gave to another wrangler,

Robert Murphy (1806–43), to develop a mathematical treatment of electricity that was suitable for undergraduate reading. His most famous pupil, however, was the mathematical physicist George Gabriel Stokes.[30] On the medical side, Cumming evidently gave great encouragement to George Kemp (1808–85) to investigate the chemistry of biliary secretions.[31] Despite his love of chemical experimentation as a youth and his older brother's enthusiasm for Cumming's course, Charles Darwin did not attend Cumming's classes when he was an undergraduate between 1827 and 1831. The only science course he took was that of John Henslow in botany. Probably he had his fill of chemistry classes during his brief medical training at the University of Edinburgh before coming to Cambridge.

Showmanship and humanising anecdotes were essential to attract students when attendance at professorial lectures was voluntary. Like his fellow professors Henslow, Sedgwick, Farish and Farish's successor Willis, James Cumming drummed up undergraduate auditors by making his lectures popular. Erasmus Darwin enthused over Cumming's showing some beautiful crystals that were supposedly artificial diamonds.

> The way he made them was, by putting a few drops of Sulphuret of Carbon at the bottom of a well-stopped bottle, & the liquid, being the most volatile of any, crystallised against the sides of the vessel. They were far more brilliant than any diamonds I ever saw, but if air was admitted they melted away.[32]

Why Darwin (or was it Cumming?) should have thought the crystals to be diamonds is unclear. A former student's annotated copy of this 1825 syllabus in Cambridge University Library reveals how Cumming sprinkled the lectures with anecdotes and historical references. For example, when demonstrating hydrogen's inflammability, he evidently pointed to the laboratory ceiling where Farish had blown a hole in an earlier demonstration. Nitrous oxide led to a digression on Southey's account of its exhilarating effects. Dulong had lost the sight in one eye from experimenting with chloride of nitrogen and Davy had been equally 'much injured'. And the perniciousness of selenic acid had caused Berzelius to cough for a fortnight. There are reports that Cumming literally shocked audiences with galvanic apparatus and frequently executed a cat with an electric shock. Little wonder, then, that in a gentle satire on Cambridge science in 1834, Cumming was placed in the company of William Clark, the anatomist, as well as Henslow and Babbage:

> And indeed I am not humming
> Thus to sing of CL-KE and C-MMING.
> Who all the universe surpasses,
> In *cutting up* and making *gasses*;

With *anatomy* and *chemics*,
Metaphysics and polemics.
Analysing and chirurgery.
And scientific surgery.[33]

In his early years Cumming would have faced intense competition from the mineralogical lecture course given by Edward D. Clarke, who took great pains to amuse as well as instruct. (We saw earlier that Herschel went to Clarke for mineralogical instruction.) Much of Clarke's course had to do with chemical analysis, something Cumming was never to provide. Clarke died in 1822, but fortunately for Cumming, apart from Henslow (1822–8), Clarke's successors Whewell (1828–32) and William Hallowes Miller (1832–80) tended to be too mathematical to be popular.[34] Unlike some of his predecessors, Cumming was also left untroubled by competition from the Jacksonian professor. Since 1813 Farish usually concentrated on natural philosophy in his lectures, while his successor, Robert Willis (1800–75), lectured exclusively on applied mechanics and engineering. For the first 25 years of his professorship, then, Cumming seems to have enjoyed good audiences. Populist tactics failed, however, in the 1840s, when undergraduate audiences plummeted. There were several reasons for this.

The principal problem lay in the serious competition from the curriculum taught by the College tutors. A Previous Examination ('Little Go'), taken in an undergraduate's fifth term of residence, was introduced in 1822 in order to make students take their study of classics, moral philosophy and mathematics more seriously. In the same year, Christopher Wordsworth and others introduced a 'postgraduate' Classical Tripos to be taken after the Mathematical Tripos for students who wanted to take honours in classics as well as mathematics.[35] This preserved the Mathematical Tripos as the most prestigious route for honours. Nevertheless, the majority of students were 'poll men', who took a lower standard pass degree without honours. Whatever their course, undergraduates had no need or obligation to attend professorial lectures, since their mathematical or classical tuition was provided entirely within the college system. However, in the 1840s Bishops began to complain that candidates for ordination lacked theological knowledge and to put this right, the theological content of the Previous Examination was tightened. In consequence, students and College tutors advised honours students against professorial lectures, however entertaining they might be.

This might have been offset by better recruitment of medical students. In 1819 the new Regius Professor of Physic, John Haviland (1785–1851), who was only too aware that Cambridge medical students were forced to complete their education elsewhere, mainly at the London hospitals, began to give regular anatomical lectures. One of the several consequences of the Apothecaries Act of 1815 was also that knowledge of practical chemistry had become obligatory for a licence to practise medicine issued by the Society of Apothecaries. This factor alone would have encouraged medical students to attend Cumming's lectures. Haviland, Cumming and Henslow further strengthened medical education and examinations in 1829 when they persuaded the Senate to agree that medical students be required to attend the lectures of the Downing Professor, as well as those of the professors of anatomy, botany and chemistry. The latter three professors were also obliged to assist the Regius Professor of Physic in the conduct of medical examinations.[36] As a consequence, in February 1831, a syndicate was established to consider the accommodation needed for the expected expansion of medical and scientific teaching. Cumming was on this committee, which reported in February 1832 that facilities could be cheaply improved by adding a wing to the Botany garden building that had been erected in 1784 and shared since then by the Jacksonian professor and the professors of chemistry and botany.[37] Despite a row over costs, the wing was erected during 1832–3 and for the next thirty years it was used by the professors of anatomy, botany, chemistry and applied mechanics. Figure 6.4 shows the early development of the New Museums site, drawn up by G. Smith.[38]

Unfortunately for Cambridge students, in 1835, in response to criticism from London and provincial doctors, the Royal College of Physicians repealed its rule that Fellows of the College had to be Oxbridge graduates. This meant that there was no longer any real advantage in taking a medical degree at Cambridge when there were better facilities elsewhere. In 1851, when Cumming gave evidence before the Graham Commission inquiring into the state, discipline, studies and revenues of the University and its colleges, he confessed to never having had more than four or five medical students attend his lectures in the last fifteen years of teaching. 'The study of chemistry has not only been neglected', he told the Commissioners, 'but discouraged in the University, as diverting the attention of pupils from what have been considered their proper academical studies.'[39] Because the numbers of students taking chemistry gradually declined, in 1845 Cumming abandoned offering 50 lectures and went back to 28.

Despite the improvement in professorial accommodation in 1833, Cumming's facilities remained basic and sufficient only for lecture demonstrations. There was no competition with London's teaching hospitals, the Royal College of Chemistry (founded 1845) or University College's Birkbeck Laboratory (founded 1846), all of which allowed students hands-on training in analytical chemistry. In 1851, Henry Bond, by then the Regius Professor of Physic, told the Graham commissioners that there ought to be more chemistry teaching at Cambridge to bring the medical teaching in line with other medical schools.

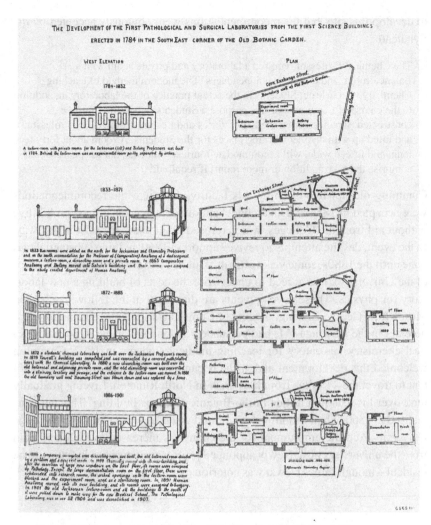

Figure 6.4. Early development of the New Museums site in Cambridge.

Cumming indicated his full support for better facilities to provide expanded teaching – meaning practical analytical training.

A new opportunity for chemistry teaching had come about with the establishment of the Natural Sciences Tripos (NST) as a 'postgraduate' degree in 1848. The Botanic Garden, its soil worn out, had been moved between 1846 and 1852, leaving valuable building space between Free School Lane and Corn Exchange Street.[40] Perhaps not surprisingly, the 76-year old Cumming was not a member of the Syndicate appointed in February 1853 to decide how best to develop this site, but his recommendations were obviously accepted by the Syndicate:

> The Chemical Professor required a laboratory and private room with a Balance-room, and other usual appendages. The modern method of teaching Chemistry is to instruct Students in the actual practice of the laboratory, in addition to the course of Lectures. For this purpose a Students' laboratory must be constructed, separated from the Laboratories and Lecture-Room of the Professor, and fitted up with distinct working places for the Students. A room about forty feet long, and fifteen wide, will accommodate fourteen Students, but the plans should comprise the easy addition of more rooms if required.[41]

Cumming did not live to enjoy this facility. Although the recommendations were accepted in 1854 and an architect appointed, the costs proved prohibitive without aid from the Colleges, which refused point blank to assist in any way. In the event, despite attempts to revive support in 1860, nothing substantial was done until the 1880s, some years after the Duke of Devonshire (then Chancellor of the University), had agreed to underwrite the costs of building a new laboratory for physics. These developments are discussed in the following chapter. Meanwhile, George Liveing, the first of Cumming's students to gain first-class honours in the NST in 1851, had persuaded St John's College to open a practical chemistry laboratory for teaching medical students in 1852. Cumming welcomed this development and, as he became increasingly infirm and reluctant to travel the 44 miles from his parish in North Runcton, Liveing gradually took over his teaching duties in the Botanic Gardens building. This was formalised in 1860, when Cumming was 83 and when, as a result of the Cambridge University Act, the emolument for the chair was increased to £300 and paid from the university chest. Newly appointed professors were also, in future, to be resident – though James Dewar was notoriously to circumvent this condition.[42]

The Natural Science Tripos

It had long been Whewell's ambition to force the Colleges to make greater use of the university professoriate and thereby extend an undergraduate's liberal education. In his book *On the Principles of English University Education* (1837), revised a few years later as *Of a Liberal Education* (1845; revised 1850), Whewell laid down his ideas for reforming the examination system and encouraging attendance at professorial lectures. Whewell viewed the classics and mathematics (including Newtonian mechanics and optics) as the fundamental basis of education because their contents were fixed and largely deductive in character. The experimental sciences were, on the other hand, inductive and progressive in the sense that new knowledge was being continually added, giving them an uncertainty lacking in classics and mathematics. A truly liberal education needed acquaintance with both the fixed and progressive sciences, but the former had to be learned before the latter. The result of Whewell's long and often acrimonious educational campaign, in which Adam Sedgwick, the Woodwardian chair of geology, joined him, was the introduction of the Natural Sciences Tripos (NST) in 1848 as an honours degree taken after the Maths Tripos. Although Cumming, like the other science professors, must have been continually consulted over these deliberations, as a non-resident and rather elderly professor he was content to follow where others led.[43] The NST examined a student in six natural sciences over a four-day period in four-hour bursts. Comparative anatomy and geology were tested on the first day, physiology and botany on the second, chemistry and mineralogy on the third, while the fourth day was devoted to a general paper containing all of these subjects, together with questions on the history and philosophy of science. The maximum marks possible were 6300, but few in the early years scored above 1500 and examiners usually gave distinctions (via vivas) in not more than two subjects.

The Board of Natural Science Studies confidently informed the Graham Commissioners in 1852 that:

> The science of Chemistry is rapidly extending its ramifications into all arts of life, and a knowledge not merely of its principles, but also of the practical power of applying them, is becoming daily more and more important as a part of general as well as professional education. The present venerable Professor of this science has lived to see the day, after many years of comparative indifference and neglect, when his lecture-room is again well attended by students; an effect partly produced by the recent legislation in the University, and partly by the interest which the results of chemistry have created amongst all classes of the community.
>
> But the science is much too extensive for the teaching of one Professor: and we venture to hope that the University will eventually add a second who may divide with the present Professor the vast range of subjects contained in this extensive science.[44]

In 1851, when six candidates first sat the new examination, Cumming told Whewell that he would set questions on Daubeny's *Introduction to the Atomic Theory* (2nd edn) and George Fownes's *Manual of Chemistry*, both of which (he noted) had good tables of contents. These were the two texts he had always recommended since the 1830s.[45] He proposed setting a dozen questions, four or five of which would relate to heat, electricity and general principles, three to organic chemistry, and the remainder to inorganic chemistry. Candidates, it must be remembered, were expected to tackle as many questions and subjects as they liked (or were able), the marks being simply aggregated for credit.

This system did not change substantially until 1861, when the NST became an undergraduate first degree (subject to candidates having first passed 'Little Go'); but this change was inherited by Cumming's successor, George Liveing.

The examination papers for 1856–8 seem typical of Cumming's general textbook approach, but like his earlier syllabuses they show that age had not prevented him from keeping up with developments in the subject.[46] Typical questions were:

> Shew by examples that the theory of chemical equivalents is a necessary result from Richter's law of the mutual decomposition of two neutral salts.

> What is the formula for the precipitate from common phosphate of soda by nitrate of silver? (This form of question was repeated every year.)

> Explain the doctrine of substitution, and shew that, in organic chemistry, the substitution of one element for another, even where the type is retained, is not limited by the electrical character of the elements.

> The analysis of an organic compound gave, in 100 parts, 20 C, 14.6 N, 6.7 H and 26.7 O; determine its formula.

> It has been assumed that equal volumes of different gases contain an equal number of atoms [*sic*]: state the arguments for and against this hypothesis.

The latter question was set in 1857, the year before Cannizzaro re-introduced chemists to Avogadro's hypothesis. Its reference to atoms, rather than molecules, is a certain indicator that Cumming's interest in the subject came from Daubeny's *Introduction to the Atomic Theory.*[47]

In the NST's final general paper there were 24 questions on all possible subjects. Cumming set three questions, typically historical in character:

> How and by whom was potassium discovered? Describe its chief properties and those of any two of its compounds.

> Write a brief history of the theory of acids.

> What is meant by the electric energy of chemical elements? Describe and explain Davy's method of protecting the copper sheaving of ships. Wherein does it fail?

Overall, it seems clear that Cumming did not change his syllabus and his style of teaching when the NST was introduced; what was new was that his lectures were now formally examined. Roberts has noted that chemistry teaching remained in the spirit of liberal education and was in no way specialised. The surviving syllabuses and examination questions all show how Cumming viewed chemistry and physics as a single subject and that he taught a general course that stressed facts rather than theories. Theories might be contentious, but Cumming was content that students heard both sides of a case rather than reaching a decision one way or the other. Like Daubeny, his opposite number at Oxford, Cumming did not think it was the purpose of undergraduate education to study science with the aim of applying it, but rather in order to understand the natural world. He would have agreed with Daubeny that, like classics, chemistry had the power to discipline the powers of the mind and to provide a truly liberal education. As Roberts has put it:

> In terms of skills, students were expected to 'know' and to 'understand' by being able to 'explain', 'describe', or 'prove' both theories and phenomena. They did not have to demonstrate ability to reason or to perform chemical operations.[48]

Had the NST been introduced in 1822 along with the Classical Tripos, at the time when Cumming was research-oriented, things might have been very different. But given the financial need to combine teaching with church preferment, and the laboratory facilities at Cumming's command, it is hard to see how his approach could have been any different.

Cumming's choice of textbooks by Fownes and Gregory[49] provides a clue as to how he kept up to date living, as he did for most of the year, far from scientific centres. No doubt his membership of the Chemical Society helped by providing him with its *Journal*, but the continually revised texts of Fownes and Gregory, together with the more philosophical approach taken by Daubeny, would have provided Cumming with all the latest information he needed.

Partly because of the Whewellian range of knowledge demanded by the examination if high honours were to be obtained, and partly because of its 'postgraduate' nature, the numbers of NST candidates between 1851 and 1861 never rose above half a dozen each year. Expansion of numbers came only in Liveing's time when the NST had become an honours degree independent of the Classical Tripos.

Scenes from clerical life

In 1819 Cumming left Cambridge on his appointment to the Rectorship of North Runcton, three miles to the south east of King's Lynn in Norfolk, a position formerly held by the Astronomer-Royal, Nevil Maskelyne. The small parish of 400 people was in the gift of Trinity College and the Quaker banker, Daniel Gurney (1791–1880), and worth £700 per annum.[50] There were fifteen acres of church land and the same amount as glebe. It was altogether a rich benefice. With his annual honorarium from the Cambridge chair of £100, Cumming was earning £800 p.a., the equivalent of £27,000 in today's values.[51] He could therefore afford to marry Sarah Humfrey (1798–1872), the daughter of a Cambridge businessman, in February 1820. From then onwards Cumming divided his time coming and going between Cambridge, where he lived in father-in-law's home in St Andrew's Close, and the Rectory at North Runcton. His life therefore revolved around conscientiously performing the role of village priest rather than a man of science. When resident in Cambridge during the Easter term (March onwards) to deliver his lectures, he seized the opportunity to socialise with colleagues and Romilly's diaries contain several mentions of dinners and parties.[52] At various periods of his rectorship, Cumming employed a curate so that he did not have to make the journey back to King's Lynn every weekend. During the years 1835–8 this was his nephew, the Revd. Joseph George Cumming (1812–68), a keen geologist who left North Runcton in 1838 to make his name as deputy headmaster of King William's College on the Isle of Man. In the 1850s Cumming's curate was the Revd. William Hay Gurney, son of Daniel Gurney, who was to succeed Cumming as rector in 1861.

All Saint's Church, shown in Figure 6.5, had been rebuilt in 1713 (its tower had collapsed in 1701) by a distinguished King's Lynn architect, Henry Bell, using local carstone and cement-rendered brickwork.[53] Its square tower and spirelet were a local landmark. The Cummings lived across the main village road in the Rectory or Parsonage. When the adjacent former manor house was transformed by Daniel Gurney into a huge 43-room mansion in 1834, the road was diverted westwards by about 300 metres and renamed 'New Road', effectively enclosing the Rectory within the church land.[54] Cumming's parish included the neighbouring farming hamlet of Hardwick and the village of Setch (formerly a market town, Great Setchley, on the river Nar, but long decayed so that in 1851 there were only 96 inhabitants). Here Cumming and Gurney built a small chapel-at-ease in 1844, where he held an additional Sunday afternoon service. The North Runcton parish incorporated altogether 2230 acres of agricultural land whose chief crops were wheat, barley, beans and turnips. Seventy-five per cent of the population were children and in 1833, like other Rectors in such rural communities, Cumming and Gurney (a rich philanthropist until his bank collapsed in 1865) raised funds for the building of a boys' and girls' National School adjacent to the church.[55] Cumming would have been responsible for the daily religious education of the schoolchildren. There is no evidence that he followed his pupil Henslow in giving scientific object lessons to the children.[56] According to Edleston, Cumming had a laboratory at the Rectory where he worked 'nearly all day'.

After the intellectual life of Cambridge, life in North Runcton must have been a humble, even dull existence. Cumming's parishioners were farmers, publicans, a blacksmith, shoemaker, carpenter and a schoolmaster and mistress. Of the three gentry in the village, the Cummings' chief intellectual companion

Figure 6.5. Nineteenth century photograph of All Saints' Church, North Runcton, where James Cumming was rector from 1819 until 1861.

was Gurney, whose older sister, Elizabeth Fry – the prison reformer – they would have met occasionally. We must remember, though, that there were well-educated clergy in the many surrounding villages, as well as in Lynn itself.

The Cummings had seven children, three of whom survived their childhoods – a son and two daughters. The son, James John Cumming (1824–1915), was educated at Caius College, Cambridge and ordained at Winchester Cathedral in 1848. He spent most of his life as Rector of East Carlton in Norfolk before retiring to Norwich in 1907. He appears to have had no interest in science.[57] A daughter, Isabella, married Henry Bacchus of Littlington Manor in Warwick, and Hannah Sophia married Cumming's obituarist, the Revd. Joseph Edleston, vicar of Gainsford in Durham.

Edleston claimed that Cumming continued to work in his laboratory despite bad health, and that at the time of his death at North Runcton on 10 November 1861 he was working on an *experimenta crucis* in physical optics.[58] He died in office, although by this time Liveing had already assumed many of the duties of the 1702 Professor. Cumming was buried in the churchyard on 14 November 1861. He left £7000 to his wife (the equivalent of £300 000 today), but his will made no mention of scientific apparatus, though Liveing stated later that this was auctioned by Cumming's executors.[59] Cumming's widow retired to King's Lynn and died in December 1872. In 1890, his surviving three children erected a memorial plaque to their parents in the chancel of North Runcton church, where it remains.

Becoming a researcher

Cumming was the first 1702 Chairholder to publish any serious research, though by later standards his output was meagre. Curiously, the responsibilities of a parish, marriage and parenthood, far from inhibiting his scientific activities, seem actually to have increased them. The period 1820 to 1834, during which his seven children were born, was the only time in Cumming's life when he communicated his scientific knowledge on paper to the wider world. The necessary stimulus for this reticent philosopher to blossom was undoubtedly the foundation of the Cambridge Philosophical Society (CPS) in 1819.[60] The idea of such a society had originated with Sedgwick and Henslow during a geological tour of the Isle of Wight during the Easter vacation in 1819. They then put the idea to the mineralogist Edward D. Clarke. The three men sounded out another thirty Cambridge dons and at a public meeting on 2 November 1819, the Society was formally created, with Farish, the Jacksonian professor, as its first president. Other founding members included Cumming, George Peacock and William Whewell. The CPS provided Cumming with a platform and persuaded this modest and unassuming man to reveal some of his own original experimental work. In return, the Society expressed its confidence in him by electing him its fourth president in the period 1825–7.

Cumming's research related more to the study of electricity than to chemistry, though we must bear in mind that imponderable bodies (as static and galvanic electricities were still defined in the 1820s) did not disappear from chemistry until the development of thermodynamics in the 1840s. Only then did the discipline of physics emerge from the shadow of chemistry. Moreover, although Whewell had rendered the French term *physicien* as 'physicist' in 1840s, the term was not popular until the end of the Victorian period.[61] Cumming was deeply interested in electricity, but he would not have seen himself as a physicist, just as Hans Oersted, the Danish discoverer of electrodynamics, never regarded himself as anything but a chemist. In the 1820s, in a series of papers given to the CPS, Cumming explored the ramifications of Oersted's discovery of electromagnetism in the spring of 1820, which (like Faraday) he learned about from the report in Thomson's *Annals of Philosophy* in that year.[62] Like Ampère, but independently of him, Cumming immediately saw that the effect of an electric current on a magnet could be used as an instrument for detecting, measuring and comparing the strengths of currents. The doubling or multiplying effect must have been obvious to anyone who understood Oersted's work. As Chapman has observed:

> If a compass needle was deflected clockwise when the wire of a particular voltaic cell lay *above* it in the magnetic meridian, the same needle would *also* be deflected clockwise if the wire was turned end-to-end and placed *below* the compass needle, without changing the rest of the circuit. Anyone perceiving this fact might deduce, as a matter of logic, that if the wire of the circuit was first passed above the needle, in the magnetic meridian, then folded and returned in a parallel path *below* the needle, the deflecting effect on the needle would be repeated, and a more sensitive indicator would result . . .[63]

Ampère called such a device a 'galvanometer',[64] and within twelve months Cumming as well as J. S. C. Schweigger and J. Poggendorff in Germany had published accounts of building and using such an instrument. Following Ampère, Cumming referred to the vertical single-wire system as a 'galvanometer' and the multiple-turn instrument as a 'galvanoscope'. In the former, the vertical slide was moved until a 'standard' deflection was obtained on the compass below. The relative 'strength' of a circuit was then read off the calibration scale (see Figure 6.6).

Cumming learned of these independent continental developments from Oersted, but made only a feeble attempt to establish his own credentials in *Annals of Philosophy*. However, he employed his own galvanoscope extensively. Using a cell with plates of zinc and copper he was able to detect currents in oxalic acid and hydriodic acids that electrochemists had not previously detected and to explore the effects on current intensity when the electrodes were brought nearer or separated. He also built a vertical galvanometer for determining the relative strengths of two voltaic cells using a slide-rule device. Both these devices were demonstrated to the CPS in April and May 1821 prior to the publications of Schweigger and Poggendorff, but history gave them the credit for the invention.[65]

Cumming had also been intrigued by Thomas Seebeck's announcement in 1822 of thermoelectricity (or thermomagnetism as the German discoverer

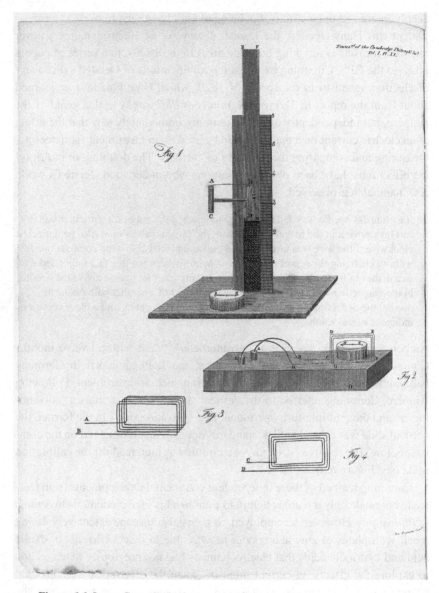

Figure 6.6. James Cumming's electromagnetic instruments.

named it).[66] He showed the CPS how the unequal heating of different conductors generated electric currents that were detectable using his galvanoscope.[67] On this occasion he introduced the astatic principle, whereby the effects of terrestrial magnetism were neutralised by 'placing a powerful magnet north and south on a line with its centre; and another, which is much weaker, east and west at some distance above it'. When Cumming varied the lengths of the connecting wire of the circuit he found that the needle deflections were almost in reciprocal proportion. The way that he writes of the deflection as being inversely proportional to the 'conducting power of the wire' suggests that he was close to hitting upon the resistance law announced by Georg Ohm in 1826. He showed that the Seebeck effect occurred even if two pieces of the same metal were joined and reasoned that the effect was due to heat, not to the junction of dissimilar metals. This led him to speculate that the diurnal variation of the earth's magnetic field might be due to the sun's passage across the earth. Cumming also reported in *Annals of Philosophy* on how the thermoelectric order of metals varied with temperature, noting that this bore little relation to the order of metals in the electromotive series or to their conductivity. He was the first person to note an inversion effect at high temperatures with couples like copper-iron, a phenomenon explained thermodynamically much later in the century.[68]

The three CPS papers reveal Cumming as a skilled experimentalist able to build his own apparatus for his lecture demonstrations. Encouraged by their warm reception, and by his election as President of the Society in 1825, Cumming planned to review the explosion of electrical research since Oersted's announcement in a book that would incorporate his own work. Whewell, who knew everything, informed him that he had been forestalled by one of Ampère's pupils, the mathematician Jean Baptiste Firmin Demonferrand (1795–1844). Whewell suggested that an English translation of Demonferrand's treatise, published in 1823, would be useful and Cumming readily agreed. The result was *A Manual of Electro Dynamics* (1827), a translation, but much-expanded by the inclusion of Cumming's own original experiments and ideas.[69] The first four parts of the book (occupying 170 pages) were a comprehensive treatment in mathematical language of Ampère's theory of electromagnetism; the remaining third of the book (pp. 172–292) described all of Cumming's own work on galvanism, galvanometers and thermoelectricity. This was the only English text on the principles of electrodynamics until Henry Noad and Thomas Tate published their largely non-mathematical treatises over a decade later. There was not another mathematical exposition in English until Maxwell's great treatise appeared in 1873.[70]

Cumming's publications ceased as abruptly as they had begun. He was an active participant in the second meeting of the British Association held at Oxford in 1832 and at Cambridge the following year, but never seems to have attended a meeting again.[71] He must also have periodically visited London at this time, since he served on the Council of the Royal Society 1832–34, and refereed a few papers for them. However, he withdrew completely from the Royal Society, the Geological Society and the British Association after 1834, though he did join the Chemical Society when it was formed in 1841, but never attended a meeting. Contemporaries blamed ill health for the paucity of his research, but, if true, this never prevented him from delivering his annual lecture course. It is also clear that Cumming went on with his research in private. According to Liveing, after Deville had perfected a method for preparing pure aluminium on a large scale in 1854, Cumming prepared the metal and investigated its electrical properties. He did not publish his findings.[72]

Conclusion

Up until the mid-1970s it was assumed that Cambridge science was unproductive before the reforms of the 1850s. Cambridge was seen as 'an institution ensnared in pre-modern intellectual ideals of liberal education dedicated to the study of barren mathematics and classical languages to the detriment of science'.[73] Professors were essentially 'superfluous figures' engaged in voluntary teaching. Real teaching was done in colleges by tutors and Fellows and for examination purposes there was no need for undergraduates to attend professorial lectures. Historians of science now view Cambridge differently, seeing reform in action continuously from the early 1800s. Its professors were active in research, lecturing, giving experimental or field-work classes, creating a scientific forum in the CPS, and networking with other members of the scientific community through the Royal Society, British Association and government departments, or through private correspondence. Above all, the professoriate brought on and promoted any promising students who attended their classes, as the cases of Darwin and Stokes demonstrate. Cumming lacked the strident and ambitious voice of Whewell, Henslow, Sedgwick, Willis and others, but quietly and efficiently worked with the reforms they helped to introduce.

Cumming, the ninth holder of the 1702 Chair, was the last of Britain's clergy-chemists. Of course, scientists, including chemists, continued to be Christians, and many chemists like W. H. Perkin and Charles Coulson have been lay preachers. However, the introduction of improved stipends in the 1860s, together with the rise of scientific naturalism, put paid to the notion that science teaching and research was anything other than a full-time avocation. It would be wrong to criticise Cumming for the fitful character of his research – indeed, contemporaries respected him for independence of thought, his unostentatious character, and the fact that he put church and family before scientific reputation. Although he published nothing on inorganic or organic chemistry and credit for his electrical research went to others, his annual chemistry lectures, taken mainly by medical students, were always kept up to date. In his sixties, he enthusiastically supported the restructuring of the Cambridge undergraduate degrees and took warmly to the possibilities of liberal and professional education opened up by the creation of the Natural Science Tripos examination. Modest, unpretentious, unambitious and venerable, Cumming was a significant representative of the class of conscientious clergymen-scientists who were active as teachers and researchers in the first half of the nineteenth century.

Acknowledgements

I am grateful to the following for sharing their knowledge of Cumming with me: Roy Humby of North Runcton, David Bates of Buxton, Christopher F. Lindsey, author of the *Oxford DNB* (2004) entry on Cumming for the correct date of his birth; and Jonathan Smith, archivist of Trinity College, Cambridge.

Notes and References

1. Grossmith, G. & W. [1892] (1969), Reprint of *Diary of a Nobody*, London: Folio Society, p. 31. The diary had first appeared as a serial in *Punch*. While preparing this paper, many of my colleagues could not resist asking how 'Cumming was going?'
2. The date of birth is given as 26 October (not 24 October, as in old *DNB*) in the St James's church parish register.
3. For Joseph George Cumming, see *Oxford DNB* (2004), and Stenning, Revd. E. H. (1920), 'The life and work of the Reverend J. G. Cumming, M.A., F.G.S.', *The Isle of Man Natural History & Antiquarian Society* **2**, p. 402.
4. The school was probably Marlborough Grammar School and not to be confused with Marlborough College, which was not founded until 1843. His tutor or headmaster was a Mr Davis. Cumming is not mentioned in Stedman, A. R. (1945), *A History of Marlborough School*, Marlborough.
5. Rouse Ball, W. W. and Venn, J. A. (1911), *Admissions to Trinity College Cambridge, 1701–1800*, Cambridge: Cambridge University Press, vol. 2, p. 361.
6. Wordsworth, William in Maxwell, J. C. (ed.) (1971), *The Prelude. A Parallel Text*, Harmondsworth: Penguin Education, Book 3, lines 535–49. For boorish, uncouth dons, see Winstanley, D. A. (1935), *Unreformed Cambridge*, Cambridge: Cambridge University Press, p. 218.
7. On Porter, see Venn, J. A. (1940–54), *Alumni Cantabrigiensis*, 6 vols., Cambridge: Cambridge University Press; Ball & Venn (1911), *Admissions* (note 5), vol. 2, p. 361.
8. Trevelyan, G. M. (1943), *Trinity College. An Historical Sketch*, Cambridge: Cambridge University Press, p. 75; Winstanley, D. A. (1940), *Early Victorian Cambridge*, Cambridge: Cambridge University Press, pp. 241–55.
9. Winstanley (1940), *Early Victorian Cambridge* (note 8), chapter 2.
10. Diary of Joseph Romilly, quoted by Searby, Peter (1997), *A History of the University of Cambridge*, vol. 3, *1750–1870*, Cambridge: Cambridge University Press, p. 253.
11. [Edleston, Revd. Joseph], *Cambridge Independent Press*, 16 November 1861; reprinted *Lynn Advertiser and West Norfolk Herald*, 23 November 1861. Edleston, Fellow and Bursar of Trinity College, married Cumming's daughter, Harriet Sophia Cumming, in 1863. His identity is given in the index to Cambridge University, UA CUR 39.11. This memorial was the principal source for an obituary (1863), *Quart. J. Geol. Soc.* **19**, p. xxxi, and the Cumming entry by Bettany, G. T. (1885–1900) in old *DNB*. For Edleston, see Venn (1940–54), *Alumni* (note 7).

12. Winstanley (1940), *Early Victorian Cambridge* (note 8), p. 59. The Proctorship is recorded by Venn (1940–54), *Alumni* (note 7).
13. Winstanley (1935), *Unreformed Cambridge*, note 6, pp. 261–2.
14. Liveing, George (1862), 'James Cumming', *J. Chem. Soc.* **15**, pp. 493–4. See also Mills, W. H. (1953), 'Schools of chemistry in Great Britain and Ireland – VI. The University of Cambridge', *J. Roy. Inst. Chem.* **77**, pp. 423–31 and pp. 467–73, esp. pp. 467–9.
15. Winstanley (1935), *Unreformed Cambridge*, note 6, p. 147.
16. Votes cast in the election of S. Tennant, Cambridge University Archives CUL, UA O.XIV.17. Alexander Marcet informed Berzelius of Cumming's aspirations. *Jac. Berzelius Bref*, ed. Söderbaum, H. G. (1913), **1**, part 3, Uppsala: Almqvist & Wiksells, p. 118.
17. I am grateful to Mel Usselman (University of Western Ontario) for this information and for transcriptions of Wollaston's letters to an unknown correspondent (27 February 1815), to Henry Hasted (28 February 1815) and to E. D. Clarke (28 February 1815).
18. The *locus classicus* for the 'Cambridge Network' is Cannon, Susan Faye (1978), *Science in Culture. The Early Victorian Period*, New York: Science History; but see also Becher, Harvey W. (1986), 'Voluntary science in nineteenth century Cambridge University to the 1850s', *Br. J. Hist. Sci.* **19**, pp. 57–87.
19. Votes cast in the election of J. Cumming, Cambridge University Archives, CUL, UA O.XIV.18. See also UA CUR 39.11, a Registrary volume of assorted papers relating to the chair, 1702–1924.
20. Buttermann, Günther (1970), *The Shadow of the Telescope: A Biography of John Herschel*, p. 17, New York: Charles Scribner's Sons.
21. [Cumming, J.] (1816), *A Plan of a Course of Chemical Lectures* (28 pp), Cambridge.
22. Cumming, J. (1825), *A Syllabus of a Course of Chemical Lectures* (57 pp), Cambridge. A duplicate copy in Cambridge University Library [Cam.c.82517] has some pencilled annotations by a former student. It has to be borne in mind that Cumming would have printed at least 100 copies in 1825 and that students in succeeding years would have purchased the syllabus until it was exhausted. Thus the annotated copy may be later than 1825.
23. Cumming, J. (1821–2), 'Notice of a large human calculus in the library of Trinity College', *Trans. Cam. Phil. Soc.* **1**, 347–50.
24. Cumming, J. (1834), *A Syllabus of a Course of Chemical Lectures* (42 pp), Cambridge. Sub-titled Part I.
25. Royal Society, EC/1815/19.
26. Willis, Robert and Clark, John Willis [1886] (1988), Reprint of *The Architectural History of the University of Cambridge*, 3 vols., Cambridge: Cambridge University Press, vol. 3, p. 193.
27. Erasmus Darwin to Charles Darwin, Cambridge, 14 November 1822. Burkhardt, F. (1985), *The Correspondence of Charles Darwin*, Cambridge: Cambridge University Press, vol. 1, pp. 3–4.
28. Walters, S. M. and Stow, E. A. (2001), *Darwin's Mentor. John Stevens Henslow, 1796–1861*. Cambridge: Cambridge University Press; Russell-Gebbett, Jean (1977), *Henslow of Hitcham*, Lavenham: Terence Dalton, p. 15.
29. Ellis, R. L. (1865), Biographical memoir, in Walton, William (ed.), *The Mathematical Writings of Duncan Farquharson Gregory*, Cambridge: Deighton & Bell, p. xiii. See Becher (1986), Voluntary science (note 18), p. 6.
30. Murphy, Robert (1833), *Elementary Principles of the Theories of Electricity, Heat and Molecular Actions*, Cambridge: Pitt Press. Only the first part on electricity was

published. Stokes made extensive notes at Cumming's lectures between February and May 1844. CUL Stokes Collection Add. MSS 7656PA29-PA31. See Becher (1986), Voluntary science (note 18), pp. 66–7.

31. Weatherall, Mark W. (2000), *Gentlemen, Scientists, and Doctors: Medicine and Cambridge, 1800–1940*, Woodridge, NY: Boydell Press, p. 79. Kemp later held the chemistry chair at Queen's College, Birmingham, before taking up general practice.

32. Erasmus Darwin to Charles Darwin, Cambridge, 8 December 1822, *Correspondence of Darwin* (note 27), p. 5. Note, Cumming was usually to lecture in the Easter Term.

33. Weatherall (2000), *Gentlemen, Scientists, and Doctors* (note 31), pp. 77–9. He quotes from the Whig magazine, *Punch in Cambridge*, 28 January 1834.

34. Clarke's showmanship is noted in Winstanley (1940), *Early Victorian Cambridge* (note 8), p. 31.

35. Winstanley (1940), *Early Victorian Cambridge* (note 8), pp. 167–8.

36. Weatherall (2000), *Gentlemen, Scientists, and Doctors* (note 31); Hodgkinson, Ruth (1971), 'Medical education in Cambridge in the 19th century', in *Cambridge and its Contribution to Medicine*, ed. Rook, Arthur, London: Wellcome Institute History of Medicine, pp. 79–106; Winstanley (1940), *Early Victorian Cambridge* (note 8), pp. 160–5.

37. Willis and Clark (1886), *Architectural History* (note 26), vol. 3, p. 155.

38. Smith, G. (1936), Plans and notes on the development of the New Museums and Downing Street Sites 1574–1936, Cambridge University Archives ref. UA PVIII [3].

39. Report of Her Majesty's Commissioners [John Graham, Bishop of Chester, George Peacock, John Romilly, Adam Sedgwick and John Herschel] Appointed to Inquire into the State, Discipline, Studies and Revenues of the University and Colleges of Cambridge, *Parliamentary Papers* 1852–3, XLIV–I, 1. Searby (1997), *History of the University of Cambridge* (note 10). An American Trinity undergraduate noted that in 1841 he had audited Cumming's lectures with just two other students. Bristed, C. A. (1872), *Five Years in an English University*, New York: G. P. Putnam, p. 128.

40. Willis and Clark (1886), *Architectural History* (note 26), chapter 2 on the museums and lecture rooms for science.

41. Willis and Clark (1886), *Architectural History* (note 26), p. 163.

42. Meeting of Congregation 8 December 1857 and 8 May 1861, see Cambridge University, UA CUR 39.11.

43. MacLeod, Roy and Moseley, Russell (1982), 'Breaking the circle of the sciences: The Natural Science Tripos and the "examination revolution"', in MacLeod, R. M. (ed.), *Days of Judgement: Science, Examination, and the Organization of Knowledge in Late Victorian Britain*, Driffield: Nafferton Books, pp. 189–212; MacLeod, Roy and Moseley, Russell (1982), 'The 'Naturals' and Victorian Cambridge: reflections on the anatomy of an elite', *Oxford Review of Education* **6**, pp. 177–95. On Whewell's campaign, see Winstanley (1940), *Early Victorian Cambridge* (note 8), pp. 178, 198–233.

44. Graham Commission (note 38), p. 102.

45. Cumming to Whewell, 12 February 1851. Whewell papers, Trinity College Add Ms a.59[3]. There are three letters from Cumming in this collection, dated 1 June 1849, 27 November 1850 and 2 April 1851. See *Cambridge University Calendar for the Year 1858*, p. 160.

46. I have looked at papers for 1856, 1857 and 1858. *Cambridge Examination Papers: Being a Supplement to the University Calendar for the Year 1856*, Natural Science Tripos 28 January 1856, pp. 98–104. The examiners were Bond, Clark, Cumming, Sedgwick, Henslow and Hort.

47. Daubeny (1795–1867) was professor of chemistry and botany at the University of Oxford. Daubeny, C. G. B. (1831), *An Introduction to the Atomic Theory*, Oxford: Collingwood / Murray, (2nd edn, 1850).

48. Roberts, Gerrylynn K. (1980), 'The liberally-educated chemist: chemistry in the Cambridge Natural Science Tripos, 1850–1914', *Hist. Stud. Phys. Sci.* 11, pp. 157–83.

49. Fownes, George (1844), *Manual of Chemistry*, London: J. Churchill. Later editions 1848, 1850 (revised by Bence Jones, H. and Hofmann, A. W.), 1852, 1854, 1856, 1858, 1861; Gregory, William (1845), *Handbook of Chemistry*, London: Taylor, Walton & Moberly (later editions 1847, 1852, 1856). Prior to 1844, Cumming probably prepared his lectures using Thomson, Thomas (1802), *A System of Chemistry*, 4 vols., Edinburgh: Bell & Bradfute. Subsequently 4 vols. (1804); 5 vols. (1807); 5 vols. (1810); 4 vols (London, 1817); 4 vols. (1820); 2 vols. (1831); and Turner, Edward (1827), *Elements of Chemistry*. Edinburgh: William Tait. Subsequently London, 1828; 1831, 1833, 1834, 1842 (revised by Liebig, J. and Gregory, W.), 1847.

50. *Crockford's Clerical Directory* (1860), 147, entry 15; Messent, Claude J. W. (1936), *The Parish Churches of Norfolk and Norwich*, p. 200. Norwich: H. W. Hunt; Anon (1836), *Commercial Directory of the County of Norfolk*, Nottingham: Craven & Co, p. 43; Kelly, E. R. (1865), *The Post Office Directory of Norfolk*, London: Kelly, p. 361.

51. The Bank of England Inflation Chart suggests £1 in 1820 ≡ £34 in 2003. A later source of income was 'Professorial Ticket Money'. After 1848 students who took only ordinary degrees (poll men) were obliged to attend one professorial class at £3 or £5 in order to raise about £1000 p.a. This, as the Graham Commission learned, was 'divided into 16 parts, two of which are assigned each of the professors of Chemistry, Anatomy, and the Jacksonian Professor, in consideration of the special expenses to which they are subjected in preparing their lectures, and one part to each of the Professors of Civil Law, Physic, History, Botany, Geology, Mineralogy, Moral Philosophy, Political Economy, and the Downing Professor of Law'. Quoted in Bury, M. E. and Pickles, J. D. (eds.) (2000), *Romilly's Cambridge Diary 1848–1864*, Cambridge: Cambridge University Press, p. 84. Thus Cumming was paid about £125 p.a. from this source.

52. Bury, P. T. (ed.) (1967), *Romilly's Cambridge Diary, 1832–42*, Cambridge: Cambridge University Press, pp. 23, 36, 57, 75, 122, 135, 194, 225; Bury, M. E. and Pickles, J. D. (eds.) (1994), *Romilly's Cambridge Diary, 1842–1847*, Cambridge: Cambridge University Press, pp. 5, 27, 62, 117, 139; Bury, M. E. and Pickles, J. D. (eds.) (2000), *Romilly's Cambridge Diary 1848–1856*, Cambridge: Cambridge University Press, pp. 60, 84. Romilly was particularly fond of Cumming's children and was their godfather. The Cummings lost a one-year-old son to scarlet fever in June 1835.

53. Beloe, Edward Milligen (1899), *Our Borough: Our Churches (King's Lynn, Norfolk)*, Cambridge: Macmillan & Bowes, pp. 182–89. Photograph reproduced from this source.

54. Information from White, Francis (1836 and 1854), *History, Gazetteer and Directory of Norfolk*. This is available on the North Runcton website, *www.runctonweb.co.uk* (as of Aug. 2004). Runcton Hall was demolished in 1967; a housing estate was built on the site.

55. This is currently used by the villagers as a day nursery. On Gurney, see Hare, Augustus (1895), *The Gurneys of Earlham Hall*, London: G. Allan; Gurney, Daniel (1848–58), *Record of the House of Gurney*, London: privately printed.

56. On Henslow's important science teaching, see Russell-Gebbett (1977), *Henslow* (note 28).
57. For James John Cumming, see Venn (1940–54), *Alumni* (note 7).
58. Cambridge Independent Press (note 11).
59. Will drawn 5 April 1851 and proved 28 March 1862 (London Probate Office). George Liveing, To members of the Senate, 3 June 1886, in Cambridge University, UA CUR 39.11.
60. Hall, A. Rupert (1969), *The Cambridge Philosophical Society. A History, 1819–1969*. Cambridge: Cambridge University Press. See also Clark, J. W. (1891), 'The foundation and early years of the society', *Proc. Cam. Phil. Soc.* **7**, pp. i–xlviii.
61. Ross, Sydney (1962), 'Scientist: the story of the word', *Ann. Sci.* **18**, pp. 65–85.
62. Oersted, H. C. (1820), 'Experiments on the effect of a current of electricity on the magnetic needle', *Ann. Phil.* **16**, pp. 273–6.
63. Chipman, Robert A. (1966), 'The earliest electromagnetic instruments', *United States National Museum Bulletin* **240**, 121–36. See also Stock, John T. (1976), 'Cumming: a pioneer in electrical instrumentation', *J. Chem. Ed.* **53**, pp. 29–30.
64. Ampère, A.-M. (1820), 'Mémoire sur l'action naturelle de deux courants électrique', *Ann. Chim. Phys.* **15**, pp. 59–76.
65. Cumming, J. (1821–2), 'On the connexion of galvanism and magnetism', *Trans. Cam. Phil. Soc.* **1**, pp. 269–79, with plate; Cumming, J. (1821–22), 'On the application of magnetism as a measure of electricity', *Trans. Cam. Phil. Soc.* **1**, pp. 281–6. See Oersted, H. C. (1823), 'On M. Schweigger's electromagnetic multiplier', *Ann. Phil.* **21**, pp. 436–45 and the response by Cumming (1823), 'Description of the galvanoscope', *Ann. Phil.* **22**, pp. 288–9.
66. Cumming, J. (1823), 'On the development of electro-magnetism by heat', *Trans. Cam. Phil. Soc.* **2**, pp. 47–76 (read 28 April 1823).
67. Seebeck, Thomas (1822–23), 'Magnetische Polarisation der Metalle und Erze durch Temperatur-Differenz', *Abhandl. Preuss. Akad. Wiss.*, pp. 265–373. Cumming read of this in *Ann. Phil.* 1822.
68. Cumming, J. (1823), 'A list of substances arranged according to their thermoelectric relations', *Ann. Phil.* **22**, pp. 177–80; Cumming, J. (1823), 'On some anomalous appearances occurring in the thermoelectric series', *Ann. Phil.* **22**, pp. 321–23.
69. Demonferrand, J. F. (1823), *Manuel d'électricité dynamique*, Paris: Bachelier; Cumming, J. (1827), *A Manual of Electro Dynamics*, Cambridge: J. Smith.
70. Tate, Thomas (1854), *On Magnetism, Voltaic Electricity, and Electro-Dynamics, for the Use of Beginners*, London: G. R. Gleig; Noad, Henry Mitchin (1839), *A Course of Eight Lectures on Electricity, Galvanism, Magnetism, and Electro-Magnetism*. London: Scott, Webster & Geary; Maxwell, James Clerk (1873), *A Treatise on Electricity and Magnetism*. Oxford: Oxford University Press. See also Murphy (1833), note 30.
71. Cumming, J. (1832), 'Report on thermoelectricity', *Brit. Assoc. Rep.* **2**, pp. 301–8; Cumming, J. (1833), 'An instrument for measuring the total heating effect of the sun's rays for a given time', *Brit. Assoc. Rep.* **3**, p. 418. For requests for Cumming to 'perform' at the Oxford and Cambridge meetings, see Morrell, Jack and Thackray, Arnold (1981), *Gentlemen of Science*, Oxford: Clarendon Press, pp. 123, 431, 475–7, 487. It is not clear whether Cumming was present at the Edinburgh meeting (1834) when he was elected to a committee to report on chemical symbolism.
72. [Edelston] (1861), note 11; Liveing (1862), note 14; George Pryme (1870), *Autobiographic Recollections*, Cambridge: Deighton, Bell, p. 117.
73. Becher (1986), 'Voluntary science' (note 18), p. 58.

7

Chemistry at Cambridge under George Downing Liveing

John Shorter

Department of Chemistry, University of Hull

During George Downing Liveing's long tenure of the 1702 Chair, the pursuit of chemistry at Cambridge developed from a minor activity to a major position in the life of the University, involving a large chemistry building and an important role in the education of students through the Natural Sciences Tripos and in medical training. This development was essentially masterminded by Liveing, who is thus a key figure in the progression from 1702 to 2002 and may be regarded as the first recognisably modern Professor of Chemistry at Cambridge, constituting a bridge from the old style to the new.

In preparing this account, I have been particularly indebted to several articles that bear on Liveing and various aspects of Cambridge in his day, by W. C. D. Dampier,[1] Dampier as revised by Frank James,[2] W. H. Mills[3] and Gerrylynn Roberts.[4]

The early years

George Downing Liveing (Figure 7.1) was born on 21 December 1827 at Nayland, Suffolk. He was the eldest son of Edward Liveing, surgeon, and his wife Catherine, the daughter of George Downing, barrister of Lincoln's Inn. He went up to St John's College, Cambridge in 1847 and was eleventh wrangler in the Mathematical Tripos in 1850. The Natural Sciences Tripos had been instituted in 1848 as a postgraduate activity for those who had already taken one of the other triposes; Liveing embarked on this in 1850–1, the first year it was available. He was placed top of the six successful candidates, with distinction in chemistry and mineralogy. Practical chemistry had not been regarded as

The 1702 Chair of Chemistry at Cambridge: Transformation and Change, ed. Mary D. Archer and Christopher D. Haley. Published by Cambridge University Press. © Cambridge University Press 2005.

Figure 7.1. George Downing Liveing (1827–1924).

an important part of the course, and Liveing evidently realised that he lacked experimental skills, so he went on to study at the Royal College of Chemistry in London under August Hofmann for some months during 1851–2.[5] In the summer of 1852 he went to work with Karl Rammelsberg, teacher of chemistry and mineralogy at the Königliche Gewerbeschule (later the Technische Hochschule) in Berlin.[6]

On his return to Cambridge later in that year, Liveing came to an agreement with the Regius Professor of Physic that he would hold a class in practical chemistry for medical students. He was the first to run such a class in Cambridge and for this purpose he rented a cottage in Slaughterhouse Lane (today's Corn Exchange Street) and fitted up a primitive laboratory at his own expense.

St John's College then came to his aid, elected him as a fellow and lecturer in 1853, built him a laboratory behind their New Court and paid him a stipend to direct it. Liveing appears to have been an active and courageous junior fellow: in 1857 he published a pamphlet attacking the existing system of government of the college and advocating reforms which foreshadowed those introduced many years later. He was associated with the supporters of science education reform in Cambridge, including the botanist J. D. Hooker, with whom he is said to have visited Charles Darwin to encourage him to hasten publication of *The Origin of Species*.[2]

Towards the end of the decade, the pace of Liveing's life quickened. James Cumming, Professor of Chemistry, was now over 80 and in 1859 Liveing was appointed to be his Deputy. In August 1860 Liveing married Catherine, daughter of Rowland Ingram, rector of Little Ellingham, Norfolk. On marriage he had to vacate his fellowship at St John's, but he retained his lectureship until 1865.[7] In 1860 Liveing acquired the additional posts of professor of chemistry at the Staff College, Camberley and at the Royal Military College, Sandhurst. When Cumming died in 1861, Liveing was elected to the 1702 Chair. The salary was £100 per annum; this was increased to £500 per annum in one jump during the late 1860s.

Early struggles to secure university provision for chemistry

When Liveing became professor, he saw that one of his urgent tasks was to secure adequate provision for practical chemistry under the auspices of the university. One reason for this was the expansion of the Natural Sciences Tripos, which was extended to undergraduates just prior to Liveing's election in 1861. At that time, the facilities available were meagre. As detailed in the previous two chapters, around 1784 some accommodation had been provided for the Professor of Botany and for the Jacksonian Professor of Natural Philosophy in a single-storey building in the Botanic Garden, just north of Downing Street, on the southeast corner of what is now the New Museums Site.[8] For reasons detailed earlier, these rooms became the primary location in which chemistry was taught, and around 1832 they were expanded northwards to provide additional space for James Cumming (see Figure 6.4 in the previous chapter).

Two decades later, however, the need for accommodation was again being felt. Brock has outlined the abortive plans made during the 1850s to re-develop part of the Botanic Garden site for this purpose. Slight progress seems to have been made in the following decade, although the various accounts of this period appear a little confused. According to Dampier, in 1863 after much controversy, the University began an expansion of laboratories, thus initiating the great development in experimental science which transformed Cambridge. He writes that during '1864 and 1865 accommodation was provided successively for zoology, anatomy, chemistry, mineralogy, and botany. In 1865 Liveing began to announce regular experimental courses in chemistry'.[1] This account is supported by a nineteenth century source which comments that 'in 1865 a University laboratory, though of a somewhat makeshift sort, was established'.[9] Liveing himself, in an obituary of one of his colleagues, remarked that 'in 1866 . . . the University provided some rooms for a students' laboratory'.[10]

It therefore seems clear that some university facilities, albeit quite provisional, were made available to Liveing at this time. Their location is not entirely clear, but it is likely that this accommodation was not purpose-built, but was provided from an existing building near the 1784 block which was vacated by another professor. A key event here was the completion in 1865 of the two-storey Salvin building, located in the centre of what is now the New Museums Site.[11] This was constructed specifically for the scientific and mathematical professors, and although chemistry itself was not accommodated there, the subject seems to have benefited from the resulting reshuffling of accommodation.[12] The provisional nature of this 1865 laboratory probably explains why it does not feature in some other accounts of this time.

The next development took place in 1872. By that time, as described by Mills, the university laboratories were becoming overcrowded and Liveing persuaded the university to construct an upper storey on the 1832 extension.[13] According to Mills, a well-lit laboratory for about 50 students was thereby formed.[14] After this 1872 expansion, there appears to have been no additional university provision for chemistry until the construction in the late 1880s of the purpose-built Pembroke Street laboratory, discussed below. This was partly because, in Liveing's words, the university was 'as poor as a rat', and so could not afford further expansion, and partly because Liveing himself resisted further attempts at a piece-meal extension of his facilities, preferring to 'wait until we could get something really suitable'.[15] However, we should also note that a dual provision of facilities for chemistry persisted throughout Liveing's tenure of the 1702 Chair. As already mentioned, St John's had built its own laboratory in 1853 (Figure 7.2) and eventually five further colleges followed suit: Gonville and Caius opened a laboratory in 1873, with Sidney Sussex, Girton and Newnham following in the late 1870s, and Downing establishing its own in the late 1890s.[16] Thus after 1872 and through the remaining 1870s the total provision for chemistry in Cambridge was gradually increased via the colleges. These facilities played an important part in chemistry education;

Figure 7.2. The laboratory of St John's College, Cambridge around the turn of the century.

even as Professor, Liveing himself still maintained a foothold in the St John's laboratory,[10] and for many years after the construction of the Pembroke Street facilities, it was possible for students to choose between university or college facilities.[16] Nevertheless, by the early 1880s, this combined provision was once again seriously inadequate.

James Dewar's role in chemistry at Cambridge

We shall digress slightly at this point to say something about the part played in chemistry at Cambridge during the second half of the nineteenth century and the beginning of the twentieth by James Dewar, Jacksonian Professor of Natural Philosophy from 1875 to 1923. As mentioned in Chapter 4, the Jacksonian Chair was founded in 1783 under the will of the Revd. Richard Jackson, Fellow of Trinity, who had specified both natural experimental philosophy and chemistry as the subjects to be covered.[3] This combination had proved difficult to implement and by the middle of the nineteenth century the connection with chemistry had become weak. During the first 14 years of Liveing's tenure of the Chair of Chemistry, the Jacksonian Professor was Robert Willis, who had held the position since 1837. Willis is sometimes described as a 'professor of mechanism', and this may be translated into modern terminology as 'professor of applied physics'. Willis died in 1875, and Liveing and others in the university saw this as an opportunity to recapture the Jacksonian Chair for chemistry. Mills gives a detailed account of how Liveing set about this.[3]

The Council of the Senate initially assumed that the next professor should have interests similar to those of Willis. The vacancy was declared and to attract suitable candidates the salary was fixed at £500 per annum. However, some within the university expressed a strong feeling that 'mechanism' lay outside the terms of the founder's will and that the next professor should be an organic or a physiological chemist; in this, the opinions of the medical school were undoubtedly a powerful factor. Liveing issued a pamphlet supporting this view and insisting that chemistry was too large a subject to be represented adequately by one professor.

Among the applicants were distinguished chemists. The election had to be made by the electoral roll – a body of over 300 senior members of the university, really quite unsuitable for this purpose. Liveing convened a meeting of the electors who thought that the next Jacksonian Professor should be a chemist, and after the meeting a notice was sent out to all the electors: 'That this meeting recommends as a fit and proper person to be elected to the Jacksonian Professorship Mr James Dewar, demonstrator of chemistry in the University of Edinburgh.' His advocacy prevailed: the two candidates offering 'mechanism' withdrew, and Dewar was elected to the Jacksonian Chair *nem. con.*

Dewar came to be Liveing's most important scientific collaborator, so it is appropriate to give a brief account of his early life.[17] He was born on 20 September 1842 at Kincardine-on-Forth in Scotland, and educated at Dollar Academy and from 1859 at the University of Edinburgh, where he studied under J. D. Forbes, Professor of Natural Philosophy, and Lyon Playfair, Professor of Chemistry. He later acted for some years as Playfair's assistant and as assistant to Crum Brown, who succeeded Playfair in 1868. Dewar spent the summer of 1867 in Kekulé's laboratory in Ghent. In 1869 he was appointed lecturer in chemistry at the Royal (Dick) Veterinary College in Edinburgh. Much of his research in Edinburgh was in organic chemistry, for example on what became known as the Dewar formula for benzene, but he was also interested in various other topics.

Dewar was thus well qualified for the Jacksonian chair and high hopes were raised by his election, but he rapidly became disappointed in Cambridge.[17] The reasons for this have been much discussed[18,19] and will not be repeated here. Suffice to say that after only two years, he accepted appointment as Fullerian Professor of Chemistry at the Royal Institution in London, but continued as

Jacksonian Professor, travelling to and fro between Cambridge and London as he deemed necessary. Indeed he occupied the Jacksonian chair until his death at the age of 80 in 1923, but for 46 years the Royal Institution was the centre of his life and work. He became a public figure and acquired many honours, including a knighthood in 1904. Dewar's great interest in low temperatures and the liquefaction of gases began at the Royal Institution. His most outstanding achievement in this connection was the liquefaction of hydrogen in 1898: it was during this work that the vacuum flask was developed. He was much interested in the properties of matter at low temperatures and made many studies of specific heats and electrical resistance at low temperatures. Other topics to which Dewar contributed included the chemistry of metallic carbonyls (in collaboration with Humphrey Owen Jones, one of his Jacksonian Demonstrators at Cambridge), the behaviour of soap films and soap bubbles, and the measurement of radiation from the sky. He was carrying out studies of the last-mentioned to within a few days of his death on 27 March 1923.[17]

What might be described uncharitably as Dewar's residual influence in Cambridge was, however, not negligible. He continued for many years to deliver lectures in chemistry for the Tripos, particularly in organic chemistry, but also occasionally on other topics. Further, right from the start of Dewar's tenure in 1875, Cambridge chemistry benefited from the activities of a succession of assistants, known as Jacksonian Demonstrators. I shall give some account of these later.

In spite of Liveing's undoubted disappointment at Dewar's behaviour, a life-long friendship developed between them. This seems surprising, for, according to one of their colleagues, C. T. Heycock, 'they were men of widely different temperaments and widely different ideals, and they were both quick-tempered'.[1] In 1878 they began a series of spectroscopic investigations, which continued into the next century, leading to 78 papers. I shall give some account of this work towards the end of this article.

The Pembroke Street Chemical Laboratory

We return now to the main story at the point around 1880 when the centralised university chemical facilities, even though supplemented by the college laboratories, had once again become inadequate and the need for further accommodation was urgent. Thus chemistry acquired a building of its own, further to the west than the previous accommodation, along Pembroke Street (Figure 7.3). According to Mills, 'The new Chemical Laboratory in Pembroke Street was brought into being through Liveing's influence and exertions; it was his greatest contribution to the Cambridge school.'[3] As one might expect, the story of its

Figure 7.3. Exterior of the Pembroke Street Chemical Laboratory.

birth is complicated. The new laboratory was built on the recommendation of a syndicate which was originally appointed to consider the site and plans for a new geological museum, but it was also instructed to take into account claims being made in various quarters for other new buildings. Liveing was a member of the syndicate and made sure that the claims of chemistry were heard. In 1883 a suitable site was found in the area of Pembroke Street, Free School Lane and Downing Street, in the southwest corner of what was by now known as the New Museums Site. By this time the claims of chemistry had become very audible.

Liveing visited and compiled details of the newest chemical laboratories in this country and on the continent, making a special tour to inspect the new laboratories at Strasbourg, Munich, Leipzig and Aix-la Chapelle. He then drew up outline plans for the new chemical laboratory on the southwestern part of the New Museums site, leaving space for a geological museum on the eastern side. His proposals were adopted by the building syndicate and approved by the university, Mr J. J. Stevenson was appointed as architect, the builders' contract was signed and the Financial Board was instructed to take the necessary steps to provide the funds. The building was begun in December 1885 and completed in the autumn of 1888 at a cost of £33 700. According to Mills in 1953, 'At the time of its completion the laboratory was one of the finest in the country, and the excellence of its planning has since been abundantly proved.'[3] The appendix at the end of the chapter reproduces the detailed description of the building published in the *Calendar* of the university for 1899–1900.

This new building in its turn was soon outgrown by the increasing popularity of chemistry as a subject. In the 1890s the college laboratories were still undertaking a considerable share of the chemistry teaching, while the number of students using the university laboratory grew steadily. By 1905 the university laboratory accommodation was under great strain; further, Caius and Sidney Sussex were making preparations to close their laboratories. The university recognised that an extension to the laboratory was needed and in 1908 an eastern wing of the building was added.[20] Liveing took no part in planning this extension; in his late seventies, he felt the future should be left to his younger colleagues. The appendix also gives a brief account of the extension.

The later years

We still have various aspects of chemistry at Cambridge under Liveing to explore, but this would be a suitable point to round off the story of Liveing's life.

First, a little about the part which Liveing played in national scientific affairs. He had been elected a Fellow of the Chemical Society in 1853, and was a Vice-President of the Society in 1883–6 and 1898–1901. In 1882 he was President of the Chemical Section of the British Association for the Advancement of Science. He was elected a Fellow of the Royal Society in 1879 and served on its Council in 1891–2 and 1893–4. He was awarded the Society's Davy Medal in 1901. In 1904 he played a leading role in the Royal Society's radium investigation committee.

Among university matters connected with chemistry, Dampier mentions that in 1888 Liveing 'arranged a course of lectures on agricultural chemistry, thus inaugurating activities which ultimately developed into the successful Cambridge School of Agriculture. For many years that school owed much to Liveing's help and support'.[1] In the early 1880s Liveing played a prominent role in the revision of the University Statutes. He also took part in local affairs, as a county and borough magistrate, and was a member of the rifle corps.

In 1908, at the age of 81, Liveing resigned from the Chair of Chemistry, although to the end of his life he remained in touch with the laboratory. After his 47 years in the Chair, chemistry at Cambridge was certainly in a very different state from what it had been in 1861. In dealing with Liveing's departure from the Chair, Dampier saw fit to include the following comment: 'Throughout his tenure of the chair, he took full financial responsibility for the maintenance of the laboratory, which in its early years must have caused a heavy drain on his private income.'[1] If 'private income' is taken to mean income from investments, and not to include university or college salary, this suggests that Liveing's private means were considerable. Similar comments indicating that Liveing put money into chemistry at Cambridge may be found elsewhere. His estate at his death in 1923 was valued at £19 908/12/2, a substantial sum.[2]

Liveing's wife Catherine had died in 1888. They had no children. In retirement Liveing continued to lead a busy life between his house and garden in Newnham, the laboratory and St John's College. Dampier wrote that 'His character seemed to mellow with age, his asperities softened, and the patriarch of 90 seemed easier of access than the professor of 50 or 60. His memories of days long past were of historic interest, both to chemists and to other members of the University.'[1] Coulton's autobiography includes the following anecdote:

> Professor G. D. Liveing of St John's . . . was the yearly guest of honour at the Fellows' Dinner on his birthday, December 21. The Master always made a short speech. Somewhere about 1922 he reminded us how the Professor had witnessed the introduction of railways, electric telegraph, motors, aeroplanes, etc., etc.: a formidable list. He added drily, "And we hope he will live to see baths at St John's."[21]

The college bathhouse was in fact opened later in 1922 on the site of Liveing's old college laboratory.[22] The following year, at the age of 96, Liveing gave up his house at Newnham, and after a short stay at the University Arms Hotel, he moved to 10 Maid's Causeway. His friends confidently expected him to live to 100, but on 11 October 1924, while walking to the laboratory, he was knocked down by a lady cyclist from Girton, and two and a half months later, on 26 December 1924, he died of his injuries.

The development of chemistry in Britain in the nineteenth century and its impact on Cambridge

In presenting chemistry at Cambridge under Liveing in the context of a short biography, I have emphasised his almost continual struggles to secure adequate accommodation for the ever-increasing number of students who needed to do practical chemistry. The initial impetus for the expansion was provided by the ambitions of the Medical School to include more chemistry in the training of its students. As we have seen, Liveing had made a start in this direction in 1852, with the support of the Regius Professor of Physic, and about seven years later practical chemistry became a required part of medical training. The other cause of the pressure on accommodation was, as mentioned, the expansion of the Natural Sciences Tripos examinations from 1861, with chemistry as a very popular new undergraduate subject.

At the beginning of this chapter I mentioned the long article by Gerry Roberts.[4] In this she traced in great detail the development of chemistry as a subject in the Tripos between 1851 and 1914 by surveying the syllabuses and recommended textbooks, the lectures listed, the examination papers, the reports of the examiners, the numbers of students involved, and the general nature of their subsequent careers. In the space available here I cannot deal with most of her interesting findings, even in summary, but I will mention a few main points that came out of her study.

The second half of the nineteenth century was a period of intense conceptual development for chemistry, in particular in the structural theory of organic chemistry and the emergence of physical chemistry. Most of the really significant work was being done in western continental Europe, while in Britain chemistry became the basis of a profession in its own right. During the first half of the nineteenth century, chemistry was increasingly important in many industries, for example, calico-printing and dyestuffs, the production of coal gas, soap making, brewing, the manufacture of chemicals, the metal industries, and so on.[23] This applied chemistry was initially largely in the hands of people whose chemical education had been haphazard. In the second half of the century, there emerged a new breed of 'practising chemists', whose role was to supply the expertise required in the industrial development. These practising chemists were more formally educated than their predecessors and employment opportunities for such trained people were increasing. The training of practising chemists was done by colleges, especially those with a strong technical bias, in London and the provinces, particularly the north of England and in Scotland. In 1877 the practising chemists acquired a qualifying association, the Institute of Chemistry.[24]

Cambridge, however, did not seek to produce people who intended to practise chemistry. The chemical component of the Tripos was as much a product of the Cambridge system and of Cambridge personalities, as a response to developments in chemistry. At Cambridge, chemistry was part of a liberal education for those who cared to include it, with the exception of its ancillary role for medicine. This is very much borne out by the careers followed by graduates of the Tripos. Work in some aspect of education was easily the most popular career, with teaching in the so-called public schools being especially attractive to the liberally educated chemist. Only a small minority went on to practise chemistry in industry. The aim of a Cambridge chemical education through the Tripos was to impart a knowledge of chemistry, not to train chemists.

Liveing as a teacher of chemistry

We must now say something about those who did the teaching of chemistry at Cambridge, starting with Liveing himself. In his early days as a professor he had few chemical colleagues and no doubt had to be prepared to teach any part of chemistry deemed to be needed by the students concerned. Even as late as the 1880s Liveing was giving very broad lecture courses to elementary students, including medics. According to Dampier, Liveing's elementary lectures were attended 'by men conspicuous more for light-heartedness than love of learning. The lectures were illustrated with experiments, and his impatience with the laboratory attendants when the experiments went wrong was eagerly watched for by his youthful class.'[1] From the 1880s, however, Liveing was also teaching spectroscopy to advanced students, this being an offshoot of his spectroscopic investigations in collaboration with Dewar.

In the 1880s he also delivered, on at least one occasion, a course on chemical equilibrium, which he turned into a small book entitled *Chemical Equilibrium, the Result of the Dissipation of Energy*, published in 1885.[25] This book was an attempt to present the ideas of Clausius, Kelvin, Clerk Maxwell and Gibbs for students of chemistry, but Liveing's own grasp of thermodynamics had certain flaws. Broadly speaking, however, his description of equilibrium, both stable and metastable, was the received wisdom of 1885, and indeed largely of our day also. Liveing also presented speculations on the nature of chemical combination, based on the vortex theory of atoms and the role of molecular vibrations. John Rowlinson concludes that 'Cambridge undergraduates must have found this a difficult course of lectures'.[26] According to Dampier, 'His advanced students found him somewhat difficult to approach; but when the approach was made he took great trouble and gave them individual attention: many distinguished chemists owe much to his teaching.'[1] (Perhaps the 'advanced students found him somewhat difficult to approach' because they found his material so difficult that they found it difficult to formulate the questions they needed to ask!)

Liveing's chemist colleagues

The space available to me here permits mention of only a selected few of Liveing's chemist colleagues in Cambridge, but this will also be an appropriate point to introduce a word which I have so far avoided using: research. The provision of facilities for research was only a minor force in the drive for more accommodation for chemistry, but nevertheless most of the chemistry staff undertook research as well as teaching. So in introducing a few of Liveing's colleagues, I shall mention briefly the research they conducted.

Information about any colleagues Liveing had in his first decade in the Chair is not readily accessible, except for one who, in an extraordinary way, became a very junior colleague from the start: William James Sell (1847–1915). Sell's obituary notice, which is the source of this note, was written by Liveing and makes fascinating reading.[10] Sell was born in Cambridge and had his early education at one of the local primary schools. His scientific education began at the chemical laboratory of St John's College in 1861, where he was employed by Liveing, who later wrote of this period:

> At that time this laboratory, maintained by the College and with great liberality, placed at the disposal of the University Professor, was the only place in Cambridge where undergraduates could get any instruction in practical chemistry.
> Subsequently in 1866, when the University provided some rooms for a students' laboratory, Sell became lecture assistant to the Professor of Chemistry.[10]

Sell married in 1870, studied to matriculate at Christ's College, and by part-time study obtained first-class honours for a B.A. in Natural Science in 1876. Within a few years he had become Principal Demonstrator in the University Laboratory, a post which he retained to the end of his life. According to Liveing, the University never properly recognised the valuable contribution that Sell made to the development of chemistry at Cambridge. In addition to exemplary attention to his teaching duties, Sell found time to do research, mainly in organic chemistry, either solo or in collaboration with one of the junior demonstrators. Perhaps his most important work was a long series of investigations on pyridine derivatives. He was elected FRS in 1900.

One of Liveing's colleagues, who became very closely associated with Sell, was Henry John Horstman Fenton (1854–1929). Fenton was born in Ealing and first studied chemistry at King's College, London. He then arrived at Cambridge, on a recently founded Clothworkers' Company exhibition in physical science, starting in the Lent Term of 1875.[27] By this time Fenton was already 21, and quite knowledgeable and experienced. After a period as a non-collegiate student, he became a scholar of Christ's College. While he was still an undergraduate, Fenton's talents were recognised by Liveing, who made him an assistant demonstrator. Fenton took a first in the Natural Sciences Tripos in 1877, and shortly afterwards became Additional Assistant Demonstrator in the University Laboratory, the Principal Demonstrator being Sell. These two largely ran the University Laboratory for many years – Fenton's lectures in general and physical chemistry, and the accompanying practical classes, being an outstanding feature of the instruction given. His research interests, however, were in organic chemistry. Much of his original work stemmed from the accidental discovery early in his career of the oxidising properties of a mixture of hydrogen peroxide and a ferrous salt, which is still known as Fenton's reagent. This was found to be of particular value for the smooth oxidation of di- or poly-hydroxy alcohols having vicinal OH groups. Fenton was elected FRS in 1899. He wrote several books, of which the best known were his *Notes on Qualitative Analysis*[28] and *Outlines of Chemistry.*[29]

From the 1870s, Liveing was well aware of the importance of gaining the cooperation of those responsible for the college laboratories in coordinating the provision of chemistry throughout the university.[4] Some of the chemistry staff who were associated with one particular college laboratory, rather than the University Laboratory, were thus among his most important colleagues. This group included Matthew Moncrieff Pattison Muir (1848–1931). Muir was born and raised in Glasgow and studied at the High School, the University of Glasgow and Tübingen University.[30] After short periods as a demonstrator at Anderson's College, Glasgow (under Sir Edward Thorpe) and Owens College, Manchester (under Sir Henry Roscoe), he was appointed Praelector in Chemistry at Gonville and Caius College, Cambridge in 1877 and elected a Fellow of the college in 1881. He ultimately became head of the Caius laboratory until his retirement in 1908. The college always had a large number of medical students and they learnt their chemistry under Muir. The laboratory also gave courses for the Natural Sciences Tripos, and provided facilities for some undergraduates from other colleges.

Between 1876 and 1888, Muir, either alone or with students, carried out research on bismuth compounds, leading to the publication of eighteen papers in the *Journal of the Chemical Society*. He always encouraged young graduates to undertake chemical research, although at that time there were no postgraduate studentships in chemistry at Cambridge. Muir was, however, more successful as a writer than as a research worker. He wrote various textbooks and treatises and contributed papers on the history of chemistry and alchemy. His greatest success was a treatise on *The Principles of Chemistry*, published in 1884, which went to a second edition in 1889. In 1907 he published *A History of Chemical Theories and Laws*. He was elected F.R.S. (Edinburgh) in 1873.

Charles Thomas Heycock (1858–1931) seems to have had difficulty in establishing himself as a lecturer or tutor in the Cambridge school of chemistry. He was born in Oakham, came to Cambridge as an exhibitioner at King's College in 1877, and took the Natural Sciences Tripos in 1880.[31] He then appears to have been a teacher for Cambridge examinations in chemistry, physics and mineralogy without a proper college or university appointment until 1895, when he was elected a Fellow of King's, becoming a College Lecturer and Natural Sciences Tutor in the following year. Heycock was an excellent lecturer and in his classes he sustained an interest in inorganic chemistry during a period when that subject seemed in danger of eclipse by the rapid advance of organic chemistry. During the period 1880–1895, in collaboration with his lifelong friend

Francis Neville of Sidney Sussex, he began (in the laboratory of that college) a comprehensive study of the phase diagrams of metals and their alloys.[32] This work led to the election of Heycock and Neville to the Royal Society in 1895 and 1897, respectively, and continued thereafter for many years. From 1908 to 1928 Heycock was Goldsmiths' Reader in Metallurgy and he was awarded the Davy Medal of the Royal Society in 1920. Liveing must have appreciated the steady growth of Heycock's research reputation from 1880 onwards. Some provision of metallurgical furnaces was included in the Pembroke Street building from its start, but full provision was not made until the extension was built around 1908, and the Sidney laboratory was about to close. Heycock wrote a very sympathetic obituary of Liveing for the Royal Society.[33]

One of Liveing's colleagues whose career involved a long and close association with him was John Edward Purvis (1862–1930). Purvis was born in Heaton Norris near Stockport[34] and went up to St John's College, Cambridge, to read for the Natural Sciences Tripos in 1889, when he was 27. Cambridge became his home for the rest of his life. He had already studied at Owens College and at the Royal College of Science, Dublin, where W. N. Hartley inspired him with an interest in spectroscopy. After the Tripos, Purvis was able to pursue this interest as Assistant to Professor Liveing, but later he also acquired an interest in the application of chemistry to public health. Thereafter his teaching was in this area. After Liveing retired in 1908, Purvis became Lecturer in Chemistry and Physics in their application to Hygiene and Preventive Medicine. He was also Director of Studies in chemistry at Corpus Christi. His main research interests remained in spectroscopy, particularly in the absorption spectra of organic compounds in the gaseous state. Purvis spent much time and energy, especially in the latter part of his life, in the service of the borough council, his connection with that body beginning in 1908 when the representatives of the colleges and halls elected him a councillor. He was elected Mayor in 1928, but became seriously ill; he struggled to carry on to the end of his term of office, but he died the following year.

No account of chemistry at Cambridge under Liveing, nor any account of Liveing's colleagues, would be complete without a mention of the pioneers who taught chemistry in the women's colleges. Girton and Newnham both opened laboratories not long after they were founded. In Cambridge there was considerable disapproval of women studying chemistry, as of their studying anything else.[35,36] Little indication of Liveing's personal views on this appears to survive, although he was in many respects more liberal than most of his peers. He petitioned against the need for bachelorhood and celibacy among Fellows, for instance, even before he was required to vacate his own fellowship by reason of marriage.[37] Nevertheless, the position of women at Cambridge was anomalous in one respect or another until 1948 and no doubt a bias against female students continued in the minds of many male members of the university, consciously or unconsciously, for years longer. In this context, I cannot resist quoting the closing words of Mills's second article, written in 1953, when the Lensfield Road laboratory was under construction.[3] 'With a spacious modern laboratory for its accommodation, bright prospects lie before the school, and the university can look to it with confidence for a steadily growing output of trained men [*sic!*] and scientific knowledge.'

The principal female chemistry teachers in Liveing's time were Ida Freund (1863–1914) and Mary Beatrice Thomas (1873–1954). Ida Freund was born in Austria in 1863 and brought up by maternal grandparents after her mother's death.[38] She came to England under the care of an uncle and read for the Natural Sciences Tripos at Girton from 1882 to 1886, taking a first in both parts. She taught briefly at the Cambridge Training College for Women (later known as Hughes Hall) and was then appointed demonstrator in chemistry at Newnham. Ida Freund was Lecturer at Newnham and in charge of the College Laboratory from 1891 until her retirement in 1913, when the laboratory closed. She concentrated very much on teaching her students at Newnham and never engaged in experimental research. She did, however, publish one substantial book: *The Study of Chemical Composition*.[39] She was greatly interested in the women's suffrage movement, and in the work of the Women's University Settlement in Southwark.

Mary Beatrice Thomas was born in Birmingham in 1873.[40] She read for the Natural Sciences Tripos at Newnham under Ida Freund from 1894 to 1898, taking a first in Part I and a second in Part II. After a brief period as a demonstrator at Royal Holloway College, she became demonstrator at Girton in 1902, and was Lecturer and Director of Studies in Natural Science of the college from 1906 to 1935. In her early years at Girton, she carried out some research in collaboration with H. O. Jones, leading to two papers in the *Journal of the Chemical Society*. During World War I, she participated in development work on anti-gas respirators. In 1920 she acted as editor (in association with Arthur Hutchinson) of Ida Freund's previously unpublished writings on *The Experimental Basis of Chemistry*.[41] On her retirement in 1935, the Girton laboratory closed, the last of the Cambridge college laboratories to go. The *Girton Review* of 1954 contains an entertaining account of her life, including her extra-mural activities of various kinds.[42] Shortly after the end of World War I she adopted a small nephew, who grew up under her care, attending the Perse School and later St John's College. She interested herself in matters connected with science teaching in girls' schools and, before World War I, in the women's suffrage movement.

The Jacksonian demonstrators

I have mentioned that one of the most valuable of Dewar's contributions to chemistry at Cambridge was through the Jacksonian demonstrators. In its origin and purpose, the post of Jacksonian demonstrator had nothing to do with enabling Dewar to spend most of his time in London and continue to hold the Chair in Cambridge, although from about 1905 onwards it probably did play a role in this respect. Positions of assistant to a professor were quite common in many universities and colleges. The term did not imply that the holder was the professor's dogsbody; assistants were sometimes quite senior men and carried out very important duties, albeit under the authority of the professor.

Five men in all occupied the post of Jacksonian demonstrator during Dewar's tenure of the Jacksonian chair. Alexander Scott (1853–1947) held the position twice: from 1875 to 1884, and again from 1892 to 1896.[43] He then joined Dewar at the Royal Institution until 1910. His main research interest was the accurate determination of atomic weights, but he had a long second career after World War I as a pioneering chemist at the British Museum.

Siegfried Ruhemann (1859–1943) was Jacksonian demonstrator from 1885 to 1891.[44] He had studied in Berlin and when Dewar had asked August Hofmann to recommend an organic chemist, Ruhemann was selected. After some years of excellent service by Ruhemann to Cambridge chemistry, differences arose between the Jacksonian professor and his demonstrator, which led to Ruhemann's transfer to a university lectureship and his moving, together with his research students, from the new University Chemical Laboratory in Pembroke Street to the Caius laboratory. When this closed in 1908, Ruhemann returned to the University Laboratory. He had taken British nationality some years earlier, but during World War I, he became a victim of extreme anti-German feeling, and in 1915 he resigned his lectureship.[45] In 1919 he returned to his native Germany for twenty years, but the political situation there in the 1930s drove him back to England just before the outbreak of war in 1939.

William Thomas Newton Spivey (1868–1901) held the post of Jacksonian demonstrator from 1896 to 1901, when he died as a result of injuries sustained in a laboratory accident with carbon disulphide.[46] His replacement was Humphrey Owen Jones (1878–1912), a graduate first of the University of Wales in 1897 (University College, Aberystwyth), and then of the Natural Sciences Tripos in Cambridge (1900). Unfortunately Jones, too, met an untimely death: he was killed in a mountaineering accident while on his honeymoon in the Alps in 1912. (His bride Muriel – a chemist educated at UCNW Bangor and Newnham – also died.)[47] Jones had become a Fellow of Clare College in 1902, and college lecturer in chemistry and physics the following year. Apart from quite considerable work with Dewar on metal carbonyls, his main research interest lay in the resolution of asymmetric quaternary ammonium salts.

To round off the story, although it takes us beyond our specified period, the last of Dewar's Jacksonian demonstrators was William Hobson Mills (1873–1959), who held the post from 1912 until it lapsed with Dewar's death in 1923.[48] He continued at Cambridge in other positions and became well known for his work on stereochemistry.

All Dewar's Jacksonian demonstrators, apart from the unfortunate Spivey, became Fellows of the Royal Society. Two of them, Scott and Mills, became President of the Chemical Society.

Dewar and Liveing's spectroscopic investigations

As already mentioned, Dewar and Liveing conducted research in spectroscopy from 1878 until about 1900.[1] Until he came into contact with Dewar, Liveing seems not to have had any particular interest in doing research himself. Why they began their work and how the collaboration operated is not clear to me, but it may be supposed that the experimental work was done mainly in Cambridge, and probably mainly by Liveing, with Dewar participating when he was available. The apparatus they used for the photographic recording of line spectra of the elements in the visible region and later in the ultraviolet, was simple, but well made and capable of giving reasonably accurate results. The work was published mainly in various 'Proceedings': of the *Cambridge Philosophical Society*, the *Royal Society*, or the *Royal Institution*; or in the *Philosophical Transactions of the Royal Society*. In 1915 Cambridge University Press published the *Collected Papers on Spectroscopy of G.D. Liveing and Sir J. Dewar*, a volume of around 570 pages, and containing 78 papers.[49]

What sort of spectroscopic topics did they investigate? We have space to give just a few examples. Eight of the papers from 1878 to 1881 dealt with 'The Reversal of the Lines of Metallic Vapours'; in other words, the comparison of the absorption spectrum of a given element with its emission spectrum. Other papers were simply concerned with the careful mapping of the emission spectra of a large number of elements, initially in the visible, later in the ultraviolet regions. There were also papers concerned with improvements to spectroscopic apparatus, and one-off topics, such as 'The Spectrum of Water', 'Sun-Spots and Terrestrial Elements in the Sun', 'The Spectral Lines of the Metals developed by Exploding Gases', 'Notes on the Absorption of Ultra-Violet Rays by Various Substances', 'The Spectroscopic Properties of Dust'. Towards the end of the work there were papers stemming from Dewar's other interests; for example 'On the Refraction and Dispersion of Liquid Oxygen and the Absorption Spectrum of Liquid Air'.

It seems appropriate to conclude this brief account of a great deal of work, by quoting Dampier.[1] 'If these 78 joint papers cannot be said to disclose any epoch-making discovery, they certainly chronicle careful, exact, and useful contributions to knowledge.'

Farewell to the Liveing era

When Liveing resigned from the Chair in 1908, he was succeeded by William Jackson Pope (1870–1939), who had been the head of the chemistry department and professor of chemistry at the Municipal School and Faculty of Technology of the University of Manchester (the predecessor of UMIST).[50] Pope's scientific career had begun at the Central Institution in London (predecessor of the City and Guilds' College of the Imperial College of Science and Technology) under H. E. Armstrong, and had continued in the Goldsmiths' Institute at New Cross, where he became head of the chemistry department at the age of 27. He was a research man from the off, working on the reactions and structures of camphor derivatives with Armstrong and then with F. S. Kipping. At Goldsmiths', Pope used the optically active camphorsulphonic acids as resolving agents, in particular for the resolution of asymmetric quaternary ammonium salts and compounds of sulphur and selenium. He had continued stereochemical work at Manchester, in association with W. H. Perkin, particularly on dissymmetric molecules that did not possess asymmetric carbon atoms.

With the coming of Pope to the Cambridge Chair, the winds of change were undoubtedly going to blow very strongly along Pembroke Street. According to Mills, the scientific activities of the laboratory were redirected largely towards stereochemistry.[3] No-one could have foreseen, however, that within six years the winds of change would blow at hurricane force under the demands of a World War.

Appendix
The Pembroke Street Chemical Laboratory

The Cambridge University Calendar for 1899–1900, pp. 656–7, describes the new Chemical Laboratory as follows:

> The new Chemical Laboratory is a large building in Pembroke Street, facing part of Pembroke College, erected in 1887 from the design of Mr J. J. Stevenson, upon a plan proposed by Professor Liveing. The principal entrance is from Pembroke

Street, but the students' laboratories are also accessible at the back from the court of the Museums. The ground floor contains the laboratories for elementary work, namely a large room for qualitative analysis fitted for 84 students, a laboratory for elementary quantitative work fitted for 38 students, with a room between them for large operations, a balance room directly accessible from the quantitative room, and in the south-west corner, an octagonal projection with windows on all sides, where operations with chlorine or bromine can be conducted in a thorough draft (*sic*). There are draft chambers in all the windows, and a separate small, well-ventilated room for work with sulphuretted hydrogen. On the same floor are three lecture rooms. The large one, seating 240, is reached by separate entrances from the street and from the rear: the smaller, each seating 60, are accessible from the principal staircase. Between the lecture rooms are the preparation room and the specimen room, and, on the mezzanine, the private laboratory of the Professor. On the mezzanine on the principal staircase are the private room of the Professor of Chemistry and a room for organic analysis.

On the first floor are the advanced students laboratories, and the laboratory for organic chemistry, balance room, lecture room, and private room for the Jacksonian Professor. Higher up are the library and one or two rooms for special researches. In the attic are rooms for the housekeeper and assistants, and for a spectroscope with a concave grating. Over the large laboratory is a flat roof, where work can be carried on out of doors. In the basement, under the large laboratory, is a classroom for demonstrations, and two rooms entirely fitted with uninflammable materials, where operations with easily inflammable substances can be safely conducted. Under the quantitative laboratory is the machinery room, containing two small steam engines, dynamo-electric machine, large air pumps and other machinery, and also the storage battery. Under the lecture rooms are unpacking and store rooms, rooms for gas analysis and for pharmaceutical and sanitary chemistry, and a room fitted with metallurgical furnaces.

The whole building is warmed by steam, and steam is laid on to all the rooms for use in chemical operations. Tubes from the airpumps are led into the different rooms, some giving low exhaustion for filtering are distributed to the working benches, while those giving the highest exhaustion are carried to the balance rooms, lecture rooms, and a few other points. The large lecture room, and the low rooms on the mezzanine floors, are lighted by electricity and there are leads from the dynamo and storage battery to all the lecture rooms for experimental purposes. A separate small storage battery can be placed in connection with any of the laboratories where an electric current is required. The whole of the working rooms are ventilated by a shaft 100 feet high, heated by the boiler furnaces.

The laboratories are open for the use of members of the University who comply with the rules, daily during term time, and during July and August, from 10 am till 6 pm. The fee for the use of the laboratory is £3/3/0 a term: large apparatus and such as can be used in common is supplied, but such small apparatus as each student requires for his own use he is expected to provide for himself, and all chemicals used and all damage done to the apparatus of the laboratory must be paid for. No one is admitted to work in the advanced students laboratories until he has proved his competence to the satisfaction of the Professor of Chemistry or the

Jacksonian Professor; nor are any experiments or researches allowed which are not approved by one or the other of those Professors.

The Assistant to the Professor of Chemistry was named as J. E. Purvis, M.A. (St John's).

In the *Calendar* for 1910–11, p. 872–3, there is an amended account of the Chemical Laboratory, describing the extension and other changes of 1908. A few quotations and notes may be of interest:

> . . . in 1908 an eastern wing was added from the design of Messrs. Stevenson and Redfern. The principal entrance is from Pembroke Street, but access can also be obtained from the Museums Court and Free School Lane; a large bust of Professor Liveing adorns the entrance hall.

The elementary laboratory then accommodated 130 students, with an auxiliary laboratory holding a further 80. There were also a large operations room, two balance rooms, and two rooms for noxious gas operations. The advanced laboratory accommodated 65 students, with an auxiliary laboratory holding a further 30.

> . . . on the mezzanine floors at the east end are 4 smaller laboratories for the use of the lecturers in chemistry . . . At the east end, communicating directly with the advanced laboratory, are the private laboratories of the Professor of Chemistry and a large physico-chemical laboratory.

In the basement were 'the metallurgical laboratory, furnace room and balance room occupied by Goldsmiths' Reader in Metallurgy . . . further east lie a large research laboratory connected with a balance room and a lecture room, to seat 150, with its accompanying preparation room.'

The fees were still £3/3/0 per term. The Assistant to the Professor of Chemistry, William Jackson Pope, was then John Read Ph.D. (Zurich).

Notes and References

1. Dampier, W. C. D. (1937), 'Liveing, George Downing', in Weaver, J. R. H. (ed.), *Dictionary of National Biography 1922–1930*, London: Oxford University Press, pp. 510–12.
2. Dampier, W. C. D. revised by James, F. A. J. L. (2004), 'Liveing, George Downing', in *Oxford Dictionary of National Biography*, Oxford: Oxford University Press.
3. Mills, W. H. (1953), 'Schools of Chemistry in Great Britain and Ireland – VI: The University of Cambridge', *J. Roy. Inst. Chem.* **77**, pp. 423–31 and 467–473.
4. Roberts, G. K. (1980), 'The liberally-educated chemist: chemistry in the Cambridge Natural Sciences Tripos, 1851–1914', *Hist. Stud. Phys. Sci.* **11**, pp. 157–83.
5. For a brief account of the Royal College of Chemistry, see Russell, C. A., Coley, N. G. and Roberts, G. K. (1977), *Chemists by Profession*, Milton Keynes: Open University Press, in association with the Royal Institute of Chemistry, London, pp. 75–81.

6. Karl Friedrich Rammelsberg was born in Berlin in 1813. After an initial training in pharmacy, he studied science at the University of Berlin. Over the course of about forty years he proceeded through the usual stages of a German academic career. By 1874 he was *ordentlicher* Professor in Inorganic Chemistry and Director of the Second Chemical Institute of the University of Berlin, a position from which he retired in 1891. He investigated the crystal forms and preparation of salts, the allotropy of sulphur, selenium, and tellurium, and other topics. He died in Berlin in 1899. See entry in Pötsch, W. R., Fischer, A., Müller, W. and Cassebaum H. (eds.) (1989), *Lexikon bedeutender Chemiker*, Thun: Verlag Harri Deutsch.

7. *St John's College Eagle*, **ii** (1861), p. 205.

8. University of Cambridge Archives, Minutes of Syndicates 1778–1803, reproduced in Willis, R. and Clark, R. W. (1886), *The Architectural History of the University of Cambridge*, Cambridge: Cambridge University Press, vol. 3, p. 153.

9. Anonymous (1899), [Obituary of Main, Liveing's successor as Superintendent of the St John's College Laboratory], *St. John's College Eagle* **xx**, p. 716.

10. Liveing, G. D. (1916), [Obituary Notice of Sell], *Proc. Roy. Soc.* **92A**, pp. xiv–xvi.

11. 'In 1865 Salvin's two-storied building, erected to accommodate all the mathematical and scientific professors, except those of Anatomy and Chemistry and the Jacksonian professor who remained in the old building and the professor of Geology, was completed. The north wing accommodated mineralogy above and botany on the ground floor, with mechanism and a lecture room at the angle. The south wing contained two rooms with an archway between them for philosophical apparatus and rooms for the mathematical professors above' [Smith, G. (1936), 'Plans and notes on development of New Museums and Downing Street sites, 1574–1936', Cambridge University Archives UA.P. VIII.1–10, plan 5].

12. Mills (1953), p. 469; Willis and Clark (1886), vol. 3, p. 153.

13. Smith, G. (1936), UA.P. VIII.1–10, especially plan 3.

14. Mills (1953), p. 469; University Grace, 6 June 1872, reported in Willis and Clark (1886), p. 184.

15. Liveing, G. D. (1922), 'Address to the President', *St John's College Eagle* **xlii**, p. 167–8.

16. Mann, F. G. (1957), 'The place of chemistry – II: at Cambridge', *Proc. Chem. Soc.*, p. 191.

17. Ross, H. M. (1937), 'Dewar, James' in Weaver, J. R. H. (ed.), *Dictionary of National Biography 1922–1930*, London: Oxford University Press, pp. 255–7.

18. Armstrong, H. E. (1926), 'Sir James Dewar, 1842–1923', *Proc. Roy. Soc. A*, **111**, xiii–xxiii.

19. Findlay, A. and Mills, W. H. (eds.) (1947), *British Chemists*, London: The Chemical Society, pp. 30–57.

20. See also Haley, C. D. (2002), *Boltheads and Crucibles: a Brief History of the 1702 Chair of Chemistry at Cambridge*, Cambridge: Cambridge University Press, p. 27. I am greatly indebted to Dr Haley for his help in trying to sort out a coherent story regarding the provision of accommodation for chemistry.

21. Coulton, G. G. (1943), *Fourscore Years: an Autobiography*, London: Cambridge University Press, pp. 97–8.

22. Cook, A. C. (1978), *Penrose to Cripps: a Century of Building in the College of St John the Evangelist, Cambridge*, Cambridge: St John's College, p. 29.

23. Russell, Coley and Roberts (1977), chapter 3.

24. Much of Russell, Coley and Roberts (1977) is devoted to the founding and history of the (Royal) Institute of Chemistry.

25. Liveing, G. D. (1885), *Chemical Equilibrium, the Result of the Dissipation of Energy*, Cambridge: Deighton, Bell & Co. I am grateful to Professor John Rowlinson for taking a look at this book for me.
26. Rowlinson, J. S., personal communication.
27. Mills, W. H. (1930), [Obituary Notice of Fenton], *J. Chem. Soc.*, pp. 889–94.
28. Fenton, H. J. H. (1906), *Notes on Qualitative Analysis: Concise and Explanatory*, Cambridge: Cambridge University Press.
29. Fenton, H. J. H. (1910), *Outlines of Chemistry with Practical Work. Part 1*, Cambridge: Cambridge University Press.
30. Morrell, R. S. (1932), [Obituary Notice of Muir], *J. Chem. Soc.*, pp. 1330–4.
31. Pope, W. J. (1931), [Obituary Notice of Heycock], *J. Chem. Soc.*, pp. 3368–71.
32. I have so far been unable to find an Obituary Notice for Francis Henry Neville (1847–1915).
33. Heycock, C. T. (1925), [Obituary Notice of Liveing], *Proc. Roy. Soc. A* **109**, pp. xxviii–xxix.
34. Mills, W. H. (1931), [Obituary Notice of Purvis], *J. Chem. Soc.*, pp. 3380–3.
35. McWilliams-Tullberg, R. (1975), *Women at Cambridge. A Men's University – Though of a Mixed Type*, London: Victor Gollancz Ltd.
36. MacLeod, R. and Moseley, R. (1979), 'Fathers and daughters: reflections on women, Science and Victorian Cambridge', *History of Education* **8**, pp. 321–33.
37. Liveing, G. D. (1857), 'Letter to the Governing Body of St John's College, Cambridge', St John's College library ref. 12.12.8[11].
38. Newnham College (1979), *Newnham College Register* vol. 1, Cambridge: Newnham College, Staff section for 1891, p. 7.
39. Freund, I. (1904), *The Study of Chemical Composition*, Cambridge: Cambridge University Press.
40. Newnham College (1979), *Newnham College Register* vol. 1, Cambridge: Newnham College, student section for 1894, p. 127.
41. Freund, I. (ed. Hutchinson, A. and Thomas, M. B.)(1920), *The Experimental Basis of Chemistry: Suggestions for a Series of Experiments Illustrative of the Fundamental Principles of Chemistry*, Cambridge: Cambridge University Press.
42. 'D. M. P.' (1954), 'In memoriam Mary Beatrice Thomas, 1873–1954', *Girton Review*, Michaelmas 1954, pp. 16–24.
43. Robertson, R. (1950), [Obituary Notice of Scott], *J. Chem. Soc.*, pp. 762–7.
44. Morrell, R. S. (1944), [Obituary Notice of Ruhemann], *J. Chem. Soc.*, pp. 46–8.
45. Saltzman, M. D. (2000), 'Is science a brotherhood? The case of Siegfried Ruhemann', *Bull. Hist. Chem.* **25**, pp. 116–21.
46. 'T. B. W.' (1902), [Obituary Notice of Spivey], *J. Chem. Soc.*, pp. 635–6.
47. Shorter, J. (1979), 'Humphrey Owen Jones, F. R. S. (1878–1912). Chemist and Mountaineer', *Notes Rec. Roy. Soc.* **33**, pp. 261–77.
48. Peacock, D. H. (1960), [Obituary Notice of Mills], *Proc. Chem. Soc.*, pp. 371–83.
49. Liveing, G. D. and Dewar, J. (1915), *Collected Papers on Spectroscopy*, Cambridge: Cambridge University Press.
50. Gibson, C. S. (1949), 'Pope, William Jackson', in Wickham Legg, L. G. (ed.), *Dictionary of National Biography 1931–1940*, London: Oxford University Press, pp. 716–17.

8

The rise and fall of the 'Papal State'

Arnold Thackray and Mary Ellen Bowden

Chemical Heritage Foundation, Philadelphia

Our subject is the era that begins with William Jackson Pope, Professor of Chemistry from 1908, and ends with the arrival of his successor, Alexander Robertus Todd, in 1944. Many readers may be familiar with Todd's pithy description of the state of the University Chemical Laboratory in 1944 as

> ... a disgrace to any university ... given a stuffed crocodile to hang from the roof, the professor's private laboratory could be more appropriately located in the Museum of the History of Science.[1]

What readers may not realise is the deliberate sting incorporated in the description, for William Jackson Pope was a noted collector of alchemical paintings and prints. To the authors' delight his collection now resides in Philadelphia, at the headquarters of the Chemical Heritage Foundation. One of them is reproduced on the dust cover of this book and also as the frontispiece. However, our subject is not 'Pope's paintings' (a topic we hope will one day be addressed by a leading supporter of the celebration of the tercentenary of the 1702 Chair, the great art connoisseur and brilliant chemist, Alfred Bader) but rather the 'Papal State'. This volume, and the symposium from which it emerged, both have the title 'transformation and change'. The words were no doubt chosen because transformation and change are not only synonyms for chemistry, but also describe today's realities in the Cambridge Department of Chemistry. However, as professional historians are acutely aware, it is also true that nothing changes more fully and completely than the immutable past. Our historical understanding of the years from 1908 to 1944 is still rudimentary, opaque and unsettled: subject to transformation and change. In short, nothing has yet crystallised.

The 1702 Chair of Chemistry at Cambridge: Transformation and Change, ed. Mary D. Archer and Christopher D. Haley. Published by Cambridge University Press. © Cambridge University Press 2005.

In the hope of at least beginning the crystallisation process, we shall first outline the bare, chronological facts of the 1702 Chair in the period 1908–44; second, point to some positive, and some negative, aspects of 'Cambridge and War'; and third, take a wider, more reflective view on the enduring significance of the era from George Downing Liveing to Todd.

Chronology

The election of William Jackson Pope (Figure 8.1) to the 1702 Chair in 1908 signalled a dramatic break with the past.[2] By no stretch of the imagination a 'liberally-educated chemist',[3] Pope held no university degrees of any kind, let alone an Oxbridge qualification. The first chair holder since Vigani not to have come through the Cambridge system, Pope was instead a Londoner, reared in the 'City and Guilds' tradition. After assisting his mentor, Henry Edward Armstrong, at City and Guilds, he became chief of the chemistry department at Goldsmiths' Institute. In later years, he maintained a close connection with their world of technical issues, offering advice related to the production and use of precious metals.

More revealing is the fact that, by the age of thirty, Pope had made an immortal name for himself as a 'brilliant crystallographer'. Working with a series of collaborators, he advanced stereochemistry beyond Jacobus van't Hoff and Joseph Le Bel by resolving a series of asymmetric, optically active compounds of nitrogen, sulphur, tin and selenium. Stereochemistry was thus greatly expanded beyond carbon compounds. The reward was appointment to Manchester's third and most junior chair of chemistry in 1901.[4] Here, Pope and William Barlow attempted to develop 'the valency volume theory of crystal structure', which held that the space that an atom occupies in a crystalline structure is proportional to its valence. For certain organic compounds, this theory worked pretty well. It ultimately failed in the case of inorganic compounds, but not before Pope had been elected F.R.S. in 1902 and called to Cambridge in 1908, at the early age of 38. Interestingly the electors bypassed the holder of the second and more senior Manchester chair, William Henry Perkin, Jr, who instead went to Oxford in 1912.

In similar competitive mode, we might note that Pope's eventual successor, Alexander Todd, made it to the 1702 Chair when one year younger than Pope, from a similar Manchester professorship (in Todd's case, the most senior chair). And – as with Todd, some 36 years later – part of the price and lure of this change was the opportunity for Pope to create what was virtually a new Department out of Liveing's Pembroke Street laboratory (Figure 8.2).

Figure 8.1. William Jackson Pope (courtesy of the Fisher Collection, Chemical Heritage Foundation).

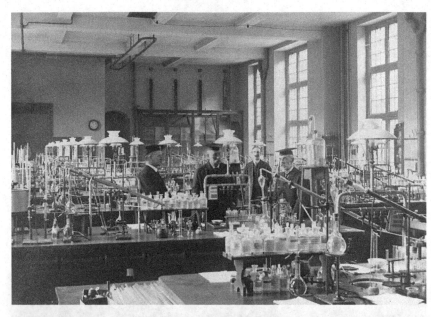

Figure 8.2. University Chemical Laboratory in Pembroke Street.

Despite not being educated through the Cambridge system, Pope took to the Cambridge milieu, which in those days was made for bachelors like himself. Early in his Cambridge career, he quite literally 'created a stink', and gave inadvertent offence to townspeople by an experiment with alkyl selenides on the roof of the Pembroke Street Laboratory (Figure 8.3). In later years he was known as a charming host; a graceful speaker in English, French, German

Figure 8.2. (*cont.*)

and Italian; an accomplished violinist; a discriminating collector of alchemical pictures; and a noted chemical bibliophile – if all this was not enough, 'in conversation he was brilliant and witty, and few could tell a good story better than he'.[5]

Pope's first objectives were to modernise Cambridge's anarchic approach to science, gain better control over scarce resources, and thereby increase his standing and prestige among the metropolitan and international elite of research-oriented chemists. These aims were partially achieved by the centralisation of chemistry teaching in the University Chemical Laboratory and the closing down of the various college laboratories that had grown up in Liveing's era, listed in Table 8.1.[6] In this strategy, he was far more successful than his Manchester-to-Oxford colleague, Perkin, who faced similar challenges.

To enable the needed further extension of the University Chemical Laboratory, Pope at the same time displayed his close links to the new commercial realities that were shaping the chemical sciences in other less tradition-bound centres and obtained major funds from industry to supplement the reserves provided by the University. Efforts engineered by Pope and efforts wholly remote from Pope led to the creation of no fewer than five additional chairs (see Figure 8.4) by the start of the 1930s, enriching and extending the various chemistries available at Cambridge. When Pope arrived in Cambridge, there

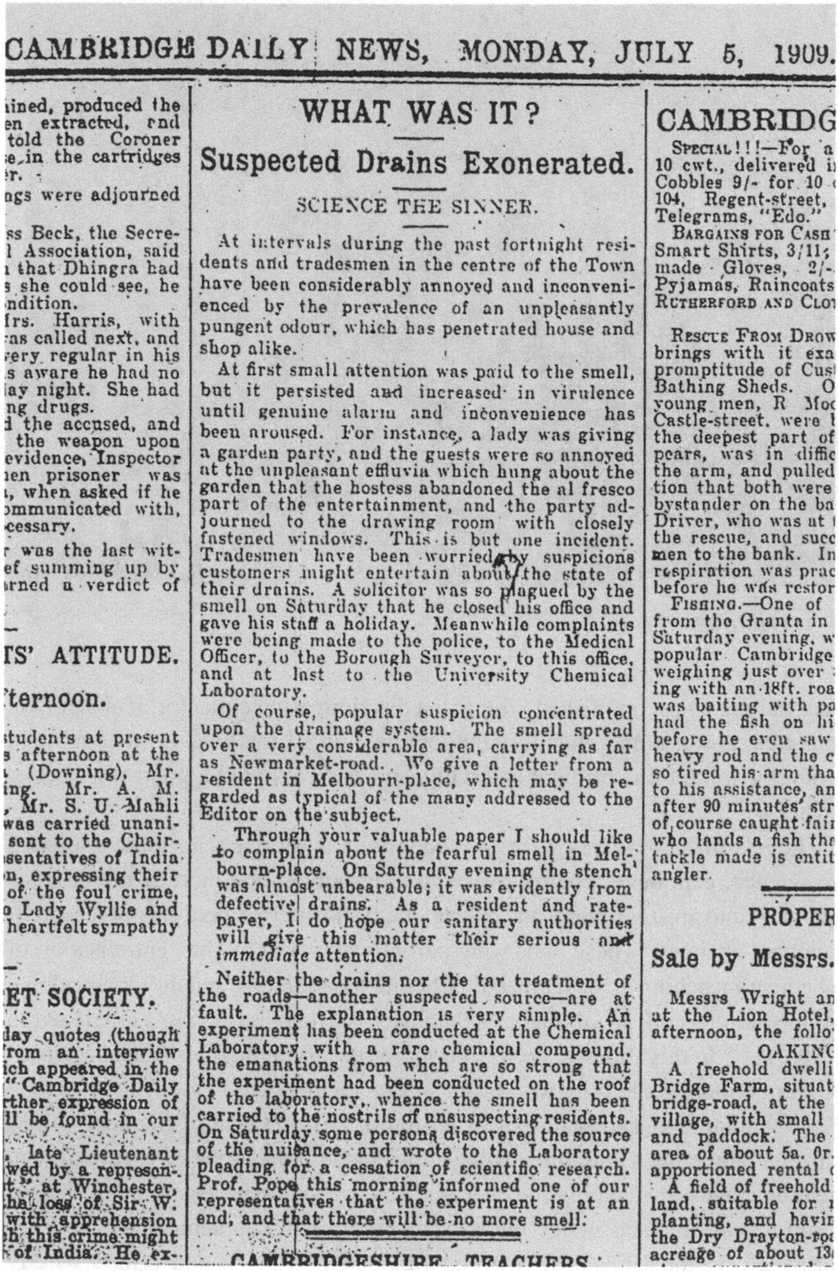

CAMBRIDGE DAILY NEWS, MONDAY, JULY 5, 1909.

ined, produced the
en extracted, and
told the Coroner
ie in the cartridges
r.
ngs were adjourned

ss Beck, the Secre-
l Association, said
i that Dhingra had
s she could see, he
ndition.
lrs. Harris, with
as called next, and
very regular in his
s aware he had no
lay night. She had
ng drugs.
d the accused, and
the weapon upon
evidence, Inspector
len prisoner was
t, when asked if he
ommunicated with,
cessary.
r was the last wit-
ef summing up by
irned a verdict of

IS' ATTITUDE.

ternoon.

students at present
s afternoon at the
i (Downing), Mr.
ing. Mr. A. M.
, Mr. S. U. Mahli
was carried unani-
sent to the Chair-
sentatives of India
n, expressing their
of the foul crime,
o Lady Wyllie and
heartfelt sympathy

ET SOCIETY.

lay quotes (though
rom an interview
ich appeared in the
"Cambridge Daily
rther expression of
ll be found in our

late Lieutenant
wed by a represen-
t at Winchester,
he loss of Sir W.
with apprehension
h this crime might
of India. He ex-

WHAT WAS IT?

Suspected Drains Exonerated.

SCIENCE THE SINNER.

At intervals during the past fortnight residents and tradesmen in the centre of the Town have been considerably annoyed and inconvenienced by the prevalence of an unpleasantly pungent odour, which has penetrated house and shop alike.

At first small attention was paid to the smell, but it persisted and increased in virulence until genuine alarm and inconvenience has been aroused. For instance, a lady was giving a garden party, and the guests were so annoyed at the unpleasant effluvia which hung about the garden that the hostess abandoned the al fresco part of the entertainment, and the party adjourned to the drawing room with closely fastened windows. This is but one incident. Tradesmen have been worried by suspicions customers might entertain about the state of their drains. A solicitor was so plagued by the smell on Saturday that he closed his office and gave his staff a holiday. Meanwhile complaints were being made to the police, to the Medical Officer, to the Borough Surveyor, to this office, and at last to the University Chemical Laboratory.

Of course, popular suspicion concentrated upon the drainage system. The smell spread over a very considerable area, carrying as far as Newmarket-road. We give a letter from a resident in Melbourn-place, which may be regarded as typical of the many addressed to the Editor on the subject.

Through your valuable paper I should like to complain about the fearful smell in Melbourn-place. On Saturday evening the stench was almost unbearable; it was evidently from defective drains. As a resident and ratepayer, I do hope our sanitary authorities will give this matter their serious and *immediate* attention.

Neither the drains nor the tar treatment of the roads—another suspected source—are at fault. The explanation is very simple. An experiment has been conducted at the Chemical Laboratory with a rare chemical compound, the emanations from which are so strong that the experiment had been conducted on the roof of the laboratory, whence the smell has been carried to the nostrils of unsuspecting residents. On Saturday some persons discovered the source of the nuisance, and wrote to the Laboratory pleading for a cessation of scientific research. Prof. Pope this morning informed one of our representatives that the experiment is at an end, and that there will be no more smell.

CAMBRIDGESHIRE TEACHERS

CAMBRIDG

SPECIAL!!!—For a
10 cwt., delivered i
Cobbles 9/- for 10
104, Regent-street.
Telegrams, "Edo."
BARGAINS FOR CASH
Smart Shirts, 3/11;
made Gloves, 2/-
Pyjamas, Raincoats
RUTHERFORD AND CLOT

RESCUE FROM DROW
brings with it exa
promptitude of Cus
Bathing Sheds. O
young men, R Mo
Castle-street, were
the deepest part of
pears, was in diffic
the arm, and pulled
tion that both were
bystander on the ba
Driver, who was at
the rescue, and succ
men to the bank. In
respiration was prac
before he was restor
FISHING.—One of
from the Granta in
Saturday evening, w
popular Cambridge
weighing just over
ing with an 18ft. roa
was baiting with pa
had the fish on hi
before he even saw
heavy rod and the c
so tired his arm tha
to his assistance, an
after 90 minutes' str
of course caught fair
who lands a fish tha
tackle made is entit
angler.

PROPER

Sale by Messrs.

Messrs Wright an
at the Lion Hotel,
afternoon, the follo
OAKING
A freehold dwelli
Bridge Farm, situat
bridge-road, at the
village, with small
and paddock. The
area of about 5a. 0r.
apportioned rental
A field of freehold
land, suitable for
planting, and havin
the Dry Drayton-roa
acreage of about 13

Figure 8.3. 'What was it?', Cambridge Daily News, 5 July 1909.

Table 8.1. *Chemical laboratories in Cambridge and Oxford colleges*

Cambridge		Oxford	
St. John's	1853–1914	Christ Church	1860s–1941
Sidney Sussex	late 1870s–1908	Magdalen (Daubeny)	1848–1923
Gonville and Caius	1873–1908	Balliol-Trinity	1853–1941
Girton	1877–1935	Queens'	1900–34
Newnham	1879–1912	Jesus	1907–47
Downing	1890s–1920		

Sources: F. G. Mann, 'The Place of Chemistry – II. At Cambridge', *Proc. Chem. Soc.* (July 1957), pp. 190–3; Edward Miller, *Portrait of a College: A History of the College of Saint John the Evangelist, Cambridge*, plate facing p. 102; Morrell, Jack, *Science at Oxford, 1914–39: Transforming an Arts University*, pp. 318–27.

were only two professorships filled by chemists – the 1702 Chair and the Jacksonian professorship in Natural Sciences. Since its inception, the Jacksonian chair had been shuffled between physics and chemistry and from 1875 it was filled by James Dewar. In 1877, Dewar accepted the Fullerian professorship of chemistry at the Royal Institution, where he thereafter focused his energies, while continuing to hold the Jacksonian *in absentia* until 1923 – his teaching obligations at Cambridge being carried out by a succession of demonstrators.

Pope's efforts inside Cambridge, and his major roles outside, were aided by his formidable experimental talent, his gift for public speaking and flair for

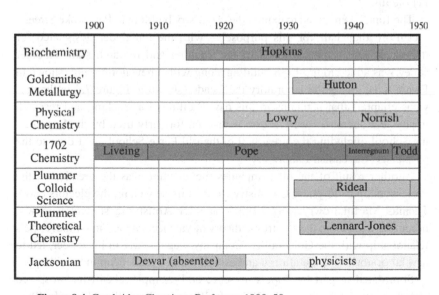

Figure 8.4. Cambridge Chemistry Professors 1900–50.

publicity, and his ability – as an individual unencumbered by family – to travel widely. For instance, the outbreak of World War I found him in Australia presiding over the chemistry section of the British Association, at its hugely successful meeting at the far reaches of empire. (He hurried home, while German chemists in attendance were promptly interned.) Almost two decades later, one obituarist still remembered how 'in his Melbourne address he depicted stereochemical investigation as an artistic pursuit . . . and in this scientific artistry he was a master unsurpassed'.[7]

On a more sombre note, the greatest of Pope's many contributions to the war was a method for preparing mustard gas, which became the standard process. Not entirely coincidentally, he was the only chemist to emerge from World War I with a knighthood. Further evidence of the esteem in which he was held came via the presidency of the Chemical Society (1917–19), his service as first president of the International Union of Pure and Applied Chemistry (1922–4), and as first president of the Solvay Chemistry Conferences (1922–36).[8]

Closer to home, Pope's wartime links and work led to the munificent gift of £210 000 to Cambridge chemistry from a consortium of British oil companies in May 1919.[9] (Conservatively estimated, this gift would be worth £6 300 000 at the time of writing.)[10] The 'Papal State' thus became the best-funded chemistry department in the United Kingdom, with support from Royal Dutch Shell as well as the Burma Oil Company, the Anglo-Persian Company and the Anglo-Saxon Petroleum Company, which would all eventually become part of British Petroleum.

The funds were given to extend the chemistry building in Pembroke Street – 'not more than half for this purpose' – with the remainder designated for its upkeep and for the salaries of the teaching and research staff. Another storey was added to the 1888 building along with a new north wing (shown in Figure 8.5), giving the laboratory the facade (shown in Figure 7.3 in the previous chapter) that it retains to this day. Facilities for teaching and research in physical chemistry were set up in an area formerly used by the Engineering School, including the great hall of the old Perse School, next door to the physicists' Cavendish Laboratory.

Another result of the oil companies' benefaction was the creation of the Professorship of Physical Chemistry. To fill this new chair, the electors chose Thomas Martin Lowry (1874–1936), another Armstrong student who was becoming a leader in the electronic theory of valence. At the same time, Pope's relationship with the Goldsmiths' Company, begun early in his career, led to new laboratories for metallurgy and assaying and a readership in metallurgy.

Stimulated in part by Pope's presence and example, chemistry prospered elsewhere in Cambridge in the 1920s. In 1924, the Sir William Dunn Institute

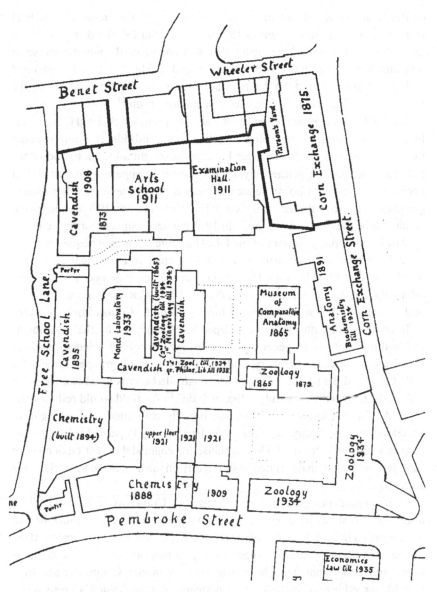

Figure 8.5. Pembroke Street Laboratory. Pope inherited the responsibility of designing and equipping the 1909 extension of the laboratory built in 1888 under Liveing's supervision. The 1921 additions were funded by the 1919 gift from the British oil companies.

of Biochemistry was founded in the medical faculty for Frederick Gowland Hopkins of vitamin fame.[11] From 1927 to 1937, John Desmond [J. D.] Bernal took up residence in the Cavendish Laboratory and regularly taught a course on 'chemical crystallography'.[12] He was succeeded by Max Perutz, who remained with the physicists until 1962 when the Medical Research Council built the Laboratory of Molecular Biology for him and his group.[13]

The 1930s saw the foundation of several more chemical chairs at Cambridge. In its efforts to upgrade the biological sciences at Cambridge, the International Education Fund of the Rockefeller Foundation sought a bridge to one of the physical sciences, but neither physics nor chemistry appeared interested. But once a lectureship in colloid science was created among the Rockefeller-funded positions and filled with chemistry's own Eric Rideal (1890–1974), cooperation seemed to be the order of the day.[14] In 1931 the university authorities, encouraged by Pope, applied portions of the John Humphrey Plummer bequest to make the colloid position permanent and to fund a chair in theoretical chemistry, which was first filled by John Edward Lennard-Jones, a pioneer in molecular orbital theory. And in the same year, the Goldsmiths' Readership was made a professorship by means of additional funding from the Goldsmiths' Company.

In comparison with the fierce independence exhibited by the incumbents of Cambridge chairs in general, the professors of colloid science, physical chemistry, theoretical chemistry, and metallurgy were all grouped around the 1702 Chair and the person of Pope, who controlled access to laboratory space and funds. Chemistry was truly a 'Papal State'.[15] As Todd would reflect years later, 'As far as I knew – and this was broadly confirmed by John Tennant Saunders (then Secretary General of the Faculties) – Pope had ruled his colleagues with a rod of iron.'[16] This centralisation enabled the creation of modern facilities, but its health depended quite directly on the health and the style of the leader.

More immediately damagingly, Cambridge University had displayed its institutional realities by dragging its feet when the creation of a Ph.D. degree was proposed in the late stages of World War I. The Senate finally approved the degree in 1919, two years later than Oxford, another 'holdout'.[17] Content in its liberal arts, undergraduate, collegiate and tutorial modes, Cambridge saw little need for something so Teutonic – or American – as a 'profession' of researcher. One Ph.D. candidate in chemistry who was later to become well known, Charles Percy [C.P.] Snow, recalled how 'a research student at that period [1928] was something of an odd-man-out in Cambridge . . . we [Snow, Patrick Maynard Stuart Blackett, and others] formed a fairly tight-knit community. We hadn't many undergraduate friends: we were rather too old for that, and we were leading a different kind of life . . .'[18]

The fact that the Ph.D. did not fit in Cambridge, and Cambridge did not care, was bad enough. Still worse was the tradition-reinforced tolerance of the idiosyncrasies of the professors. Dewar continued his absentee ways. Lowry, the new professor of physical chemistry, 'was a bit of an injustice collector', with 'a curious kind of obstinacy', who had 'got stuck with researches on optical rotation that didn't attract many pupils'.[19] Rideal, when he became professor of colloid science, attracted students away from Lowry, but was afraid of having too many students.[20] Ronald George Wreyford Norrish (1897–1978), who was elected to the chair of physical chemistry on Lowry's death in 1936, had a mixed reputation with students.[21] Lennard-Jones, on the other hand, certainly had the power to attract students to his brand-new, but then quite segregated, field.

Pope himself had little interest in the idea of a 'research school' in stereo-chemistry, of the type long familiar in Germany, and newly powerful in the United States. As John Read, his collaborator in the early Cambridge days, was to put it: 'With his research men he was always resourceful when help was called for; but after a topic had been decided upon and fully discussed he encouraged his junior collaborators to cultivate self-reliance'.[22]

An even greater handicap became apparent around 1927. Pope began to suc-cumb to an illness, which his memorialists did not identify but which appears to have progressively disabled him. He took less interest in chemistry and became almost a recluse. His advanced students were increasingly handled by William Hobson Mills (another stereochemist, recruited by Pope in 1912, officially to deputise for the absent Dewar). Mills was notorious for writing understated letters of recommendation, for being unapproachable, and for expecting 'his research men both to order and buy any chemicals they required which were not available in the very limited departmental store'.[23] Since the few, meagre government grants available from the Department of Scientific and Industrial Research covered only two years, and Cambridge required three for the Ph.D., this was yet another major disincentive. As Table 8.2 shows, there was little or no growth in the number of chemistry Ph.D. degrees awarded until after World War II, when Todd's energetic programme of recruiting graduate students took hold.

Pope died in October 1939 after a long, lingering illness, but still in harness, at the age of 69. The exigencies of war meant that the 1702 Chair languished, unfilled.

Alex Todd was well aware of Cambridge realities. Active in war work in Manchester, and well connected to Britain's scientific establishment through his marriage to the daughter of Sir Henry Dale (President of the Royal Society, 1940–5), Todd had been informally sounded out regarding the Cambridge chair soon after Pope's death. In 1940, the timing was poor; in 1943, it was better, and

Table 8.2. *Numbers of Ph.D. degrees awarded, 1926–39*

Year	Cambridge chemistry	All British universities
1925–6	9	285
1930–1	12	405
1935–6	12	532
1938–9	13	547

Sources: *Abstracts of Dissertations Approved for the Ph.D., M.Sc. and M.Litt. Degrees in the University of Cambridge for the Academical Year 1925–6, 1930–1, 1935–6, 1938–9, 1945–6* (Cambridge: Cambridge University Press); Simpson, Renate, *How the Ph.D. Came to Britain* (Guildford, Surrey: Society for Research into Higher Education, 1983), p. 165.

Todd was offered – and cannily declined – the more peripheral biochemistry chair, made newly vacant by the death of Gowland Hopkins. In 1944, with the end of the war in sight, the time was ripe to fill the 1702 Chair. Todd could name his terms. And he did – echoing Pope's early strategy by including on his list the renovation of existing facilities and the construction of a new building as soon as was feasible.[24] However, having enjoyed several years of relative autonomy during this interregnum, Rideal and Norrish respectively asserted the independence of colloid science and physical chemistry; and Rideal moved off to the Royal Institution in 1946.[25]

Cambridge and war

World War I

Contemporaries glorified the bravery of the university students and staff who heeded the clarion call in the opening days of World War I. In fact, war in its initial phases was a disaster, for both students and staff. Cambridge had well over 3000 male undergraduates on the eve of World War I. Within weeks, this number had halved to 1658; by 1915, it halved again, to 825; by 1917, to 398.[26] Cambridge men quickly volunteered, and quickly found their way to the Continent and to their deaths. For instance, John Dunlop, Assistant Lecturer in Chemistry, was commissioned a Second Lieutenant in the Royal Dublin Fusiliers, and fell in battle on 17 August 1914.[27] The category of 'reserved occupations' and the idea of conserving talents too valuable for the trenches did not come into effect until November 1915.[28]

In the early days of the war, the government proceeded without much scientific advice. Pope later bitterly recalled the beginning of British Dyestuffs:

The scheme adopted by the Government for resuscitating [the coal-tar industry] in our country . . . was launched without scientific advice; the Cabinet mouth-piece, indeed, declared that the directorate of the company was not to include men of scientific knowledge on the ground that a director who knew something about the business of the company would have an advantage over his less well-informed colleagues.[29]

Slowly, scientists began to organise themselves to provide advice in many areas. Pope was a member of a group from the Royal Society and the Chemical Society who pressed for a committee to link universities with government and industry. Their wish was granted in May 1915, with the creation of a Committee of the Privy Council on Scientific and Industrial Research, and from 1916, the separate Department of Scientific and Industrial Research.[30]

From the industrial side, an important initiator of cooperation between industry and academe was Robert Waley Cohen, a graduate of Emmanuel College who had taken both parts of the Natural Science Tripos in 1898–90. Before the war, Cohen had engineered the merger of the Royal Dutch and Shell oil companies, and was among the British oilmen who hoped to pull together all the British oil companies to compete with the American oil companies. During the war, he urged the British government to use petrochemicals as alternatives to coal-tar derivatives, then in scarce supply, in the production of explosives.[31] Early in the war, Waley Cohen sent to Cambridge a team of Dutch chemists, headed by J. E. F. ('Frits') de Kok, manager of Royal Dutch Shell's experimental laboratory in Amsterdam, to collaborate with Pope and his staff to develop processes whereby toluene distilled from petroleum imported from Borneo could be used to manufacture explosives.[32]

Soon after the Germans introduced gas warfare into the conflict, in April 1915 on the Western Front, the newly formed Ministry of Munitions began to pay certain chemistry departments to find methods of synthesising poisonous materials that could be scaled up with a good yield. Pope and his colleagues worked on methods to prepare phosgene. To Pope's chagrin, however, his was not among the first laboratories to be asked to work on mustard gas ($\beta\beta'$-dichlorodiethyl sulphide – $(C_2H_4Cl)_2S$), which was introduced into the conflict in July 1917. He complained to Harold (later Sir Harold) Hartley, Oxford chemist and then head of gas defence in the Army's Special [Gas] Service:

No chemist outside London is allowed any effective part in the work. The country is full of excellent chemists who are as capable as anyone in Germany of working out the question of what can be done in the production of unpleasant materials –

Perkin, Kipping, Wynne, Lapworth, Henderson, Patterson, Orton, McKenzie, etc. have been available all along. [British chemical warfare research] is being run in the spirit of the little village publican doing his two barrels a week.[33]

Pope's complaint was heard, and he went to work on methods of preparing mustard gas. In the end, he was successful in convincing Winston Churchill, then Minister of Munitions, of the superiority of a method based on a synthesis from sulphur monochloride and ethylene discovered by Frederick Guthrie more than fifty years before:

$$2C_2H_4 + S_2Cl_2 \rightarrow (ClCH_2CH_2)_2S + S$$

Eventually, the temperature parameters that Pope recommended for the reaction to produce its highest yield had to be changed.[34] He nonetheless carried the reputation of having solved a major military problem – and at no small cost to himself and his assistants, one of whom suffered sloughing sores over a great part of his body following a laboratory accident.[35]

The other major military contribution of Cambridge chemistry to World War I was the development of photographic sensitisers. Daring pilots in their flying machines would take photographs at dawn to determine changes in the enemy positions, but silver halide emulsions are sensitive only to the violet and blue regions, not to the rosy light of early morning. The Germans solved this difficulty before the war with cyanine compounds; the British were again in the position of playing catch-up. On a contract from the Committee on Scientific and Industrial Research, Pope and W. H. Mills achieved their objective in 1915 within a few months. The University Chemical Laboratory henceforth became the principal site for the manufacture of cyanine dyestuffs used by the Allies during the war.[36]

World War II

World War II was less disruptive for Cambridge than World War I. The colleges, classrooms and laboratories were not deserted as before; rather, many students were put on a fast track to obtain undergraduate degrees in two years. The difficulties caused by this speed-up were compounded by the evacuation from London of entire institutions: the chemistry building was soon bursting with students from Bedford College, Queen Mary College and St Bartholomew's Hospital Medical School. The gas lighting in the building posed safety hazards, and with blackout requirements, ventilation became worse than ever. Lecture halls and laboratories were used continuously, and as one assistant recalled, there was a powerful stale smell emanating from the large lecture hall when the doors were opened to let out the last herd of students.[37]

Given the demands on university expertise, teaching staff were, not surprisingly, in short supply. Lennard-Jones, having converted the calculating machines of the Cambridge Mathematics Laboratory from working on problems in molecular orbital theory to problems in ballistics, was called to be Chief Superintendent of Armament, directing a staff of over 3000.[38] Eric Rideal remained at Cambridge, but served on a multitude of defence committees.[39] The minutes of the Faculty Board of Physical and Chemical Studies reveal the struggle to keep lectures staffed. Physical chemist Emyr Alun Moelwyn-Hughes went off to the Ministry of Supply in December 1940. Anxiously, Cambridge's Secretary General of the Faculties wrote:

> If the Government wished to maintain a supply of physical chemists, it was essential for the University to retain at least two teachers of Physical Chemistry in addition to yourself [Norrish].[40]

Two men whom Norrish was bent on keeping were Frederick Dainton and William Charles Price.

Temporary faculty had to be employed to fill the holes in the curriculum. Among the temps, James Riddick Partington, from Queen Mary College and hence in Cambridge, gave eight lectures on thermodynamics to prepare students for Part II of the Tripos.[41] Harry Julius Eméleus, then at Imperial College, gave lectures in inorganic chemistry.[42] (After the war, he was to be appointed to Cambridge's new chair of inorganic chemistry, founded at Todd's insistence.) And when, despite Norrish's request, Price went off to work for ICI, Norrish turned in desperation to Frederick Eirich, who was finally doing research for the Ministry of Supply, having earlier been interned as an enemy alien, first in England and then in Australia.[43]

Many of the trials and tribulations of university life went on as before. Young Rosalind Franklin won a fourth-year fellowship to work with Norrish in 1941, but left because of his harsh treatment of her – apparently not uncommon behaviour for him at that time in his life.[44] Dainton and his wife supported Franklin in her decision to work for the Ministry of Fuel, even though at the last minute Norrish offered to put her on 'reserved status'. She returned to Cambridge only briefly to submit a dissertation based on her war work: 'The Physical Chemistry of Solid Organic Colloids, with Special Reference to the Structure of Coal and Related Materials'.

The Cambridge Chemical Society, which hosted talks by faculty and visitors on the latest advances in chemistry, met regularly – in contrast to the six-year hiatus imposed by the exigencies of World War I.[45] Only occasionally was the topic at hand war-related; most on-going war research was kept secret. Shortly after the war, on 19 October 1945, Rideal delivered a talk to the society entitled 'Back-Room Boys at Work', which included brief discussions of several wartime research projects conducted at Cambridge.[46]

World War I is often called the 'Chemists' War', with World War II being claimed by the physicists. Nevertheless, Cambridge chemists did much, even if in the 'back rooms'. In Norrish's group, Harold Bamford was working on spontaneously inflammable gels; Morris Sugden, Douglas Axford and Raymond Andrew on the suppression of gun muzzle flash; Fred Dainton, on a photochemical switch, various smoke mixtures and the ignition of stored aircraft fuels; and Arthur Donald Walsh, on combustion in internal combustion engines.[47]

Among Rideal's group of physical chemists, some focused on explosives. For example, Hugh Campbell and Paley Johnson examined differences in nitrocellulose prepared from four different sources of cellulose (a study noted by Rideal in his talk); David Charles Pepper and Daniel Douglas Eley worked on the rheological problems of plastic explosives and cordite; Andrew John Blackford Robertson studied the thermal decomposition of a whole range of explosives; and Eric Hutchinson worked on flow problems in amatol. Meanwhile, problems in the physical characteristics of fuels and lubricants occupied James Herbert Schulman and Albert Ernest Alexander, Arthur Stuart Clark Lawrence, Alastair Cameron, Andrew Robertson and Philip George. Ronald Francis Tuckett's group solved problems with polymers for the Ministry of Aircraft Production. There was one wartime tragedy in Rideal's group – an experiment by Oliver Gatty and Charles Frank involved passing superheated steam through petroleum. The equipment set up in Grantchester Meadow overturned, and Gatty was severely burned and died a few days later.[48]

For the organic chemists, chemical warfare was again part of the picture. Bernard Charles Saunders synthesised organophosphorus compounds as potential nerve gases for the Chemical Defence Section of the Ministry of Supply.[49] Fred Mann, aside from carrying a heavy teaching load and acting as an adviser to local air-raid wardens, researched potential anti-malarials and trypanocides.[50]

When Todd arrived in Cambridge in October 1944, he was astounded by much that he saw, including a number of groups in the Pembroke Street chemistry laboratory whom he considered invaders. Among them was a small group working on the separation of uranium isotopes for the atomic bomb project. Serving with this group was future faculty member Alfred Gavin Maddock.[51] Perhaps the greatest versatility in responding to wartime needs was displayed by William Tutte. This young man was plucked from his graduate studies in chemistry at Cambridge by his undergraduate tutor to engage in deciphering German codes at Bletchley Park. Without graduate training in mathematics,

Tutte managed to deduce the structure of 'Fish', the code that the German Army High Command employed.[52]

So much for the triumphs. What of the questions? It is sobering to reflect how, in World War I, the initiative in chemistry lay with Germany. There was no British Haber to synthesise ammonia, transform munitions or launch gas warfare. Instead, there was only a slow, belated scramble to respond. Even when processes were understood on the laboratory scale, Britain in general, and Cambridge in particular, were devoid of chemical engineers. The very phrase 'chemical engineering' had been coined in Manchester, long before World War I by George E. Davis.[53] But the American Institute of Chemical Engineers, founded in Philadelphia in 1908, was not to have a British analogue until 1922.[54] For a degree in the subject one went to Cambridge, Massachusetts, not Cambridge, England. Winston Churchill half understood this when, as Minister of Munitions in World War I, he was asked by Woodrow Wilson what he most needed: he replied, 'Send us chemists', by which he meant men experienced in the design, construction and operation of plants.[55] One such man was the American Kenneth Bingham [K. B.] Quinan. British TNT production in 1914 was 20 tons a week. Summoned by telegraph in August 1914, Quinan had built a plant at Oldbury by May 1915, which eventually produced over 200 tons a week.

Reflecting on the war in his 1918 and 1919 presidential addresses to the Chemical Society, Pope warned that society's estimate of the value of science might decline precipitously once the dust of battle had settled. He predicted that scientists would no longer have ready access to the governing councils of the land, and scientific research would again be under-funded. He understood that the German chemical industry would rebound, because they had converted their standing chemical factories to war work, and could easily return them to peacetime pursuits. The British, in contrast, had hurriedly erected wartime production facilities that were not associated with any commercial infrastructure. Moreover, he foresaw that 'competition in pure and applied chemistry between Europe and America will become increasingly keener during the years to come', and feared a re-establishment of 'that parsimonious treatment of scientific effort' which had characterised Britain in the previous decades.[56]

Events were to prove his fears well-grounded. In World War II, the centre of the chemical stage was held by the synthetic rubber programme, high-octane gasoline, the chemical engineering feat known as the Manhattan Project, and the commercial production of penicillin and DDT. Metaphorically speaking, Cambridge, Massachusetts, and Urbana, Illinois, not Oxford and Cambridge, had replaced Berlin and Munich as the prime theatre for the chemical drama.

The wider view

Between 1908 and 1944 – perhaps even from the 1870s to the 1970s – Cambridge chemistry has had certain assets and certain liabilities. Its greatest asset has been its capacity to generate an elite. With this have come superior resources – of living accommodation, food and wine, sports, libraries and intellectual traditions – which meant that Cambridge also became a magnet for English-speakers overseas. This role as cradle of an elite favoured undergraduates and generalists and militated against the graduate specialism of the Ph.D., the business school and clinical medicine. Viewing chemistry as a science, the 'great tradition' goes from Boerhaave and Leyden, to Edinburgh, then on to Liebig at Giessen and the nineteenth century hegemony of the German university. The English echo, and research tradition, lay supremely in Manchester and also in London, but not in Oxbridge. Hopkins, Harvard and Illinois all heard the music of the research school more clearly than Cambridge or Oxford. The new institutions of California heard it most clearly of all – Caltech made a job offer to Alexander Todd before either Manchester or Cambridge.[57]

Cambridge chemistry was strongly centralised, by Cambridge standards, in the first half of the twentieth century. Initially, that aided Pope, and later it helped Todd as well. However, the charms of Cambridge's more customary anarchic and idiosyncratic habits were also plainly displayed in Pope's era. Gowland Hopkins' work in biochemistry owed little to central planning. Rideal, Lennard-Jones and Mann were half-accidental additions of great strength. The work of Norrish, of Archer Martin and Richard Synge, of George Porter, of C. P. Snow, of J. D. Bernal, of Joseph Needham and of Rosalind Franklin – all pursuing chemical research in Cambridge in this period – shows how talent will respond to resources in unforeseen ways.

Perhaps the chemist whose name will live longest in the annals of Cambridge chemistry, even more for his philanthropy than his science, is an organic chemist from this era who attracted little attention while in Cambridge, and who, like the authors, passed his adult life in suburban Philadelphia. To see how difficult is the task of understanding how to plan education and research, we need only mention Cambridge chemistry's greatest benefactor, Herchel Smith, whose career lay in organised industrial research, of a kind and on a continent quite remote from the Cambridge in which he obtained his education. Smith became an undergraduate at Emmanuel College in 1942, and he was awarded his doctorate in organic chemistry in 1952. His fortune was based primarily on patents concerning the synthetic steroids used in birth control pills.[58] During his lifetime, Dr Smith made benefactions to Cambridge of more than £15 million to support medicinal chemistry and intellectual property law; on his death in

2001, he set new records for Cambridge philanthropy by bequeathing more than £45 million to provide endowments for the chairs and other benefactions established during his lifetime.[59]

Cambridge chemistry in the period 1908–44 – like Cambridge in this era – was quirky and resistant to centralised direction, yet alive and full of ingenious talent and unexpected outcomes.

Notes and References

1. Todd, Alexander (1983), *A Time to Remember: the Autobiography of a Chemist*, Cambridge: Cambridge University Press, p. 68.
2. Pope's biographical details and discussions of his chemical contributions are available in Gibson, Charles S. (1941), 'Sir William Jackson Pope', *Obituary Notices of Fellows of the Royal Society, 1939–1941*, vol. 3, London: Royal Society, pp. 290–324; Mills, W. H. (1941), 'Sir William J. Pope', *J. Chem. Soc.*, pp. 697–715; Mann, Frederick George (1981), 'William Jackson Pope', *Dictionary of Scientific Biography*, vol. 11, New York: Charles Scribner's Sons, pp. 84–92.
3. For the definition of a 'liberally-educated chemist', see Roberts, Gerrylynne (1980), 'The liberally-educated chemist: chemistry in the Cambridge Natural Sciences Tripos, 1851–1914', *Historical Studies in the Physical Sciences* **11**, pp. 157–83.
4. Pope was appointed professor of applied chemistry in 1901 at the Manchester Municipal School of Technology, which would eventually become Manchester Institute of Science and Technology – the Faculty of Technology of the University of Manchester. Over at Owens College, Manchester, which would become the Arts and Sciences Faculties of the University of Manchester, Henry Roscoe had occupied the chair of chemistry since 1857, and William Henry Perkin, Jr its chair of organic chemistry since 1892.
5. Gibson, 'Sir William Jackson Pope', p. 323.
6. Read, John (1947), *Humour and Humanism in Chemistry*, London: G. Bell and Sons, p. 285.
7. Mills, 'Sir William J. Pope', p. 715.
8. See Gibson, C. S. (1946), 'Life and work of William Jackson Pope', *J. Roy. Soc. Arts* **94**, pp. 668–74; Lampitt, L. H. (1947), 'Sir William Jackson Pope: his influence on scientific organisation', *J. Roy. Soc. Arts* **95**, pp. 173–83.
9. Cambridge University Registry, Professor of Chemistry (CUR 39.11), f. 69, Cambridge University Archives. Illustrations of the facilities built with the British oil companies' endowment are included in Mann, F. G. (1928), 'The Cambridge University Chemical Laboratory', *J. Soc. Chem. Ind.* **47**, pp. 690–6.
10. The calculation of the present worth of the gift is derived from McCusker, John J. (2001), 'Comparing the purchasing power of money in Great Britain from 1264 to any other year including the present', *Economic History Services*, URL: *http://www.eh.net/hmit/ppowerbp* (as of Aug. 2004). McCusker gives values based on the retail price index.
11. Weatherall, Mark and Kamminga, Harmke (1992), *Dynamic Science: Biochemistry in Cambridge*, Cambridge: Wellcome Unit for the History of Medicine, p. 34 ff.
12. Minutes of the Faculty Board of Physical and Chemical Studies, 1937–1943, Min. v. 86, f. 26, Cambridge University Archives; and for the course's disposition during World War II, *ibid.*, 1944–49, Min. v. 87, f. 47.
13. Nobel Foundation (1964), *Nobel Lectures in Chemistry, 1942–62*, New York: Elsevier Publishing Company, pp. 674–5.

14. Kohler, Robert E. (1991), *Partners in Science: Foundations and Natural Scientists 1900–1945*, Chicago: University of Chicago Press, pp. 182 –8; Eley, D. D. (1976), 'Eric Keightley Rideal, 1890–1974', *Biographical Memoirs of Fellows of the Royal Society*, vol. 22, London: Royal Society, p. 385.
15. For reference to the 'Papal State', see Tabor, D. (1969), 'Frank Philip Bowden, 1903–1968', *Biographical Memoirs of the Fellows of the Royal Society*, vol. 15, London: Royal Society, p. 5.
16. Todd, *A Time to Remember*, p. 64.
17. Simpson, Renate (1983), *How the PhD Came to Britain: A Century of Struggle for Postgraduate Education*, Guildford, Surrey: Society for Research into Higher Education. For Cambridge's final capitulation, see pp. 140–7.
18. Tabor, 'Frank Philip Bowden', quoting Snow, C. P., p. 4.
19. *Ibid.*
20. Eley, 'Eric Keightley Rideal', p. 388 ff. For Rideal's reluctance to have too many students, see Minutes of the Faculty Board of Physical and Chemical Studies, 1944–1949, Min. v. 87, insert following f. 113.
21. Dainton, Sir Frederick and Thrush, B. A. (1981), 'Ronald George Wreyford Norrish, 1897–1978', *Biographical Memoirs of Fellows of the Royal Society*, vol. 27, London: Royal Society, p. 406 ff. And see note 43 below on Rosalind Franklin's relationship with Norrish.
22. Read, *Humour and Humanism in Chemistry*, p. 287.
23. Mann, F. G. (1960), 'William Hobson Mills, 1873–1959', *Biographical Memoirs of Fellows of the Royal Society*, vol. 6, London: Royal Society, p. 219; Millar, I. T. (1984), 'Frederick George Mann, 1897–1982', *Biographical Memoirs of Fellows of the Royal Society*, vol. 30, London: Royal Society, p. 413.
24. Todd, *A Time to Remember*, pp. 61, 64, 68–9.
25. *Ibid.*, p. 68; Eley, 'Eric Keightley Rideal', p. 387.
26. Sources vary on these headcounts. See Brooke, Christopher N. I. (1993), *A History of the University of Cambridge, Vol. IX, 1870–1990*, Cambridge: Cambridge University Press, p. 331; *Yearbook of Universities of the Empire, 1916 and 1917*, London: Herbert Jenkins Ltd, 1917, and *Yearbook of Universities of the Empire, 1918–1920*, London: G. Bell and Sons, 1920.
27. MacLeod, Roy (1993), 'The chemists go to war: the mobilisation of civilian chemists and the British war effort, 1914–1918', *Ann. Sci.* **50**, p. 459; *Yearbook of the Universities of the Empire* (1915), 'War Rolls', London: Herbert Jenkins, 1915.
28. MacLeod, 'The chemists go to war', p. 474.
29. Pope, William Jackson (1918), 'Presidential address: the future of pure and applied chemistry', *J. Chem. Soc.* **113**, p. 292. See also Pope, William J. (1917), 'Chemistry and the nation', *The Chemical News* **116**, p. 200.
30. MacLeod, 'The chemists go to war', pp. 467, 468.
31. 'Sir Robert Waley Cohen', *Dictionary of National Biography, 1951–1960* [7th Supplement], Oxford: Oxford University Press (1971), pp. 236–37; Ferrier, R. W. (1982), *The History of the British Petroleum Company*, vol. I: *The Developing Years 1901–1932*, Cambridge: Cambridge University Press, pp. 239–41.
32. Read, *Humour and Humanism in Chemistry*, p. 297.
33. Haber, L. F. (1986), *The Poisonous Cloud: Chemical Warfare in the First World War*, Oxford: Clarendon Press, p. 124, quoting from letter from Pope to Hartley, 20 August 1917. The chemists to whom Pope referred by name were William Henry Perkin, Jr, Frederic Stanley Kipping, William Palmer Wynne, Arthur Lapworth, George Gerald Henderson, Thomas Stewart Patterson, Kennedy Joseph Previté Orton and Alexander McKenzie.

34. *Ibid.*, pp. 112–13, 165–67.
35. *Report of Conference on H. S. [mustard gas] held on Thursday, 11 July 1918, at 28 Northumberland Ave., WC*, p. 3, War Office 142/274, Public Records Office.
36. Pope, W. J., 'The future of pure and applied chemistry,' p. 296; *Yearbook of the Universities of the Empire* (1918–1920), pp. 424–25.
37. Oral history interview with Edgar 'Tiger' Coxall by Haley, Christopher and Bowden, Mary Ellen, at Lode, Cambridge, 11 September 2002; tape preserved in Archive of the Department of Chemistry, Cambridge.
38. Mott, N. F. (1955), 'John Edward Lennard-Jones, 1894–1954', *Biographical Memoirs of Fellows of the Royal Society*, vol. 1, London: Royal Society, pp. 178–80.
39. Eley, 'Eric Keightley Rideal', p. 386 ff.
40. Minutes of the Faculty Board of Physical and Chemical Studies, 1937–43, Min. v. 86, f. 52.
41. *Ibid.*, f. 68.
42. *Ibid.*, f. 79.
43. *Ibid.*, f. 99; Oral history interview with Frederick Eirich by James Bohning, 18 March and 10 June 1986, pp. 11–12. Transcript in Othmer Library, Chemical Heritage Foundation.
44. Maddox, Brenda (2002), *Rosalind Franklin: the Dark Lady of DNA*, London and New York: HarperCollins Publishers, pp. 70–82.
45. See Lent Term, 1915, in Cambridge University Chemical Club/Society Minutes, Cambridge University Archives.
46. *Ibid.*, 19 October 1945.
47. Dainton and Thrush, 'Ronald George Wreyford Norrish', pp. 385–6.
48. Eley, 'Eric Keightley Rideal', p. 386.
49. Todd, *A Time to Remember*, p. 67.
50. *Ibid.*, pp. 67–68; Millar, 'Frederick George Mann', p. 413.
51. Maddock, Alfie, 'Protactinium in Cambridge', Talk to the Department of Chemistry Museum Group, 29 November 2001; transcript preserved in Archive of the Department of Chemistry, Cambridge.
52. Saxon, Wolfgang, 'William Tutte, 84, who broke German code', *New York Times*, 10 May 2002, p. C13.
53. Davis, George E. (1901–2), *Handbook of Chemical Engineering*, 2 vols., Manchester: Davis Bros. See Davis's remark in the preface that his lectures on the subject at Manchester Technical School commenced in 1887.
54. Mackie, Robin (2000), ' "But what is a chemical engineer": profiling the membership of the British Institution of Chemical Engineers, 1922–1956', *Minerva* **38**, 171–99.
55. MacLeod, 'The chemists go to war', p. 473.
56. Pope, William J. (1919), 'Presidential address: chemistry in the national service', *J. Chem. Soc.* **115**, pp. 404–5.
57. Todd, *A Time to Remember*, p. 90 ff.
58. Chang, Kenneth, 'Herchel Smith, 76, major player in developing birth control pills', *New York Times*, 23 January 2002; 'Philanthropist leaves Cambridge its largest ever legacy', *CAM: Cambridge Alumni Magazine* no. 36, Easter 2002, p. 2.
59. University of Cambridge *Reporter*, 19 June 2002.

9

Alexander Todd: a new direction in organic chemistry[1]

James Baddiley

Department of Biochemistry, University of Cambridge

Daniel M. Brown

MRC Laboratory of Molecular Biology, Cambridge

Towards the end of World War II, organic chemistry at Cambridge University had declined. Sir William Pope had died in the autumn of 1939 and the 1702 Chair had not been filled during the early war years. In fact Pope, who had a distinguished record of work on the stereochemistry of a wide range of chemical structures, had been unwell for a considerable time and had not been active in the Department of Chemistry for several years. Consequently, effective leadership in organic chemistry had become essential, as Alex Todd himself pointed out in his autobiography.[2] Physical and theoretical chemistry, on the other hand, were in a healthier state under Ronald Norrish and John Lennard-Jones respectively, so there was an imbalance in the several branches of chemistry. There was therefore a need for an active, distinguished organic chemist to fill the 1702 Chair.

Alexander Todd (Figure 9.1) met these qualifications admirably. He was born in Glasgow in 1907, educated at Allan Glen's School and graduated in chemistry at the University of Glasgow. A brief period of research under Thomas Stewart Patterson was followed by a doctorate in Frankfurt in the laboratory of Walther Borsche where he worked on steroids. He went on to work in Robert Robinson's laboratory in Oxford on anthocyanins and then to Edinburgh in the Department of Medicinal Chemistry, where George Barger introduced him to the fascinating and topical subject of the chemistry of vitamins. This was followed by a move to the Lister Institute of Preventive Medicine in London, where experience in animal experiments for vitamin research was highly developed and where one of us (JB) later worked (Figure 9.2). From there, Alex was appointed to the Sir Samuel Hall Chair of Organic Chemistry at the University of Manchester in 1938.[3] It is a remarkable testimony to his achievements

The 1702 Chair of Chemistry at Cambridge: Transformation and Change, ed. Mary D. Archer and Christopher D. Haley. Published by Cambridge University Press. © Cambridge University Press 2005.

Figure 9.1. Alexander Robertus Todd (1907–1997).

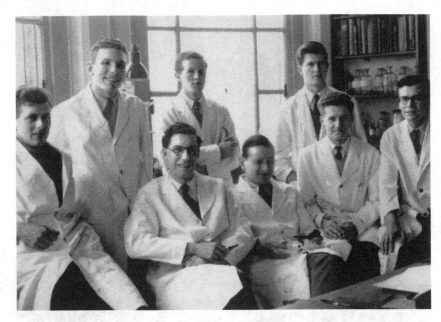

Figure 9.2. James Baddiley with his research group at the Lister Institute of Preventive Medicine in London, *c.* 1950. Seated from left: Dick Hodges, James Baddiley, Grant Buchanan, Laszlo Szabo and Malcolm Thain. Standing behind from left: Bob Handschumacher, Tony Mathias and Graham Jamieson.

that he had been appointed at the age of only 30 to a senior chair in a major university. An experimental subject such as chemistry would normally require a much longer apprenticeship in order to achieve distinction.[4] It was while he was in London that Alex married Alison Dale, daughter of the distinguished physiologist and Nobel laureate, Sir Henry Dale.

In the early 1940s, Alex had been offered the chair of biochemistry at Cambridge, recently vacant through the retirement of Sir Frederick Gowland Hopkins, but for several reasons he had declined it. The University of Manchester had an excellent reputation, especially in chemistry, physics and medicine, and Alex had built up an active and enthusiastic research group. The Manchester Department of Chemistry was working well and Alex's group was making progress on vitamins, penicillin and other natural products, especially in the exciting new areas of nucleosides and nucleotides, constituents of the co-enzymes, and giving consideration to nucleic acids.[5] In addition to these research interests, he and the group had communal experience of wartime departmental civil defence duties, in which the young professor participated, so he and his research group knew each other exceptionally well. He was a regular firewatcher and shared meals and overnight accommodation in the department with his research team. Moreover, the difference in age between him and most of the group was quite small. Those of us who joined him to research on natural products had been stimulated by attending the brilliant lecture course that he gave to final-year chemists.

Alex Todd's decision in 1944 to accept the 1702 Chair was taken with considerable reservations, but pressure from Cambridge University and the potential of the Cambridge department for future development convinced him that he should move. To some of his colleagues the reasons for this decision were less obvious. Like the chemistry building in Manchester, the Cambridge Chemical Laboratory in Pembroke Street was old. However, in Manchester the group, much later known as the Toddlers,[6] had good facilities and an environment that was conducive to pioneering advances.[2] Most of the research group were local, as was usual at that time, and had similar backgrounds in scientific training, having graduated from the same chemistry department. A further factor – more or less taken for granted – was the friendly and mutually respectful relationship between the heads of Physical and Organic Chemistry, Michael Polanyi and Alex Todd: both were ably supported by high-quality members of staff. The Department of Chemistry in Manchester worked remarkably well.

Inside the Pembroke Street laboratory

The arrival of the Todd group in Cambridge in October 1944 has been described by one of us (JB) elsewhere;[7] it is only necessary to describe here the situation facing them. Minor difficulties were expected through shortages arising from the effects of war, but as the highly effective Laboratory Superintendent Ralph Gilson was a member of the group, it was expected that delays would be minimal.

The state of the laboratories, however, was a more serious matter. The Pembroke Street buildings were ancient and gloomy and there were signs of neglect everywhere. There were few electricity sockets and fume cupboards were either absent or of an entirely unacceptable standard. The overhead lighting was dim and individual ones over the benches were gaslights! It is difficult to understand how disastrous fires were avoided, especially as flammable and toxic solvents were used then with much less care than they are today. Apparatus already in normal use elsewhere at that time, such as ground glass-jointed flasks, condensers and so on were not generally available or of obsolete design, and the total departmental supply of ground glass-jointed apparatus was kept in a small glass-faced cupboard, as if for display only. There were no effective stores and generally little, if any, organisation. This might be described as the 'cork and rubber stopper' era of organic chemistry. Most of the laboratories were of the large, open type that enforced the sharing of acidic and alkaline fumes with neighbours, thereby restricting what kind of work could be done. It also appeared that some senior members of the department did not accept standard techniques, such as column chromatography, that were in regular use elsewhere, as a means of purification of organic compounds. Crystallisation, mixed melting points and microanalysis were the traditional means of identification; other techniques were not approved.

Most of these deficiencies were overcome in a remarkably short time and Alex's persuasive ability, together with the efforts of research workers, academic and technical staff, enabled the programme to proceed with little delay. The arrival of an active, enthusiastic group changed the atmosphere in the department and this really was the beginning of modern organic chemistry in Cambridge. There was a strong sense of loyalty towards the professor, already existing with his former Manchester team, and rapidly acquired by those who joined him at that time. This was in part a result of the wartime common experiences but also because Alex was a highly effective and devoted leader of research. He discussed objectives openly and accepted comments, criticisms and alternative approaches from both senior and junior members. His ability to manage a number of inter-related programmes simultaneously without territorial embarrassments or misunderstanding was exceptional. He visited each member of his group almost daily and listened to difficulties that had been encountered, offering suggestions and useful advice for further progress. Roger Brown, one of the Australian contingent who did his Ph.D. with Alex in the 1950s, recalled of these visits:

> One of the quirks of the building was that the only direct communication between the ground floor and the first floor was a wrought iron spiral staircase. Todd was impressively and solidly tall, and it was a remarkable sight to see him spiral from above – first the gleaming shoes, then the pinstriped trousers, and finally the magisterial head.[8]

The predominantly Mancunian composition of the laboratory personnel did not last for long. Alex's presence boosted applications to work in the Cambridge department, and attracted a number of exceptional individuals. For example, Cambridge is especially indebted to the later work of one early member of the group, Herchel Smith, who invented key steps in steroid synthesis essential for the manufacture of oral contraceptives, and became a major benefactor to the present department, to the University of Cambridge and elsewhere (see Chapter 8).

Figure 9.3. Charles Dekker, Dan Brown (standing centre back) and Hugh Forrest, with Herchel Smith in front.

There was easy contact and discussion in Cambridge, not only between members of our laboratory, but also with researchers at the Molteno Institute, the Cavendish, Biochemistry and other locations; one of us (DMB) and others from Chemistry, for instance, lunched regularly with Francis Crick and Jim Watson in the Eagle. Colleges were a rather unimportant part of the landscape in this respect. In the whole laboratory, there was constant interaction, up and down the above-mentioned iron spiral staircase (over which was the bust of John Dalton shown in Figure 9.3) that joined the so-called XL laboratory to the smaller but more senior one above. Bench space was essentially randomly allocated so that there was constant cross-fertilisation. Joining the laboratory in the 50s and 60s were people from many other countries who continued afterwards to extend, *inter alia*, the nucleotide/phosphorylation fields discussed below: many from Australia and New Zealand, and others from countries such as India, Germany, Poland, France, Canada, the USA and the USSR. The international character that developed has continued and has been an important factor in the distinguished reputation that the department enjoys today.

Todd's grip on the work in the laboratory was excellent. However, although Alex was a dominating figure, he achieved this control through charisma rather than despotism. Many of his staff members and senior visitors looked after and directed the work, and having discussed the general direction of the work they were proposing to follow, trusted them implicitly. This was, in fact, necessary as, before long, demands on his time outside the laboratory grew rapidly due to increasing work in London and lecturing and advising in other countries, so that his appearances in the lab were perhaps weekly or less in order to visit each member of his research group. But his presence was reassuring to those that needed it, his memory was retentive and the system worked extremely well. It also allowed him to make some of his own most important contributions relating to his interactions with chemical industry and government, and to develop his thoughts on science education.

Before describing some of the major scientific achievements of Alex Todd and his colleagues, something should be said about the departmental structure that existed on his arrival and continued formally for a considerable time. In 1944, teaching and research in chemistry in Cambridge was carried out in two essentially independent departments: the Department of Physical Chemistry, run by Ronald Norrish, and the Department of Organic and Theoretical Chemistry, into which Alex arrived. The 1702 Chair had, on Alex's election in 1944, been officially re-styled as the Chair of Organic Chemistry, so that Alex's official role was to be both Professor of Organic Chemistry and Head of the Department of Organic and Theoretical Chemistry. John Lennard-Jones was Professor of Theoretical Chemistry. A separate Department of Colloid Science under Eric Rideal carried out excellent work that today could be described as physical chemistry and, of course, chemical work was also carried out in a number of independent departments in the university, such as Biochemistry and the Goldsmiths' Metallurgy Laboratory.

There might have been a time when the disciplines of physical and organic chemistry could be treated as separate subjects, but for a long time before 1944 it was clear that they had much in common and separate departments were inappropriate. However, Norrish and Todd did not agree on important matters and their personal relationship was cool. Consequently, the departmental structure was to remain unchanged, although some integration of teaching occurred.

A notable serious deficiency was the absence of inorganic chemistry from the formal departmental structure. The subject was taught but modern developments were not represented, despite notable advances that had been made therein. Alex was successful in persuading the university to establish a Readership in Inorganic Chemistry, to which Harry Eméleus was appointed. Shortly thereafter, Eméleus was made Professor of Inorganic Chemistry and a distinguished teaching and research group arose under his guidance. With the passage of time most of the territorial distinctions disappeared and close integration of the several branches of chemistry was achieved in the undergraduate course structure. However, it was not until 1986 that a single Department of Chemistry was formed with university approval.

The move to Lensfield Road

A factor in the development of a unified department was the geographical need to bring together the several subject areas into one building. The need for a new building to replace the outdated Pembroke Street facilities had been foreseen at the time of Alex's appointment and one had been promised in 1944.[9] However, negotiations and restrictions delayed a start until 1950 and the first phase in Lensfield Road was not completed until 1955–6. The building was officially opened in 1958 (Figure 9.4), although several additions have been made since then in line with advances in the subject and the outstanding reputation of chemistry in Cambridge today. A poem, which gently jibed at Alex's occasional egotism and perhaps illustrated a little envy in other universities of the palatial accommodation of the new laboratory, was published under the pseudonym 'pH6' (in other words, slightly acidic):

> Lord, give me leave to build a lab
> So large that when I've trod
> Its lofty aisles and nave I'll think
> I'm in thy house, O Todd.[10]

The move to the new Lensfield Road laboratory was accompanied by new members of staff and new fields of study, including rapidly developing methodologies and instrumentation. Examples were the mass spectrometric and NMR understanding brought by Dudley Williams, and the introduction of a new chair for Alan Battersby, described in Chapter 11. There is little doubt that Alex was consciously moving the new enlarged laboratory into a new era.

Research work

At the time of his arrival in Cambridge, Alex already had a record of important successes in the chemistry of natural products. Early work on steroids, anthocyanin pigments and vitamin structures, including vitamin E, was crowned by an elegant first total synthesis of vitamin B_1 (thiamine, the anti-beriberi factor) that is more or less the one used commercially today. When in Manchester, he had made important contributions to the identification and structure of the active principles of cannabis. Alex described how his application to the authorities

Figure 9.4. Alex Todd and H.R.H. Princess Margaret at the opening of the Lensfield Laboratories in 1958. In the foreground is a model of Vitamin B$_{12}$ which was constructed for Todd by Glaxo when its structure was solved.

for permission to import *Cannabis indica* resin (hashish, cannabis) for research purposes presented an unusual problem to HM Customs.[11] We know, of course, that today many thousands of unauthorised, but allegedly pleasurable, experiments are carried out daily on the pharmacological properties of cannabis by enthusiastic volunteers in many countries. Even in the 1940s, however, use or sale of this drug was strictly illegal. After much deliberation the Customs and Excise Authorities accepted Alex's request, albeit with considerable reservations. He was required to inform the authorities of any work he published. When asked how such papers would be classified the answer was 'obscene publications'!

Nucleosides, nucleotides and nucleic acids

The most exciting programme, however, was that started in Manchester and continued as the major interest in Cambridge. This was the design of synthetic methods to establish beyond doubt the structures of the known nucleosides, nucleotides and related co-enzymes for which Alex was later awarded a Nobel Prize. The contributions to nucleic acid chemistry came later.

The study of nucleosides and nucleotides came to play a vital part in the elucidation of the structure of nucleic acids, but it is important to realise that at that time the science of nucleic acids was primitive.[12] Although deoxyribonucleic acid (DNA as it is now called) was discovered by Friedrich Miescher in 1871 as a component of human pus cells and salmon sperm, its significance was unclear. Even its macromolecular nature was neither appreciated nor accepted for a considerable time. However, by the end of the 1940s, it was generally recognised to be a component of chromosomes: so it seemed most probable that it was the key material of genes and was directly involved in the inheritance of life's properties. In fact, O. T. Avery and his associates at the Rockefeller Institute for Medical Research in New York had already shown in 1943 or earlier that isolated DNA was responsible for genetic transformation of *Pneumococcus* spp. (*Streptococcus pneumoniae*). This was itself based upon an observation made in the 1920s by Fred Griffith in London that this organism could be transformed from rough (i.e. lacking an outer polysaccharide layer) to smooth (i.e. possessing this layer) strains by addition of transforming factor obtained from anti-rough immune serum. However, these first demonstrations of the genetic role of DNA were recognised only slowly, and in the late 1940s there were still a few conservative individuals who were not convinced and who believed that proteins were the carriers of genetic information.

The nature and role of ribonucleic acid, RNA, was even less clear than that of DNA. No function had been ascribed to it and analysis of its structure proved to

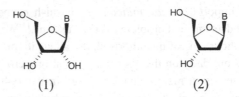

Figure 9.5. Structures of ribonucleosides (1) and 2′-deoxyribonucleosides (2).

be quite intractable. It was much more easily hydrolysed than DNA and so was much degraded – except in the case of some RNA viruses which contained RNA of high molecular weight, and which were just beginning to provide material for study. As a result, biochemical information was suspect. Nothing was known about the biosynthesis of either nucleic acid.

Synthesis of nucleosides

This uncertainty of the roles of RNA and DNA, coupled with both the difficulty in obtaining them in a pure state and the likely complexity of their structure, suggested that initially we should concentrate our researches on simpler units, the nucleosides and their phosphate esters (the nucleotides). These were known to be constituent parts of the polymeric nucleic acids. They were also components of the nucleotide coenzymes, the main subject of our interest at that time. Modern representations of nucleosides are shown in Figure 9.5, where B represents one of the four nitrogenous bases of RNA (1) or DNA (2).

There was little talk in the laboratory of nucleic acids per se, since there was little solid evidence on which to think rationally. Rather, Alex was essentially concerned with investigating methods of synthesis of mononucleotides and ensuring nucleoside structures were correct. It was envisaged that the nucleic acids would be examined at a later date.

It was agreed, therefore, that the first priority was the confirmation by unambiguous chemical synthesis of the structures assigned (mainly by Phoebus Levene) to the nucleosides and nucleotides isolated from nucleic acids, together with the development of phosphorylation techniques that could be applied to the synthesis of nucleotide co-enzymes. This work had already been in progress in Manchester when the relocation took place, and it was correctly assumed that little delay would be encountered through the transfer to Cambridge of most members of the research group.

The major problem, however, was that the chemistry of phosphate esters was so little understood. The procedures generally used at that time for chemical synthesis often included reactions carried out in organic solvents, at elevated temperature, under acidic or basic conditions and requiring highly reactive

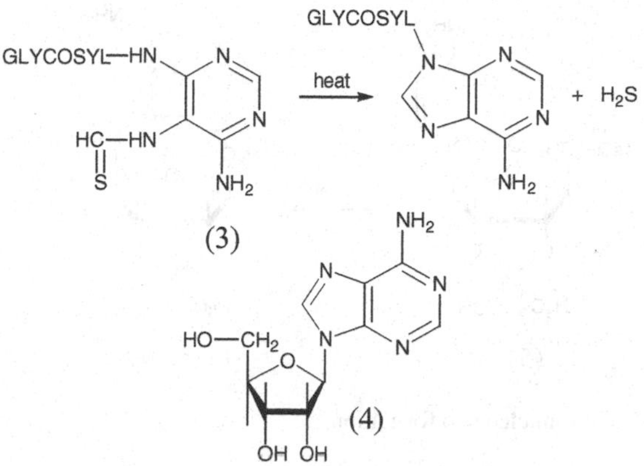

Figure 9.6. Cyclisation of a thioformamido pyrimidine (3), a step in the synthesis of adenosine (4), the nucleoside of the base adenine.

reagents. However, nucleosides and nucleotides are water-soluble and are insoluble in most organic solvents, and so differ in these respects from the kind of molecules to which most organic chemists had been accustomed. These factors had to be reckoned with in attempting synthesis in this new, difficult field.

A number of biochemists at that time expressed the view that the structures assigned to nucleotide co-enzymes would be too difficult to synthesise chemically. Alex disagreed; he retorted that living cells possess enzymes that are able to synthesise these structures, and if simple bacteria can make them, so should we be able to using pure organic chemistry. It is interesting that, at least during the time that one of us (JB) worked in the laboratory (1941–9), neither living cells nor enzymes were used in our synthetic efforts: Alex would have considered it improper to resort to such procedures. The undertaking was a formidable one, and this probably accounts for the fact that the Todd laboratory was the only place where such work was being attempted. There was little or no competition; most organic chemists would have kept clear of such labile structures. This situation contrasted with the later one when work on the nucleic acids coincided with the recognition of their extreme biological importance.

Good progress had already been made at the time of Alex's move to Cambridge.[13,14] Several phosphorylation techniques were being investigated and a route had been developed successfully by Baddiley, Basil Lythgoe, George Kenner and Todd for the structurally definitive synthesis of purine–N9 glycosides. The synthesis of these was based on the earlier vitamin B_1 synthesis using thioformylation of an amine which cyclised on heating. Using the appropriate sugars, adenosine (4) and other nucleosides were synthesised, thus confirming the location of the glycoside substituent in the natural nucleosides (Figure 9.6).

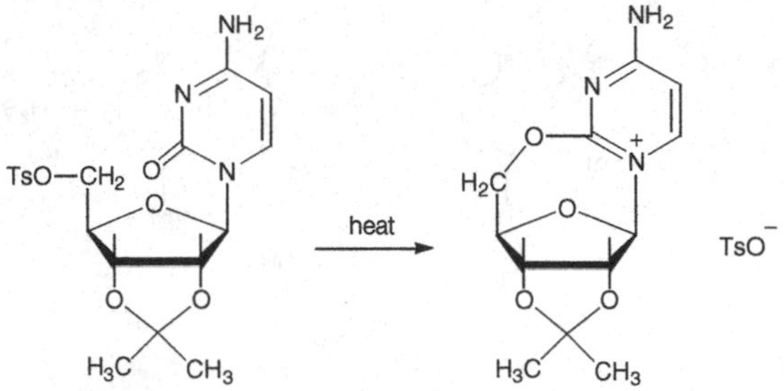

Figure 9.7. *Cyclo*nucleoside formation.

A complication in this work was the low availability of some of the raw materials required. For example, ribose is a rare sugar and none was available in the UK at that time. The prospect of preparing it, probably from yeast ribonucleic acid (RNA), was daunting. However, Alex's excellent relations with industry resulted in a generous gift of about 500 g from Hoffman-La Roche in Switzerland, without which the programme would probably have been delayed by at least a year or so.

The configuration of the glycosidic linkage in several pyrimidine and purine nucleosides was also established by the easy ring closure of 5'-0-tosyl nucleoside derivatives (5) to give *cyclo*nucleosides (6) (Figure 9.7).[15,16] Such cyclisation is only possible if the glycosidic linkage of the base to the sugar has the so-called β-configuration of structure (6).

Synthesis of AMP, ADP, ATP and nucleotide co-enzymes

The phosphorylation of nucleosides to give nucleotides could not have been considered using the procedures available at that time. These usually employed phosphorus pentoxide or phosphoryl chloride. Both were violent reagents and were unsuitable for controlled application. This was earlier mitigated to some extent by using diphenyl phosphorochloridate. However, we were aware that synthesis of complex nucleotides required protecting groups to control the direction of reactions, and to have the ability to remove these in a defined and orderly manner. This work began with the synthesis of dibenzyl phosphorochloridate.[17] The work that followed opened up an enormous amount of phosphorylation chemistry in the synthesis of co-enzymes and in nucleic acid studies. Using this reagent, the first synthesis by JB of a nucleotide co-factor of biological importance, adenosine triphosphate (ATP), the energy exchange factor present

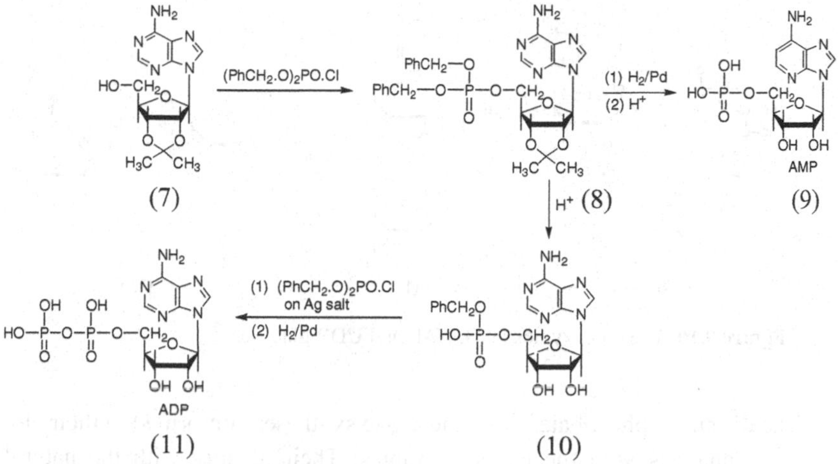

Figure 9.8. Synthesis of adenosine-5'-phosphate (AMP) and adenosine-5'-diphosphate (ADP).

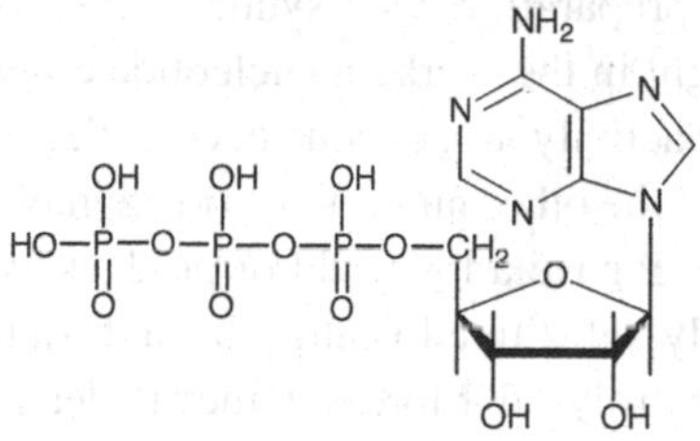

Figure 9.9. Adenosine-5'-triphosphate (ATP).

in all living cells, was achieved. This was synthesised by Baddiley and Todd following the synthesis of the mono- and di-phosphates of adenosine.

For the synthesis of these phosphates of adenosine, a suitably protected derivative of the nucleoside (7) was treated with the phosphorylating reagent to give the product (8), which, on removal of the protecting groups, gave adenosine-5'-phosphate (AMP), structure (9). Alternatively, careful treatment of (8) with dilute acid gave an intermediate (10) which, on further treatment with the phosphorylating agent and removal of protecting groups, yielded adenosine diphosphate (ADP), structure (11) in Figure 9.8.[18]

The intermediate adenosine-5'-benzyl phosphate (10), on further treatment with the phosphorylating reagent and removal of protecting groups, gave adenosine triphosphate (ATP), structure (12) in Figure 9.9.[19]

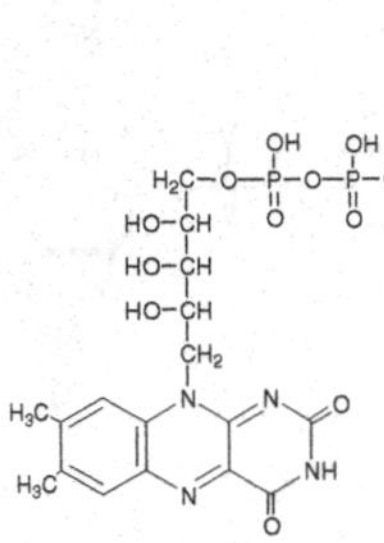

(13) 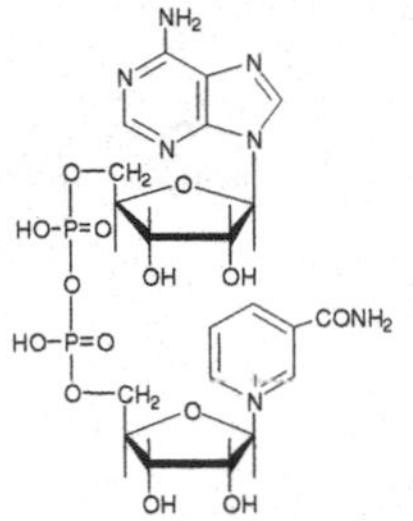   (14) 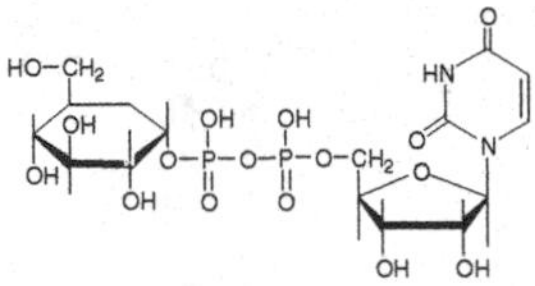   (15)

Figure 9.10. The co-enzymes FAD, NAD and UDP-glucose.

The di- and triphosphates from the above syntheses conformed in their elementary analysis with the expected values. Their identity with the natural nucleotides was convincingly established by demonstration of the biological activity of the triphosphate using an actomyosin preparation from muscle (by Kenneth Bailey and Sam Perry in the Department of Biochemistry). The contraction of the muscle preparation when synthetic ATP was added by one of us (JB) remains a highlight in the work on nucleotide co-enzymes. Such demonstrations of biological activity are, of course, essential and have been the basis for the identification of the other nucleotide co-enzymes that were synthesised.

The synthesis of ATP gave a low yield of product. We recognised two reasons for this. The fully substituted neutral di- and triphosphates were highly unstable and readily hydrolysed. Moreover, they underwent exchange reactions giving rise to a mixture of products, including symmetrical and unsymmetrical pyrophosphates. It was found that reaction between a fully substituted phosphoryl chloride and a salt of a mono-substituted phosphate gave improved yields of unsymmetrical pyrophosphates because of the stabilising effect of a charge on phosphate.[17] Application of this to the appropriate phosphomonoesters gave flavin adenine dinucleotide (13; FAD), nicotinamide adenine dinucleotide (14; NAD) and uridine diphosphate glucose (15; UDP-glucose) (Figure 9.10).[20,21]

At about this time, however, the Indian chemist Gobind Khorana was a postdoctoral worker in the laboratory. He discovered that dicyclohexyl carbodiimide (16) reacted readily with phosphomonoesters to give pyrophosphates (18) in high yields (Figure 9.11).[22] Using this elegant procedure, the above nucleotide co-enzymes and others were synthesised in good yields, but still in admixture with the two symmetrical products as the putative intermediate (17) had no useful lifetime.

However, the phosphoramidates, long studied in Todd's laboratory, had a surprising property. Their monoanions (19) were stable intermediates but

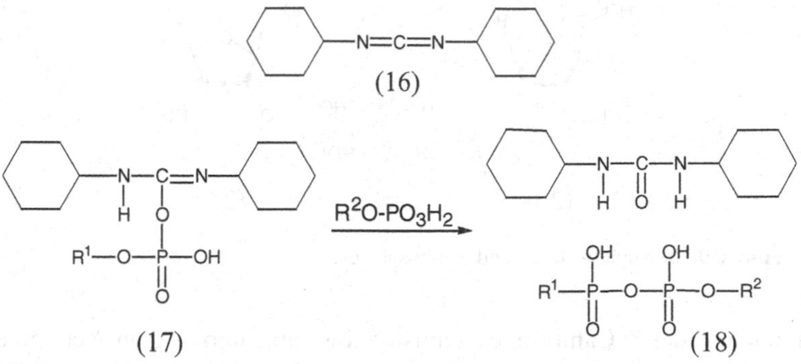

Figure 9.11. Synthesis of a pyrophosphate (18) using a carbodiimide.

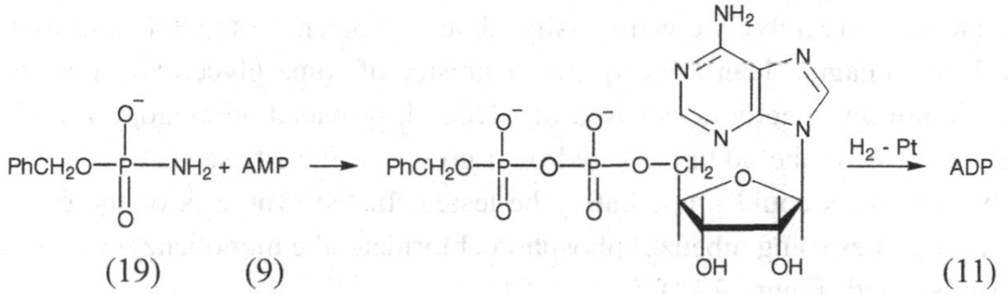

Figure 9.12. Synthesis of ADP from AMP using a phosphoramidate (19).

reacted cleanly with other phosphates, as in the synthesis of ADP from AMP (Figure 9.12).[23] This general method was extended in Cambridge and by Khorana and colleagues to the preparation of a range of co-enzymes.[24] Thus, synthesis of unsymmetrical pyrophosphates, uncontaminated by symmetrical products, the goal in the co-enzyme field, was finally achieved.

RNA and DNA structures

In 1932, Levene and Harris had reported that hydrolysis by alkali of RNA gave four ribonucleoside-3'-phosphates, corresponding to the four bases, whereas DNA was stable.[25] For some time, this was essentially the only structural evidence relating to RNA beyond that of the nucleosides themselves. Although this conclusion turned out later to be erroneous, it was explicable in an interesting way. Since the hydrolytic products were reported as only 3'-phosphates, a key question was posed: what constituted the other link in the polymer chain of the nucleic acids? There was a hiatus lasting fifteen years before a key piece of information emerged. Then, in 1949, Waldo Cohn and colleagues at Oak Ridge showed that there were actually *four pairs* of nucleotides based on the

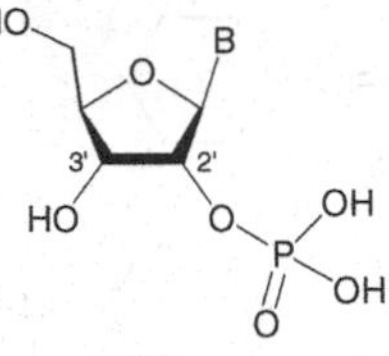

(20) 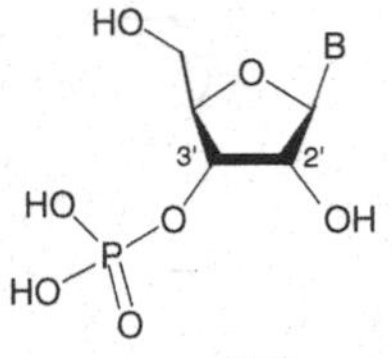   (21)

Figure 9.13. Nucleoside 2′- and 3′-phosphates.

four nucleosides.[26] Cambridge chemistry now came into its own. Very quickly the nucleotides were recognised as the isomeric 2′- and 3′-phosphates (20) and (21) (Figure 9.13). Later it was demonstrated which was which.

But were these the linkage points in the polymer? One of the linkage points had to be correct, but we were unsure about the other. Fortunately, discussing with a colleague, Jean Lecocq, the chemistry of some glycerophospholipids which are phosphodiesters based on glycerol, provided an analogy for RNA lability; it was the additional hydroxyl group on the ribose that caused this lability.[27] This could immediately be tested. In the synthesis of the 2′- and 3′-phosphates using dibenzyl phosphorochloridate, the monobenzyl esters had been isolated (Figure 9.14).[28]

These monobenzyl esters, such as (22), were to hand and sure enough, they were, like RNA, labile to alkali, losing only the benzyl group to give the 2′- and 3′-phosphates (20) and (21); by contrast the 5′-ester (10) was stable. If the ester group in (22) was considered as equivalent to the next nucleoside in the RNA polymer, the analogy with RNA was complete. It took some time to decide that the 3′-position was involved since a pseudo-symmetrical cyclic phosphate (23) was formed as an intermediate, accounting for the lability (Figure 9.14). This eliminated a 2′,3′-internucleotide linkage. The other 5′-linkage point fitted the evidence but the mechanism dictated that a 5′-nucleotide would not appear as a hydrolysis product. The very process by which the 2′-phosphate was formed was the same as the process that concealed the 5′-linkage.

As shown in Figure 9.14, the cyclic phosphate is an obligatory hydrolysis product; mechanistically it was a case of neighbouring group transesterification. Roy Markham and John Smith at the Molteno Institute were working with ribonuclease A hydrolysis of RNA. One lunchtime in 1951, one of us (DMB) and Chuck Dekker met Markham and Smith in the Bun Shop, the Pembroke Street meeting place of many biologists, to talk about the curious products they had found. By opening time the same evening, the comparison with materials synthesised by Dave Magrath had been made; they were uridine and cytidine 2′,3′-cyclic phosphates.[29] It later turned out that the simple

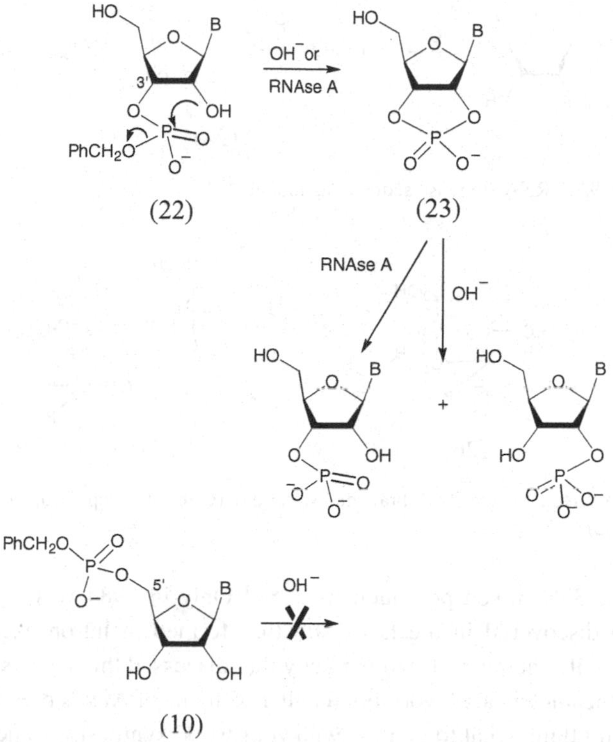

Figure 9.14. Hydrolysis of a nucleoside 3'-phosphate ester through a 2', 3'-cyclic phosphate intermediate.

3'-benzyl ester (22) (but not the 2'-) was hydrolysed by RNAses, so the linkage point in RNA was established and an important connection between RNA chemistry and biochemistry was made.[30]

One might note that, despite the apparent complexity of RNA, almost all of the basic chemistry was worked out in Alex's lab using simple synthetic compounds. This even extended to demonstrating the first method for stepwise nucleotide sequencing of RNA; in effect, converting the (terminal) diol of (10) to a dialdehyde (24), then running the so-called β-elimination (Figure 9.15).[31] The resulting benzyl phosphate (25) mimics the RNA chain less one residue. This was applied at the same time in Markham's laboratory to di- and tri- nucleotides[32] and later by others, in one case to 26 sequential steps, a process so mild that other inter-nucleotide linkages were unaffected.[33] It was also possible to discuss rationally the possibilities for chain branching in RNA, such as through phosphotriesters (26) or the 2'-position (27) (Figure 9.16) and in effect eliminate them, again using synthetic analogues,

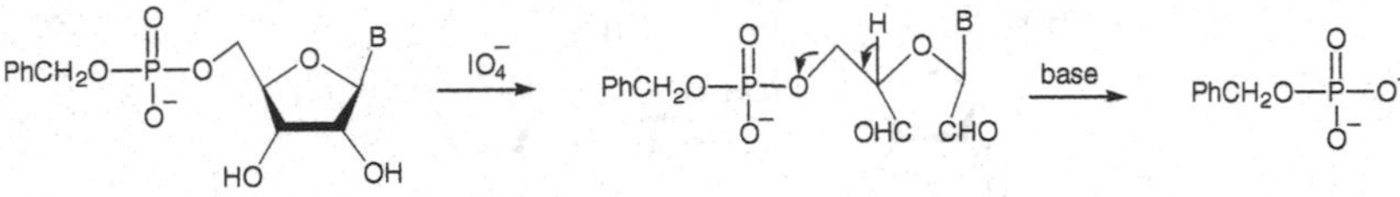

(10)   (24)   (25)

Figure 9.15. RNA stepwise sequencing model.

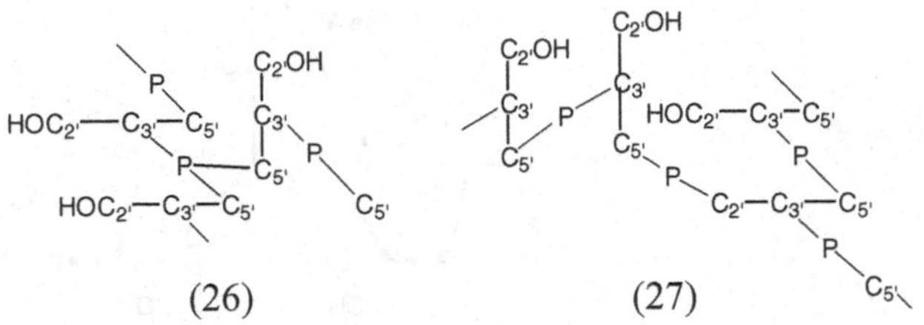

(26)   (27)

Figure 9.16. Possible RNA branched structures ($C_{2'}C_{3'}$-$C_{5'}$ represents a ribonucleoside).

leaving the $3',5'$-linked polynucleotide.[32,34] Only in 1984 was a case of $2'$-branching discovered in the lariat structure formed in intron excision from pre-mRNA. Its chemistry fitted perfectly that discussed thirty years earlier.

All of the nucleic acid work fitted rather well one of Alex's precepts, to wit, 'If you can't think what to do next with your work, synthesise something.'

The question of the DNA general structure (28) was much more straightforward than that of RNA (29); there was no $2'$-hydroxyl group, and hence no phosphate migration to confuse the issue (Figure 9.17). The structures of the deoxynucleosides were already established or were confirmed in Todd's laboratory as being β-furanosyl derivatives (2) and the internucleotide linkage positions being shown synthetically by Dekker, Mick Michelson and Todd confirming the structures of the nucleotides isolated enzymatically from DNA by others.[35]

In 1952, the 250th anniversary of the 1702 Chair, the first papers discussing the general chemistry and structure of DNA (28) and RNA (29) were published.[28] As Francis Crick later pointed out, the chemical work from Alex's laboratory was of much help in the DNA model-building in 1953 in the Cavendish Laboratory.[36] Soon after the double helical model had been constructed by Watson and Crick, Lawrence Bragg, head of the Cavendish Laboratory, asked Alex and DMB to view the DNA model before the work was submitted for publication. We were able to confirm all the essential chemical features and, when the genetic and replicative potential of the complementary double helical structure was pointed out, we were immediately convinced of its

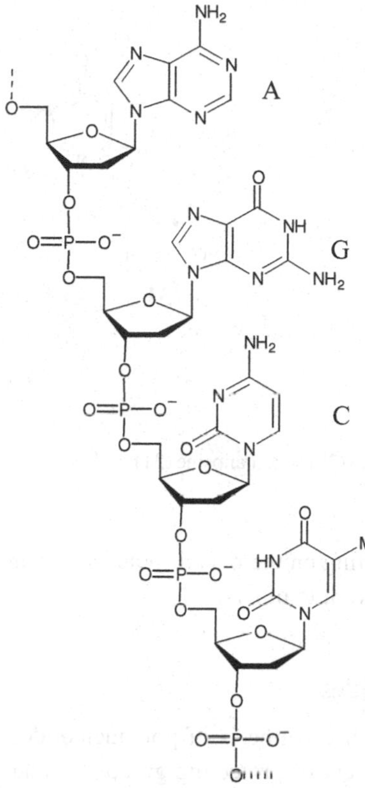

Figure 9.17. Structures of DNA (28) and RNA (29), showing the four bases of each: A = adenine; G = guanine; C = cytosine; T = thymine; U = uracil.

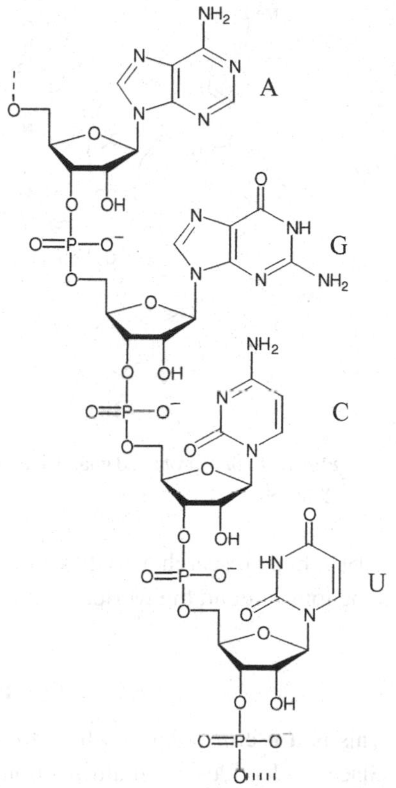

veracity. The structure and function of DNA were initially acclaimed by only a limited number of scientists, but they were 'a revelation to prepared minds', as Watson Fuller[37] has recently pointed out, and we were among those prepared minds. One of us (DMB), indeed, had given a few lectures in the Chemistry Department on nucleic acids a month or two before this, but without seeing the significance of information already in the literature that fell naturally into place in the duplex model.

Once the general chemistry of nucleic acids was understood, Alex's interest waned, although in other laboratories, using the new knowledge, activity accelerated remarkably in many directions. He was in a sense following another of his precepts: 'Choose any important subject in which few others are working; among other things it saves you time reading the literature.' But by the late

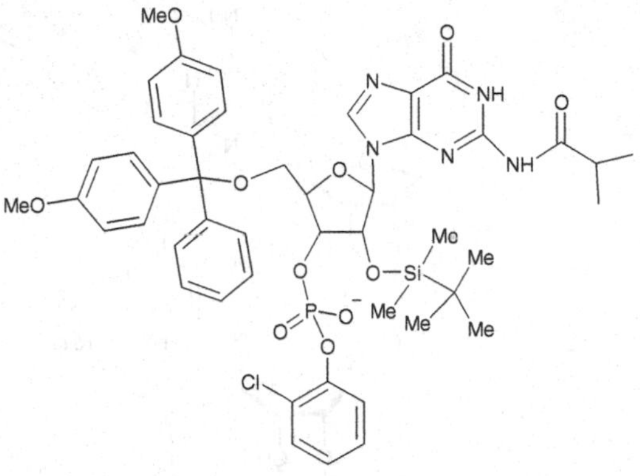

Figure 9.18. A protected guanosine intermediate (30) and guanosine (31) for RNA synthesis.

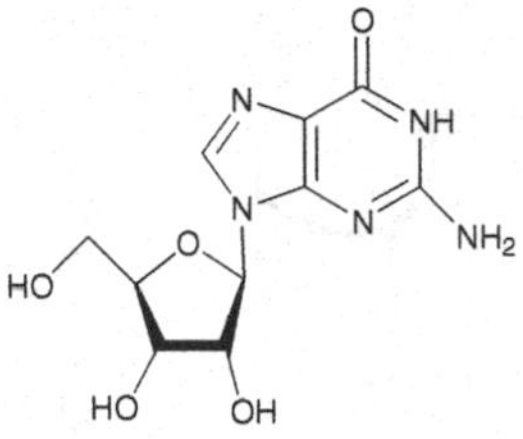

1950s, huge research activity was concentrating on the nucleic acids and their functions, around the world. 'Therefore, leave it to others!'

Polynucleotides

This is a useful point at which to mention the synthesis of polynucleotides. Since nucleotides are multifunctional, a variety of protecting groups was necessary and this was in agreement with our laboratory practice. Gobind Khorana, mentioned earlier, had established a polynucleotide synthesis using an extension of his co-enzyme coupling synthesis methodology, a method that lacked phosphate protection. It required enormous manpower to deal with purification of the polyanionic intermediates, but they succeeded and gave products needed, *inter alia,* to help in solving the genetic code.[24] Much earlier in 1956, Michelson and Todd had shown that having a protecting group on the nucleoside phosphate gave a neutral product when coupled to the next nucleoside. This was soluble in organic solvents and therefore much more tractable.[38]

Some years later, Colin Reese in Cambridge and then in London, and many others, made an in-depth attack with this principle in mind. They coupled fully protected intermediates, for example (30) derived from the nucleoside guanosine (31) as in Figure 9.18.[39] This opened the way toward present-day methods, which now utilise reactive trivalent phosphorus intermediates, by which machines can put together long, defined nucleotide sequences accurately, quickly and entirely automatically. This, the combined work of many laboratories, must rank as one of the greatest achievements in synthetic organic chemistry of the past half-century.

Mechanistic chemistry

Another matter which, in the authors' opinion, has not been given sufficient prominence is the effect of Alex's work in opening up the field of mechanistic chemistry. Frank Westheimer of Harvard said to one of us (DMB) that when Alex, at an earlier time, lectured in the USA, he strongly influenced a number of younger chemists, including Westheimer himself, to move into the mechanistic chemistry of organophosphates and by extension, enzyme mechanism. Alex's conviction that the new nucleotide field was virtually untouched, and that phosphate chemistry was unexpectedly rich, must have carried them, as it did his own colleagues. Nearly all organic mechanistic studies, which were by that time very sophisticated with great predictive power, dealt with reactions at carbon centres and much of organic chemistry rested on this foundation. Alex had interacted with many of the great exponents of the subject – Linus Pauling in Caltech, Lapworth and Polanyi in Manchester, his great friend Bob Woodward in Harvard, and he had worked with Robinson in Oxford. Yet, curiously he seemed to have little necessity for mechanistic thinking. His wide-ranging classical understanding of organic chemistry and a marvellous memory stood him in good stead. His younger colleagues – among them, Basil Lythgoe, George Kenner, Malcolm Clark, Tony Kirby, Neil Hamer, Stuart Warren and Mike Blackburn – were important to him in this connection.[40] The synthetic work helped to provide the materials and stimulus for such mechanistic studies. The whole subject blossomed enormously over the following decades in many centres.

Other research

Besides their work on nucleosides, nucleotides and co-enzymes, Alex and his group also made very significant contributions to the chemistry of a number of other natural products. As he explained in his Nobel lecture, he took the view that organic chemistry should focus upon substances found in living matter.[41] Work on vitamin B_{12} was one such major area. As mentioned earlier, before coming to Cambridge Alex had worked on vitamin B_1 (first with Barger at Edinburgh, and at the Lister Institute) and on vitamin E (at the Lister Institute and in Manchester). Indeed, it was largely through this work that he made his name as an independent researcher.[42]

Since these early days Alex had retained an interest in the so-called 'B group' of vitamins and it is noteworthy that some of the nucleotide co-enzymes discussed above incorporate B group vitamins.[43] The intervening period had seen some confusion in the field, as researchers struggled to separate the constituent parts of the group. Nevertheless, in 1948 the last of the group – the so-called 'anti-pernicious anaemia factor' – was isolated from liver. This isolation was achieved by Lester Smith at the British Glaxo laboratories and almost simultaneously by a group at the American Merck laboratories, who named it B_{12}.[44] However, soon after its isolation, the Director of Research at Glaxo, Thomas Macrae, realised that its chemical structure was 'exceedingly complex' and would be very difficult to resolve.[45] Macrae therefore turned for help to Alex, whom he knew from their undergraduate days at Glasgow.[46] Alex accepted the challenge and, with the help of his colleague Alan W. Johnson and research students, including Grant Buchanan, isolated and identified several degradation products of B_{12}.[47] One in particular, containing the central chromophoric group isolated and crystallised by Jack Cannon,[48] was of very great help to Dorothy Hodgkin, who was tackling the same problem through X-ray crystallography in Oxford. This led to the announcement of a complete structure of the vitamin in 1955.[49]

In their efforts to solve the structure of B_{12}, members of the group introduced new techniques to the Cambridge department, such as paper chromatography. Chromatography, as an analytical tool, had been successfully developed by Archer Martin and Richard Synge in the early 1940s.[50] By the late 1940s it was being used in Cambridge by Fred Sanger and by Miles Partridge. However, as mentioned earlier, within the chemistry department itself there was some resistance to the technique. Alex came around to paper chromatography when he saw its usefulness. By the early 50s the technique was widespread, leading Robert Robinson to rib the Cambridge chemists for being 'just a lot of paper hangers'.[51]

Another major area in which Alex and his group spent much time involved the pigments from the haemolymph of several species of aphids (Figure 9.19).[52] This was the kind of chemistry with which Alex had a familiarity, dating back to the period when he worked in Oxford on the synthesis of flower pigments – the anthocyanins (e.g. 32), for example, in which he first showed his experimental prowess. The aphins (e.g. 33), were much more complex but he could relate easily to the work; it was back to carbon chemistry.

Other roles

At the same time, Alex's energies became increasingly channelled into his advisory activities. In 1963 he became Master of Christ's College. He filled this position with distinction and the college derived considerable benefit from his

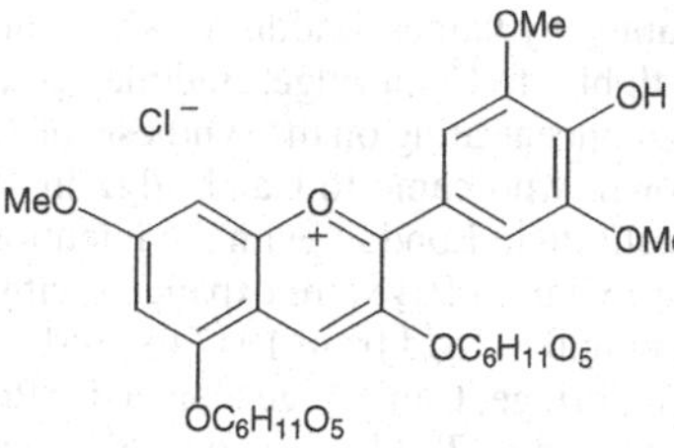

Figure 9.19. Hirsutin (32) and protoaphin (33).

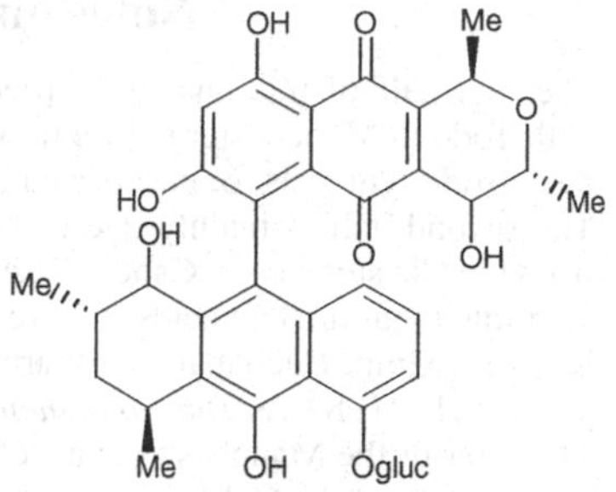

tenure. Others can testify that the strength of the college increased in many ways during his time as Master, especially in its academic standing.

His distinction in science also led to many calls on his time as an adviser and leader of scientific policy, and he chaired several national bodies concerned with scientific policy and scientific and medical education.[53] Amongst these numerous roles, special mention should be made of his presidency of the Royal Society and other learned Societies, his long association with the Nuffield Foundation (he became a trustee in 1950 and served as chairman from 1973–9) and his period as Chancellor of the University of Strathclyde. He also played an active part in the foundation of Churchill College and as an adviser to a number of chemical and pharmaceutical firms. He especially valued his advisory role with the Dyestuffs and the Pharmaceutical Divisions of ICI, a connection that lasted for 25 years. He also had long associations with Roche Products UK, Fisons (of which he was a non-executive director for 15 years from 1963) and the Croucher Foundation – the latter giving him many opportunities to visit Hong Kong. His scientific and broader activities resulted in the award of a knighthood, a life peerage, membership of the Order of Merit, about forty honorary degrees and other honours. He was a skilled lecturer and he delivered many important named lectures and addresses in his characteristic Glasgow accent.

Alex's advice was always being sought; he was a man of affairs and with his self-confidence he expected to be at the top. He once said that had he studied Christian exegesis he would have been a professor in that subject; had he done so he might have emulated some of the earlier holders of the Chair!

Alex retired from the 1702 Chair in 1971, prompted in part by a heart attack. He continued to serve as Master of Christ's College until 1978, and died in January 1997.

Notes and References

1. The first half of this chapter is predominantly by James Baddiley, who worked with Todd in Manchester before moving with him to Cambridge. Baddiley studied nucleoside synthesis for his doctorate before concentrating on the synthesis of ATP. The second half is mainly due to Dan Brown, who came to Cambridge in 1948 from the Chester Beatty Cancer Research Institute in London, with the intention of learning about nucleic acids. The remarks on vitamin B_{12} were expanded with the help of J. Grant Buchanan, a research student under Todd from 1947 to 1951.
2. Todd, A. R. (1983), *A Time to Remember*. Cambridge: Cambridge University Press.
3. The offer of the Manchester chair came along just as Todd was about to accept an appointment at the California Institute of Technology, following a two-month visit there in 1938. He often told his three children how nearly they were born American citizens (Helen Brown neé Todd, personal communication, 2004).
4. As Todd records in his introduction to *The Toddlers: 1972–1992* (privately produced in limited number by Barbara Mann, Todd's long-serving secretary, to mark the twentieth anniversary of the formation of the Toddlers): 'I took up my appointment as Sir Samuel Hall Professor of Chemistry and Head of the Department of Chemistry in the University of Manchester on 1st October 1938 and attained the ripe age of 31 the next day. I was at that point the youngest member of the staff of the Manchester Department, I. M. Heilbron, who I succeeded, having taken most of the younger members of the organic teaching staff to Imperial College with him.'
5. Nucleotides are the monomeric units of DNA and RNA. Each nucleotide is composed of a phosphate group, a sugar residue and one of four bases. In DNA, the sugar is deoxyribose and the four bases are adenine (A), guanine (G), cytosine (C) and thymine (T). In RNA, the sugar is ribose and the four bases are A, G, C and uracil (U). When the phosphate group is absent, the compound is called a nucleoside.
6. The core of the group known as the Toddlers (a name suggested by Alex Todd himself at their inaugural dinner in the Café Royal on 10 December 1971) comprised the students who moved with Todd from Manchester to Cambridge: Basil Lythgoe, Frank Atherton, Jim Baddiley, George Howard, Harold Howard, Roy Hull, George Kenner, Denis Marrian, Peter Russell, Peter Sykes, Arthur Topham and Norman Whittaker. Ralph Gilson, the laboratory steward in Manchester, Barbara Mann, who became Alex's secretary, Hal Openshaw, Cedric Hassall, John Davoll, Alan Johnson, Lex Lyons, Anthony Holland and later Herchel Smith were also elected members.
7. Baddiley, J. (1994), 'A chemical group migration: Professor Todd's move from Manchester to Cambridge fifty years ago', *Christ's College Magazine* **219**, pp. 13–18. See also Todd (1983).
8. Brown, R. F. C (2001), *Chemobiography*, ISBN 0 9579632 0 3, Royal Australian Chemical Institute, North Melbourne, p. 21.
9. Todd (1983), pp. 69 and 131.
10. Anon. (1962), *Proc. Chem. Soc.* p. 9.
11. Todd (1983), pp. 37–9.
12. Levene, P. A. and Bass, L. W. (1931), 'Nucleic acids', *J. Amer. Chem. Soc.*, Monograph Series No. 56, Catalog Co. N. Y.; Gulland, J. M. (1938), 'Lecture on nucleic acids', *J. Chem. Soc.*, pp. 1722–33.
13. Baddiley, J., Lythgoe, B., McNeil, D. and Todd, A. R. (1943), 'Experiments on the synthesis of purine nucleosides. Part I. Model experiments on the synthesis of 9-alkylpurines', *J. Chem. Soc.*, pp. 383–6.
14. Todd, A. R. (1946), 'Synthesis in the study of nucleotides' (Pedler Lecture, 1946), *J. Chem. Soc.*, pp. 647–53.

15. Clark, V. M., Todd, A. R. and Zussman, J. (1951), '*Cyclo*nucleosides. A novel rearrangement of some toluene-*p*-sulphonyl nucleosides', *J. Chem. Soc.*, pp. 2952–8.
16. Brown, D. M., Varadarajan, S. and Todd, A. R. (1958), '*Cyclo*nucleosides', in *Chemistry and Biology of Purines* (Ciba Purine Symp.), London: Churchill, pp. 108–16.
17. Atherton, F. R., Openshaw, H. T. and Todd, A. R. (1945), 'Studies on phosphorylation, Part 1. Dibenzyl chlorophosphate as a phosphorylating agent', *J. Chem. Soc.*, pp. 382–5.
18. Baddiley, J. and Todd, A. R. (1947), 'Nucleotides, Part 1. Muscle adenylic acid and adenosine diphosphate', *J. Chem. Soc.*, pp. 648–51.
19. Baddiley, J., Michelson, A. M. and Todd, A. R. (1949), 'Nucleotides, Part 2. A synthesis of adenosine triphosphate', *J. Chem. Soc.*, pp. 582–6.
20. Christie, S. M. H., Kenner, G. W. and Todd, A. R. (1952), 'Total synthesis of flavin–adenine dinucleotide', *Nature*, **170**, p. 924.
21. Todd, A. R. (1954), 'Chemistry of the nucleotides' (1954 Bakerian Lecture), *Proc. Roy. Soc. A* **226**, pp. 70–82.
22. Khorana, H. G. and Todd, A. R. (1953), 'Studies on phosphorylation, Part X. The reaction between carbodiimides and acid esters of phosphoric acid. A new method for the preparation of pyrophosphates', *J. Chem. Soc.*, pp. 2257–60.
23. Clark, V. M., Kirby, G. W. and Todd, A. R. (1957), 'Studies on phosphorylation, Part XV. The use of phosphoramidic esters in acylation', *J. Chem. Soc.* pp. 1497–501.
24. Khorana, H. G. (2001), *Chemical Biology: Selected Papers*, World Series in 20th Century Biology, Vol. 5.
25. Levene, P. A. and Harris, S. A. (1932), 'The ribosephosphoric acid from xanthylic acid II', *J. Biol. Chem.* **98**, pp. 9–16.
26. Cohn, W. E. (1950), 'The anion-exchange separation of ribonucleotides, *J. Amer. Chem. Soc.* **72**, pp. 1471–8.
27. After completion of the Cambridge work, a paper obscure to us, by Andreas Fonö (*Ark. Kemi., Min., Geol.*, 1947, **24A**, p. 1) came to light, in which the same analogy had been made. But Fonö could proceed no further with the evidence available at that time.
28. Brown, D. M. and Todd, A. R. (1952), 'Nucleotides Part IX. The synthesis of adenylic acids *a* and *b* from 5'-trityladenosine', *J. Chem. Soc.*, pp. 44–51; 'Part X. Some observations on the structure and chemical behaviour of the nucleic acids', pp. 52–8.
29. Markham, R. and Smith, J. D. (1951), 'Structure of ribonucleic acid', *Nature* **168**, pp. 406–8.
30. Brown, D. M. and Todd, A. R. (1953), 'Nucleotides Part XXI. The action of ribonuclease on some simple esters of monoribonucleotides', *J. Chem. Soc.*, pp. 2040–9.
31. Brown, D. M., Fried, M. and Todd, A. R. (1953), 'The determination of nucleotide sequence in polyribonucleotides', *Chem. and Ind.*, pp. 352–3.
32. Whitfeld, P. R. and Markham, R. (1953), 'Natural configuration of the purine nucleotides in ribonucleic acids; chemical hydrolysis of the dinucleoside phosphates', *Nature* **171**, pp. 1151–2.
33. Uziel, M. and Khym, J. X. (1969), 'Sequential degradation of nucleic acids. Degradation of *Escherichia coli* phenylalanine transfer ribonucleic acid', *Biochem.* **8**, pp. 3254–60.
34. Brown, D. M. and Todd, A. R. (1955), 'Nucleic acids', *Ann. Rev. Biochem.* **24**, pp. 311–38.

35. Dekker, C. A., Michelson, A. M. and Todd, A. R. (1953), 'Nucleotides Part XIX. Pyrimidine deoxynucleoside diphosphates', *J. Chem. Soc.*, pp. 947–51; Michelson, A. M. and Todd, A. R., 'Nucleotides Part XX. Mononucleotides derived from thymidine. Identity of thymidine-5'-phosphate and thymidylic acid', pp. 951–6.
36. Crick, F. H. C. (1995), 'DNA: a co-operative discovery', pp. 198–9; *Ann. N.Y. Acad. Sci.* **758**, 19B, p. 1.
37. Fuller, W. (2003), 'Who said helix?', *Nature* **424**, 876–8.
38. Michelson, A. M., Szabo, L. and Todd, A. R. (1956), 'Nucleotides Part XXXVI. Adenosine-5' uridine-5' phosphate', *J. Chem. Soc.*, pp. 1546–9.
39. Reese, C. B. (2002), 'The chemical synthesis of oligo- and poly-nucleotides: a personal commentary', *Tetrahedron* **58**, pp. 8893–920.
40. Kirby, A. J. and Warren, S. G. (1967), *Organic Chemistry of Phosphorus*. Amsterdam: Elsevier.
41. Todd, A. R. (1964), 'Synthesis in the study of nucleotides' (Nobel Lecture, 11 Dec. 1957) in Nobel Foundation, *Nobel Lectures: Chemistry 1942–1962*, pp. 522–36. Amsterdam: Elsevier.
42. Buchanan, J. G. (1999), 'Lord Todd 1907–1997', *Adv. Carbohydrate Chem. Biochem.* **55**, pp. 1–13.
43. Todd, A. R. (1941), 'Vitamins of the B group. The Tilden Lecture', *J. Chem. Soc.*, pp. 427–32.
44. Todd, A. R. (1955), 'Introduction', in Williams, R. T. (ed.), *Biochem. Soc. Symp. No. 13: Biochemistry of Vitamin B_{12}*, pp. 1–2. Cambridge: Cambridge University Press. The Merck group was led by Karl Folkers.
45. Macrae, T. F. (1957), 'The research work of the Glaxo Laboratories Ltd.', *Proc. Roy. Soc. B* **146**, pp. 181–93.
46. Macrae was in the year above Todd at Glasgow.
47. Buchanan, J. G., Johnson, A. W., Mills A. J. and Todd, A. R. (1950), 'Chemistry of the vitamin B_{12} group. Part I. Acid hydrolysis studies. Isolation of a phosphorus-containing degradation product', *J. Chem. Soc.*, pp. 2845–55. See also Todd (1983), p. 93.
48. Cannon, J. R., Johnson, A. W. and Todd, A. R. (1954), 'Structure of vitamin B_{12}. A crystalline nucleotide-free degradation product of vitamin B_{12}', *Nature* **174**, pp. 1168–9.
49. Hodgkin, D. C., Johnson, A. W. and Todd, A. R. (1955), 'The structure of vitamin B_{12}', in *Recent Work on Naturally-occurring Nitrogen Heterocyclic Compounds*, Schofield, K. (ed.), *Chem. Soc. Special Publ.*, **3**, pp. 109–20.
50. Martin, A. J. P. and Synge, R. L. M. (1941), 'A new form of chromatogram employing two liquid phases. 1. A theory of chromatography. 2. Application to the micro-determination of the higher monoamino-acids in proteins', *Biochem. J.* **35**, pp. 1358–68. Martin and Synge shared the 1952 Nobel Prize for their work on partition chromatography.
51. Peter Sykes, private communication; and see Birch, A. J. (1995), 'To see the obvious', in Seeman, J. I. (ed.), *Autobiographies of Eminent Chemists*, Washington, DC: American Chemical Society.
52. Cameron, D. W. and Todd, A. R. (1967), 'Aphid pigments', in Taylor, W. I. and Battersby, A. R. (eds.), *Organic Substances of Natural Origin Vol. 1 (Oxidative Coupling of Phenols)*, New York: Marcel Dekker, pp. 203–41.
53. Alex Todd's clear, down-to-earth, views on these matters are summed up in his Presidential Addresses as Appendices 1–6 in Todd (1983), pp. 205–57.

10

Ralph Alexander Raphael: organic synthesis – elegance, efficiency and the unexpected

Bill Nolan and Dudley Williams

Department of Chemistry, University of Cambridge

Robert Ramage

Albachem Ltd., East Lothian, Scotland

The early years

Ralph Alexander Raphael was born in Croydon on New Year's Day 1921, the eldest child of Jacob (Jack) Raphael and his wife Lily (*née* Woolf); there were also three daughters. The paternal family hailed from Poland; Ralph's grandfather, Solomon Raphael, had emigrated to the UK in 1864 and settled in London.

Ralph remembered his father as a sweet-natured, gentle man, moderately orthodox in the Jewish faith, and his mother as a strict disciplinarian and splendid cook. Jack Raphael was a master tailor, a perfectionist in his craft, but in the pre-war Depression days, work was hard to come by, and the family was continually moving to where the jobs were. Ralph's schooling was disjointed, encompassing several London boroughs, then Leeds, Bradford, Sunderland, Wesley College, and back to London. Life was far from easy, and Ralph remembered his father making 'suits' out of newspaper to keep his hand in. This period left a deep impression on Ralph, instilling a very strong work ethic in him and a determination to look after family.

Ralph's secondary schooling started in Wesley College, Dublin, where he held a scholarship. Prompted by strong parental leaning, his initial school education was almost completely on the arts side and it appeared that he might teach classics or become a rabbi. However, when the family moved back to London in 1936, Ralph was enrolled at Tottenham County School, which excelled in mathematics and sciences. The chemistry master, Edgar Ware, was so appalled that Ralph knew nothing whatsoever about the subject that he took young Ralph personally in hand. Edgar Ware was a truly inspiring teacher and mentor, and

The 1702 Chair of Chemistry at Cambridge: Transformation and Change, ed. Mary D. Archer and Christopher D. Haley. Published by Cambridge University Press. © Cambridge University Press 2005.

Ralph took to chemistry with enthusiasm. As he wrote some time later, 'The resulting conversion was complete and deep-rooted and the way ahead was clear.' Spurred on by his teacher's advice, Ralph obtained London Intercollegiate and Royal Scholarships, entering Imperial College, London, in October 1939.

Imperial College and wartime

The chemistry course at Imperial was compressed into two years during the war, although the normal three-year curriculum was taught; with botany as his subsidiary subject, Ralph emerged with first-class honours and the Hofmann Prize for practical chemistry. But these bland facts do not convey the realities of the time. The winter of 1940–1 was a tough time to be living in London – air raids were almost continuous from September 1940 until May 1941, and thousands were killed. In the midst of the carnage, life went on, and Ralph took half of his final papers in February 1941 and the other half in June. He was one of five students to obtain a first-class degree that year. By courtesy of Ralph's Ph.D. supervisor and later mentor at Imperial, Dr 'Tim' Jones (later Professor Sir Ewart Jones F.R.S.), we can disclose his marks (%) for public scrutiny: Inorganic, 78; General and Physical, 62; Physical, 68; Organic, 78; Course Work, 82; Practical Organic, 83 and Analytical, 85. Ralph was narrowly beaten in the competition for top position by Geoffrey Wilkinson (who was later to receive the Nobel Prize for Chemistry).

The quality of undergraduate organic chemistry teaching at Imperial was high, and it was this that orientated Ralph towards the organic branch of the subject for his Ph.D., for which he remained at Imperial. The Ph.D. course was also of two years' duration but the Heilbron – Jones school of organic chemistry was extremely active and exciting, and Ralph's brilliant research efforts were contained in five collaborative papers in the polyene area.

Ralph's first paper, co-authored with Professor Sir Ian Heilbron F.R.S.[1] and Ewart Jones stems from this period, and appeared in the *Journal of the Chemical Society* in 1943.[2] It reported the formation of a carbon–carbon bond by the attack of an alkyne anion on an aldehyde, as shown in Figure 10.1.

The work, the long-term goal of which was the synthesis of polyene alcohols related to vitamin A, makes use of the highly nucleophilic alkyne anions in the construction of carbon–carbon bonds. The seeds of a life-long interest in acetylene chemistry had already been sown in Ralph. Imperial College was justly proud of its ultraviolet spectrometer, and large pictures of the spectra grace the pages of Ralph's first paper. The young Raphael not only meticulously prepared

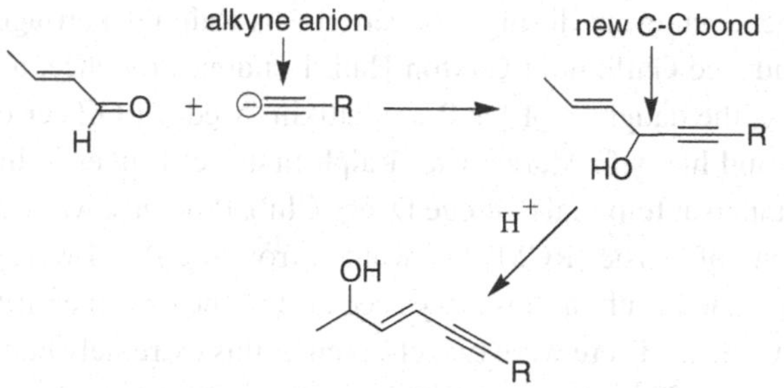

Figure 10.1. Carbon–carbon bond formation by the attack of an alkyne anion on an aldehyde.

3,5-dinitrobenzoates and α-naphthylurethanes of his alcohols and semicarbazones of his carbonyl compounds, but also showed that his alcohols really had one OH per molecule by measuring the volume of methane evolved in the Zerewitinoff determination. This period was the origin of Ralph's undying love of acetylene chemistry.

All younger citizens were subject to government direction in wartime, and following the award of his Ph.D. in 1943, Ralph was sent to May & Baker Ltd, where he spent three years as head of the Chemotherapeutic Research Unit, engaged on the collaborative UK–US research on the chemistry and synthesis of the then-new antibiotic penicillin. This work is recorded in book form in the account of penicillin research published after the war.[3] Despite his rather grandiose title, there was still time for Ralph to work at the bench with, on at least one occasion, catastrophic results. He was in the process of isolating a carboxylic acid by the time-honoured procedure of acidifying an aqueous solution of the sodium salt in dilute sodium carbonate, and then extracting the liberated acid into ether. Ralph gave the separatory funnel a couple of good shakes and was left holding the stopper whilst the funnel shot upwards and burst against the wall of the laboratory, precisely above the swing door through which the Assistant Director of Research was entering!

Ralph thought the penicillin project wonderfully stimulating and worthwhile. However, he became increasingly impatient with the research direction in May & Baker at that time, which he considered to be unimaginative and conventional, and he resolved to move on as soon as he was able.

During the war, Ralph lived with his family in Walthamstow and contributed to the war effort by serving as one of the region's part-time gas identification officers, in charge of training air-raid precaution personnel in the identification and handling of war gases (which fortunately were never used against the British). At night, Ralph and his father undertook fire-watching duties. A most important personal event during this period was Ralph's marriage on 23 May 1944 to Prudence Gaffikin at Caxton Hall, London. Prudence, a professional violinist, was the daughter of Dr P. J. Gaffikin, Medical Officer of Health for Maidstone, and his wife Marguerite. Ralph first met Prudence in 1942 while learning to tango at Imperial College Dance Club. Prudence was a student at the Royal College of Music (RCM) but women from the RCM were always welcome at the Dance Club in those days because of the extreme rarity of women at Imperial College. There were two children of this extremely happy marriage: Richard Anthony (Tony), born in 1945, and Sonia Elizabeth, born in 1952.

Imperial College and Glasgow

Taking advantage of the newly instituted ICI Research Fellowship scheme, Ralph was able to return to Imperial College in 1946 as an independent researcher. The main thrust of his research was the synthetic exploitation of acetylene chemistry, which at that time had been infrequently used. Imperial was an obvious centre of excellence in his chosen field at that time – sixteen of Ralph's contemporaries at Imperial would go on to hold major chairs in organic or biological chemistry. Ralph's days (1946–9) as an ICI Research Fellow were a happy time of great chemical activity, recognised in his receipt of the Meldola Medal of the Royal Institute of Chemistry in 1948. Basil ('Jimmy') Weedon F.R.S. (later professor of organic chemistry at Queen Mary College, and subsequently Vice-Chancellor of the University of Nottingham) was one of Ralph's colleagues. He recalled of those days that Ralph did some preliminary work, which Weedon later took over, on the reaction in which primary and secondary alcohols are oxidised to acids and ketones, respectively, in the presence of chromic acid, aqueous sulphuric acid and acetone. This reaction is now generally known as the Jones oxidation. He also recalled that, following Ralph's marriage to Prudence, it was not uncommon to enter the departmental library and find Ralph surrounded by journals, but at the same time minding the infant Tony in a pram. He added, 'This does not sound very significant these days, but I can assure you it made quite an impact on that male-dominated community of that time.'[4]

During this period, Ralph developed acetylene chemistry in a number of synthetic directions. One of these was in the controlled production of *cis*- and *trans*-double bond linkages in various molecular frameworks. Some of this work was carried out in collaboration with Ralph's great friend Franz Sondheimer (Figure 10.2).[5] Together they provided the first synthesis of the important natural product linoleic acid (1) with its methylene-interrupted conjugation.[6] This

Figure 10.2. Ralph Raphael (right) with Franz Sondheimer at Imperial College *c*. 1946 – searching for crystals.

opened the way to the synthesis of more complex but similar acids, such as linolenic acid and arachidonic acid. Figure 10.3 shows the key points of the synthesis.

Here the alkyne functionality is used to put together the molecular framework, which is then partially reduced on the surface of a catalyst to ensure the *cis*-geometry of both double bonds in the target molecule.

However, during this early period the work that rightly pleased Ralph the most was his synthesis of the antibiotic and biosynthetic intermediate penicillic acid (4), performed entirely by his own hands. This also involved some acetylene chemistry in the synthesis of the key starting material, the lactone (2).[7] The key steps in the synthesis are shown in Figure 10.4.

Ralph described the final step thus: 'On boiling an aqueous solution of (3) with excess magnesium oxide, trimethylamine is evolved and penicillic acid is produced directly. The yields in all the above stages are substantially quantitative.'[7] This synthesis, early in his career, shows his preference for 'tricky' synthetic targets – relatively small molecular systems with challenging mixtures of functionality.

In 1949, at the conclusion of his tenure as an ICI Research Fellow, Ralph was appointed to a tenured position as a lecturer at the University of Glasgow.

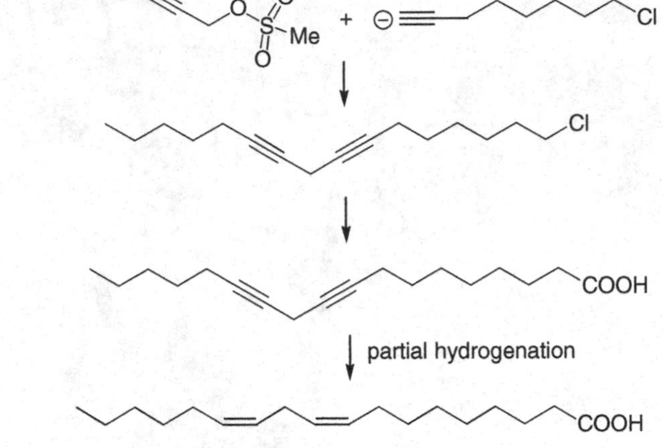

(1)

Figure 10.3. Synthesis of linoleic acid.

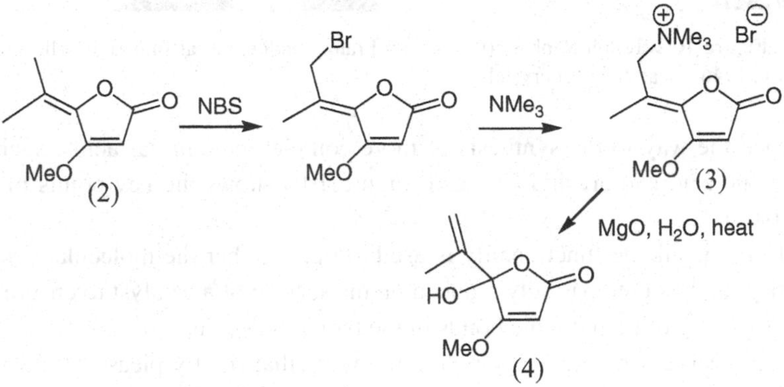

Figure 10.4. Synthesis of penicillic acid.

Here, as he noted, he 'spent the next five years learning my job as a university teacher and research supervisor'.[8] He was in consequence well equipped to contribute no less than nine of the eleven chapters to an early volume in the series *Chemistry of Carbon Compounds*.[9]

Ralph's thoughts on the chemistry department at Glasgow at that time are interesting: 'The department was run very hierarchically and with a decidedly pinch-penny attitude towards research and teaching equipment, and I was convinced (like all the junior staff) that I could run it much better given the opportunity.'[8] It was in Glasgow that Ralph earned his spurs as a lecturer, and

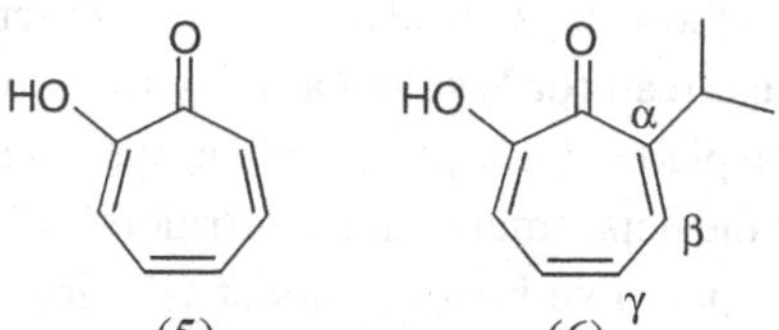

Figure 10.5. Structures of tropolone and the thujaplicins.

Figure 10.6. Ralph Raphael with Ian Scott (left) and Willie Parker (right), the first two of his research students to be appointed to chairs.

he described his students as 'warmly voluble and appreciative'[8] of his efforts. He continued his research in the acetylene field but, in collaboration with Professor J. W. (later Sir James) Cook F.R.S., he also opened up a new line on the synthesis of the quasi-aromatic tropolones, at that time very novel molecules. This work led to the first practical synthesis of tropolone (5) itself,[10] and also the first synthesis of three naturally occurring tropolones, the α-, β- and γ-thujaplicins (6).[11,12] Figure 10.5 shows the structures of these molecules.

As Professor Cook was heavily involved in administrative matters, much of the research direction of the department was left to Ralph. Ian Scott (Figure 10.6), himself to become distinguished for his work on the biosynthesis of alkaloids and of vitamin B_{12}, was among Ralph's students at this time.

His colleagues from those days recall Ralph's phenomenal powers of concentration; he could sit and read for hours in a noisy room or laboratory. Perhaps this 'concentration' explains the incident when, on a wet day, he was found walking along one of the department's long corridors still holding his umbrella over his head. Ralph spent five happy years at Glasgow, but new vistas were to be opened by his appointment in 1954 as professor of organic chemistry at Queen's University, Belfast.

Queen's University, Belfast

Ralph Raphael was the first professor of organic chemistry at Queen's University, Belfast. He found there an unduly neglected subject, a situation he worked hard to improve. He and the staff took the opportunity to overhaul both teaching and research. In this refurbishment of the subject, they were solidly backed by the Vice-Chancellor, Sir Eric (later Lord) Ashby F.R.S., whom Ralph described as the 'most supportive Vice-Chancellor I have ever worked with'.[8] Ralph revelled in the opportunity to create a department effectively from scratch, along with a new building. It was during his three years at Queen's, in 1955, that Ralph published his monograph *Acetylenic Compounds in Organic Synthesis*,[13] beautifully written, rich in experimental detail and making the subject much more accessible to the non-specialist organic chemist.

The Regius chair at Glasgow

In 1957, Ralph received an invitation from the Vice-Chancellor of the University of Glasgow to return to the University as the Regius Professor of Chemistry, in succession to Derek Barton (later Sir Derek Barton F.R.S., Nobel prize winner in 1969). Ralph later commented[14] that, in view of his grumblings as a junior lecturer there, he found it impossible to resist this invitation, to set right the many shortcomings that he had seen. His tenure of the Regius chair from 1957–72 turned out to be one of the great periods of Glasgow chemistry, producing a galaxy of organic chemical research stars from both staff and students, amongst whom are numbered Peter Doyle (Executive Director, AstraZeneca), Geoffrey Eglinton F.R.S. (Professor, Bristol), Keith James (Pfizer), Alexander (Sandy) McKillop (Professor, East Anglia), Tom McKillop (Chief Executive, AstraZeneca), James Maxwell F.R.S. (Professor, Bristol), Tom Money (Professor, British Columbia), Karl Overton (Professor, Glasgow), William Parker (Professor, Stirling), Robert Ramage F.R.S. (Professor, Edinburgh), A. Ian Scott

F.R.S. (Professor, Texas A & M University), Emanuel Vogel (Professor, Universität Köln) and Douglas Young (Professor, Sussex).

Ralph arrived at Glasgow with a ten-strong research group in July 1957 and, with the enthusiastic participation of the staff, totally remodelled every organic teaching course in the place during the following hectic three months of intensive work. Time-hallowed but largely irrelevant courses were removed and replaced by up-to-date and interesting themes. Research interests were expanded in the area of natural product synthesis and were broadened to include structure work and the study of certain mould metabolites. These changes, and the development of them, weathered extremely well over the subsequent years at Glasgow.

In his address at the memorial celebration of Ralph's life, given in Christ's College Chapel on 30 June 1998, Professor Eglinton remembered those times as follows:

> [Ralph] seemed to be everywhere at once – one met him striding energetically along the long echoing corridors, bounding up and down the great curved staircases, drawing reaction schemes for awestruck students in the teaching labs, peering short-sightedly at the latest journals in the Library, telling arcane stories in the tea room, and lost in deep contemplation of his bridge hand in the College Club. The department seemed to pulse with energy and fun.

This happy image of the department spread throughout the world and, as always with Ralph, 'non-chemical' stories abound along with the science. When he appeared on the bowling green during a departmental outing, he had to be warned for 'body-line' bowling, and he remains on the record books as the only person who ever managed to bounce a bowl right across the green, over the ditch, across the path, and into the neighbouring bowling green. The affection with which he was held in the department is conveyed by a depiction of him as Santa Claus on the departmental organic bulletin.

Ralph had a strong sense of humour – sometimes almost surrealist – and as he did not mind being undignified, the students loved it. He would get involved in daft schemes such as the pram-pushing race (for younger professors only) at Glasgow. Each professor raced a decrepit old pram around the campus loaded with a young female student in a baby-bonnet, hanging on for dear life.

Ralph also loved a 'spoof' lecture, and Dr George Buchanan remembers one given to the Alchemists' Club in Glasgow:

> His lecture was on a new type of polymer – a chain of linked rings then unknown, but now known as catenanes. He began with a credible synthesis and then moved on to describe the polymers' remarkable properties in great detail – IR, UV, mass spec. etc., all illustrated with elegant charts, which we later found out had been

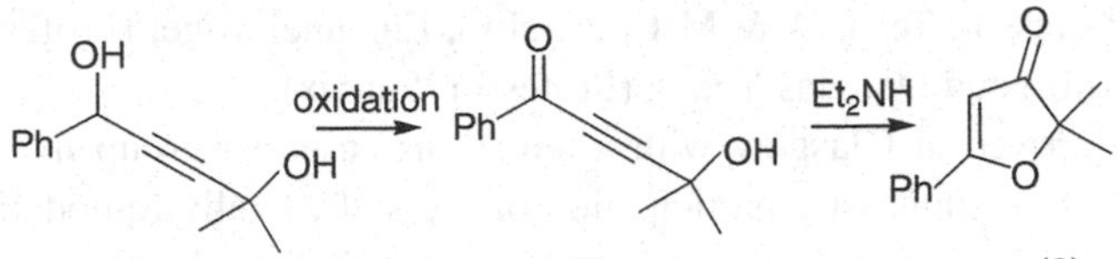

Figure 10.7. Synthesis of bullatenone.

completely forged. These pearls of wisdom were being carefully noted down by the more conscientious students. But all became imperceptibly less believable. On the front row Dr Klar[15] began to 'tut tut'. Ralph ploughed on and described how these polymers absorbed in the audio region over a broad band from A flat to F sharp with an extinction coefficient in high millidecibels. That did it – Klar was now vehemently shaking his head, all note-taking stopped and pandemonium ensued!

During this second period in Glasgow, Ralph's science blossomed as before. In 1959, he published a paper on bullatenone (9) with Parker and Wilkinson.[16] The paper determined the correct structure of the natural product and confirmed it by the first total synthesis, an outline of which is shown in Figure 10.7.

The starting diol (7) was speedily constructed using Ralph's favourite acetylene chemistry, and oxidation of the secondary alcohol gave the hydroxy-ketone (8). Addition of diethylamine, followed by distillation, unexpectedly gave the crystalline bullatenone directly. Since then bullatenone has been resynthesised many times as a favourite target for trying out alternative synthetic methods.

Other highlights of this very productive period were an important total synthesis of (-)-shikimic acid (10), an important intermediate in amino acid synthesis[17] and a neat synthesis of (±)-*trans*-chrysanthemic acid (11).[18] The latter is an important component of the natural pyrethrin insecticides and their synthetic pyrethroid relatives. Ever since his time working on penicillin, Ralph had maintained an interest in antibiotics, and this was taken up in Glasgow and continued during his time in Cambridge. The complex antifungal (±)-trichodermin (12) was the first member of the antibiotic trichothecin family to be synthesised.[19] The synthesis of the macrolide antibiotic (±)-pyrenophorin (13) was also completed by Ralph's group in Glasgow;[20] this seems to have been the first synthesis of any macrocyclic natural-product lactone. Figure 10.8 shows the structures of these molecules.

In recognition of this and much other outstanding synthetic work, Ralph was the Tilden Lecturer of the Chemical Society in 1960, and he was elected a Fellow of the Royal Society in 1962. His wider services to the chemical community in this period were also expressed through his membership of the Chemistry Committee of the Science Research Council, and his role as Council

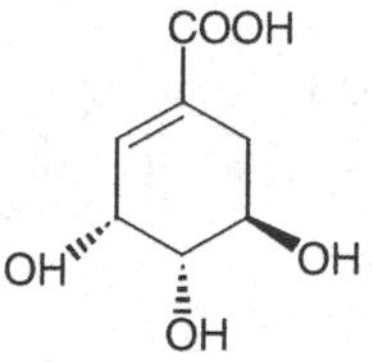

(10) shikimic acid 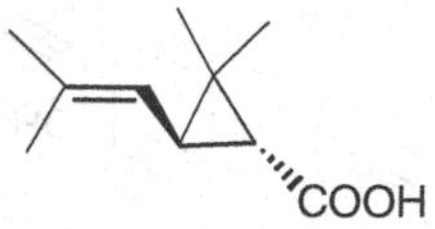   (11) chrysanthemic acid 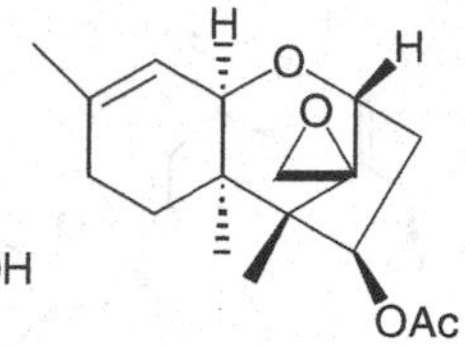   (12) trichodermin

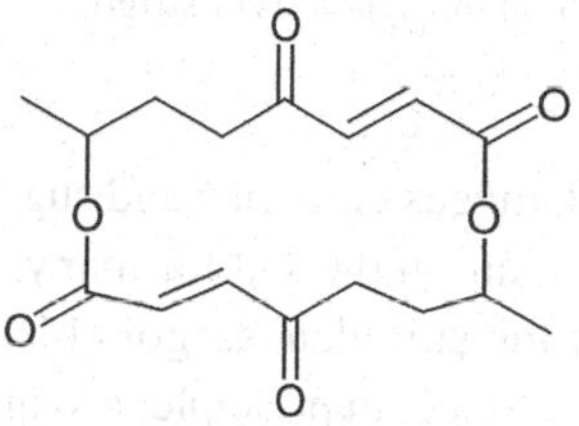

(13) pyrenophorin

Figure 10.8. Structures of some natural products synthesised by Ralph Raphael.

member (1960–3) and Vice-President (1967–70) of the Chemical Society. He undertook lecture tours to many countries including the USA, France, Germany, Belgium, Switzerland and Hungary, but he had a special affection for Glasgow, describing himself as 'a brainwashed pseudo-Scot'. He was deeply honoured when, in 1994, a new laboratory in the university chemistry department was named after him.

The Cambridge years

It was perhaps inevitable that Ralph's talents in teaching, research and administration would lead other institutions to entice him to move on. In 1972, he was elected 1702 Professor of Organic Chemistry at Cambridge, in succession to Lord Todd. This position carried with it the headship of the Department of Organic, Inorganic and Theoretical Chemistry; the separate Department of Physical Chemistry shared the same building and facilities, an arrangement which, as the current head of the Department of Chemistry, Professor Jeremy Sanders, notes in his *Dictionary of National Biography* entry on Ralph, 'did not always lead to comfortable relationships between the two Departments'.[21]

In Cambridge, Ralph carried out more elegant synthetic work and to illustrate this, we have chosen points from his syntheses of strigol, steganacin and pseudomonic acid. Strigol is the germination stimulant of witchweed, a troublesome

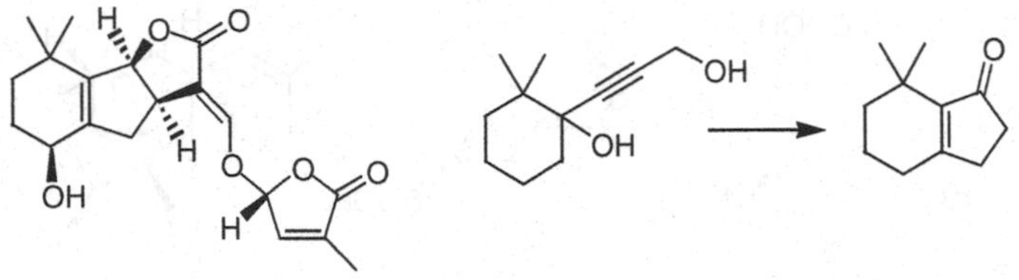

(14) strigol (15) (16)

Figure 10.9. Key reaction in the synthesis of strigol.

semi-parasitic plant that damages corn, rice and sugar cane crops. The seeds of the witchweed can lie dormant in the soil for many years; their germination is triggered by contact with the stimulant strigol (14), exuded from the roots of the growing 'victim' plant. Since, in principle, a synthetic substitute for natural strigol would render witchweed seed germination possible in the absence of standing crops, the synthesis of strigol was an important goal. The successful synthesis[22] epitomised Ralph's love of acetylene chemistry, and the vigour with which he was attacking synthesis in the early 1970s. It is economic, as can be illustrated by the one-pot conversion of the acetylenic diol (15) into the bicyclic enone (16). The conversion is concisely summarised in the paper as one 'which blends the Newman mechanism for the Rupe rearrangement with a consequent Nazarov-type conrotatory electrocyclisation'. Figure 10.9 shows the target molecule and key reaction.

One investigation that gave Ralph much personal pleasure concerned the synthesis and properties of the anti-leukaemic lignan (-)-staganacin (17). Efficiency is once more the key word in the Raphael synthesis.[23] The enamine (18) was synthesised by relatively conventional methods, and the desired 8-membered ring of the target then elegantly produced by treating this with dimethyl but-2-ynedioate. The product (19) was converted in seven further steps, via (±)-steganone (20), to (±)-steganacin. The intermediate (±)-steganone was achieved in ten steps with an overall yield of 23%, corresponding to an average yield of almost 90% per step. The key reactions are shown in Figure 10.10.

The total synthesis of (±)-pseudomonic acids (acid A, 21) was achieved in 1984 and with a characteristic elegance of strategy.[24] The lactone (22) was treated with di-*t*-butyl sodiomalonate in the presence of bis(dibenzylidene-acetonato)palladium(0) and 1,2-bis(diphenylphosphino)ethane, resulting in allylic displacement of the internal carboxylate with double inversion to give the pyranylacetic acid (23). Having ensured the *cis* stereochemistry of the 2- and 5-substituents, (23) was converted in several steps to (24), which was then *cis*-dihydroxylated from the face opposite to the 2- and 5-side chains using N-methyl-morpholine N-oxide and a catalytic amount of osmium tetroxide.

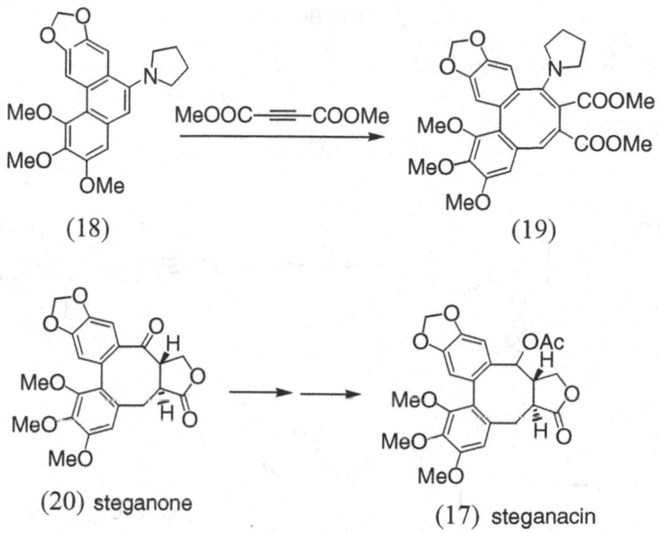

Figure 10.10. Key steps in the synthesis of steganacin.

The cyclohexylidene acetal (25) of this diol has been converted to the pseudo-monic acids in the hands of others. Figure 10.11 shows the key points in this synthesis.

Some other natural products synthesised by Ralph's group in Cambridge are shown in Figure 10.12. These include virantmycin (26),[25] an antiviral natural product of unique structure, and aaptamine (27),[26] a compound isolated from a sea sponge and possessing β-adrenoceptor blocking activity. Practical and flexible methods for the synthesis of naturally occurring indolocarbazoles (for example, staurosporinone 28)[27] were developed, and these are of value for their pharmacological interest. All these syntheses carry the Raphael hallmarks of efficiency and elegance of strategy.

Ralph's lectures to Cambridge undergraduates were clear and laced with humour and examples of the relevance of organic synthesis to society.[28] Most undergraduates are aware of Murphy's Law, one statement of which is: 'What can go wrong, will'. But Ralph frequently made them sit up in puzzlement with O'Brien's Law, concisely stated as: 'Murphy was an optimist'. Ralph also used humour to illustrate chemical points to his research group. He would often describe the acetylene anions as 'Heineken nucleophiles' since (in a parody on the words of a 1980s beer advertisement) their small size allows them to reach areas of the molecule that other nucleophiles cannot reach. Ralph was keen to promote the Society for the Abolition of Ethyl Esters – since they complicate proton NMR spectra and can sometimes be tricky to remove. Nearly all Raphael syntheses used methyl esters!

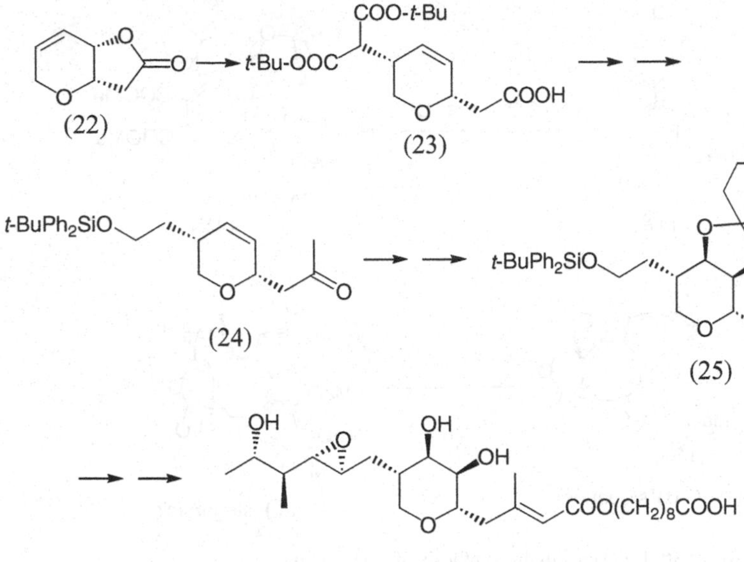

(21) pseudomonic acid A

Figure 10.11. Synthesis of pseudomonic acid A.

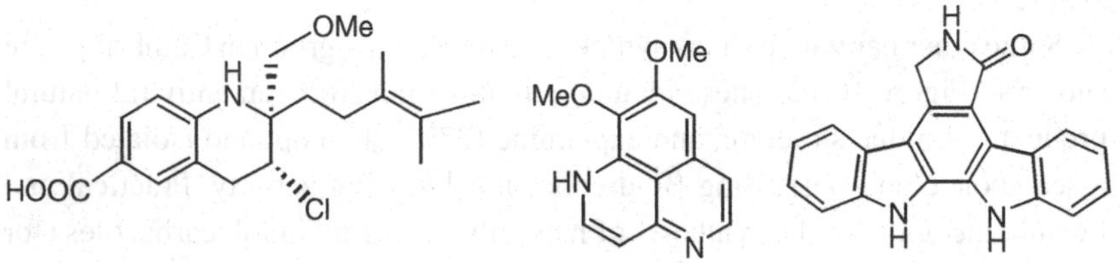

(26) virantmycin (27) aaptamine (28) staurosporinone

Figure 10.12. Structures of virantmycin, aaptamine and staurosporinone.

Occasionally Ralph would display his satirical humour in an article. After his retirement, he wrote one for Christ's College magazine in which he invented an electronic vice-chancellor for a university financed by the advertising industry on business lines, with admission restricted to juvenile delinquents. The humour he demonstrated in the lecture theatre carried over into his daily life, and particularly to the dinner table – either when he was an after-dinner speaker, or at home following Prudence's superb cuisine. He was an excellent raconteur, employing the measured delivery of his deeply resonant voice to good effect.

Research, lecturing, and maintaining a sense of humour were but a small part of Ralph's Cambridge position. In view of the large size of the department, a considerable portion of his time was inevitably spent on administration, but he had no liking for excessive paperwork and formalities; he had on his side considerable ability as a negotiator, having both charm and moderation. Staff were always listened to and their views considered in detail, and whenever possible Ralph moved to a consensus decision. He had a very characteristic administrative style: early arrival in the department, responses to paperwork first drafted out in his attractive and flamboyant handwriting and an open-door policy to anybody who felt like a chat. Ralph saw grants and individual budgets as barriers to creative science and believed that the role of a head of department was to manage people and money, freeing staff to concentrate on teaching and research. He wanted staff to produce and carry out good adventurous projects, which he would encourage unreservedly and for which he would try to find the necessary support.

Not only did Ralph cope with the administrative load almost single-handedly, but he also lent his broad shoulders to other institutions. He sat on sufficient professional appointment committees to be affectionately known as 'the Godfather' in some circles. Following his arrival in Cambridge, his talents were tapped as a member of the Council of the Royal Society (1975–7) and president of the Perkin Division of the Royal Society of Chemistry (1979–81). Additionally, he was a Pedler lecturer of the Chemical Society in 1973, winning its Ciba–Geigy Award for synthetic chemistry in 1975, and he was awarded the Davy Medal of the Royal Society in 1981. Membership of the Royal Irish Academy and honorary doctorates from the Universities of Stirling and East Anglia, and Queen's University, Belfast, were acquired along the way. All Ralph's services to chemistry were recognised when he was appointed CBE in the Queen's Birthday Honours List in 1982 (Figure 10.13).

A new experience when Ralph arrived in Cambridge was to be a Fellow of a Cambridge college. This experience can be educational, frustrating and pleasurable; and it gives exposure to the eccentricities that occasionally go with long-established tradition. On election to a Fellowship at Christ's College, Ralph found himself, although a very senior member of the university, the most junior Fellow of the college. As such, he was 'Mr Nib', with the duty to pour after-dinner drinks for other Fellows and their guests. Ralph seems to have performed his duties in suitably traditional style, for the entry in Christ's wine-book for 26 June 1972, reads: 'Professor Raphael presented a bottle of Sauternes to celebrate the first occasion he had occupied Mr Nib's chair, and to express his appreciation of the friendly forbearance of the Room'.

Ralph was intensely proud of his family, his children, his grandchildren and his first great-grandchild. Prudence and he spent fifty-four happy years together, sharing the enjoyment of music (especially opera), travel, good food and wine, good company and bridge. Meals at the Raphaels were always an experience

Figure 10.13. Ralph Raphael at Buckingham Palace with Prudence after receiving his CBE in 1982.

to be enjoyed and savoured. The conversation would range far and wide, with Ralph delving into his library to check a point raised, be it on art, sciences, music, philosophy, politics or literature. A discerningly chosen glass of wine in his hand, his eyes would light up and another anecdote or joke would illustrate the point. Prudence was a freelance professional violin and viola player, and this heightened Ralph's interest in music.

This interest was further increased by contact with the distinguished Cambridge-based *luthier* David Rubio, who was investigating the violin-making methods of Stradivarius and the Cremonese school. In a joint research project,[29] involving the scanning electron microscopic and energy-dispersive X-ray spectroscopic examination of the wood surfaces of Stradivarius stringed instruments, the 'ground' used on such instruments by Cremonese masters was identified. It was based on an Italian pozzolana volcanic ash, and a modern version was formulated. Using this 'ground', the rich appearance of the wood grain and the improvement in sound were quite remarkable for newly made instruments, eliminating the usual 'rawness' of tone.[30] In the study of violin varnishes and treatments, Ralph saw a relationship between chemistry and music, and he and Prudence developed this, giving several joint lectures. The highlight of Ralph's later years must have been the invitation to visit Washington where, during a special meeting on music at the American Academy of Sciences' annual meeting in April 1993, Prudence played a less than two-year-old David Rubio and a 1685 Nicola Amati violin, and Ralph lectured on the research into the construction and finish of instruments with remarkable qualities of sound and his work with David Rubio over the previous nine years. At this time, Ralph was a member of the Science Scholarships Committee of the Royal Commission for the Exhibition of 1851, which supported young scientists from overseas wishing to study in the UK; he also served on the Honorary Scientific Advisory Committee of the National Gallery from 1986, attending his last meeting in February 1998 only a few weeks before his death.

Ralph was a friendly man, and the half-humorous glint behind those thick bottle-glass spectacles was reassuring to even the most timid student. In speech, he had a slight impediment, but he knew how to turn it to advantage as a telling pause in talking or lecturing. Above all, he was a kind and generous man who spared no effort to help his colleagues and students. This could be important career help, or could just be the bottle left on the bench of a student who had succeeded in an important reaction, or the ice creams bought for the secretaries working hard on a hot day. As Jeremy Sanders noted in his tribute at Ralph's memorial celebration, 'Ralph defined his success not in selfish terms of his own scientific or monetary triumphs but the collective success of his Department, and above all in the success of his young appointees' (Sanders, J. K. M., personal communication, 2002).

Ralph retired from the 1702 Chair in 1988 but retained a lively interest in chemistry. On most days he could be found reading the chemical literature in the departmental library. Ralph always thought of the research interests of his colleagues and friends and many useful references were passed on via a short note in his distinctive handwriting. He was reading the latest papers enthusiastically until a few weeks before his death in Addenbrooke's Hospital from ischaemic heart disease on 27 April 1998. He left his body to the University's Department of Anatomy for medical teaching and research.

Reflecting on his time in Glasgow, Ralph wrote the following summary of what he hoped had been achieved during his time as Regius Professor:

> The department now runs on a firm but flexible organisation that can anticipate and handle problems rapidly, efficiently and informally. The general opportunity to take part for all involved has had the desired effect of engendering a friendly and lively atmosphere where intelligent criticism is a normal expectation. Thoughtful innovation in research is the key and all possible support is mustered to give worthwhile ideas the backing they deserve.[14]

That was true of Glasgow and of his time as the 1702 Professor in Cambridge. Professor Ralph Alexander Raphael was a highly distinguished synthetic chemist, and a popular administrator with unique qualities of warmth, humour and humanity. This made him one of the most successful heads of organic chemistry departments that the UK has known.[31]

Acknowledgements

This chapter is based on the personal recollections of the authors and the contributions and correspondence of the many friends and colleagues of Ralph Raphael. The authors would particularly like to thank Prudence Raphael for her substantial help in research and writing.

Notes and References

1. Professor Sir Ian Heilbron F.R.S., the 'old man' as he was known to all his students, came from a well-known Glasgow family, and held the chair of organic chemistry at Imperial College at that time. During the war, the milk bottles in which he and his staff cultured penicillin lined the underground walkway from South Kensington underground station to Imperial College. Heilbron was a gifted teacher, revered by all his students, and Prudence Raphael recalls his extraordinary care for them with two examples: Franz Sondheimer suffered from bad acne, and Heilbron sent him to a specialist and paid the bill. Similarly, when Prudence was suspected of having tuberculosis in the late 1940s, Heilbron insisted that she be sent to a Swiss sanatorium at his expense should the diagnosis be confirmed; fortunately it was not. It was Heilbron who proposed Ralph for the Royal Society.
2. Heilbron, I. M., Jones, E. R. H. and Raphael, R. A. (1943), 'Studies in the polyene series. Part IX. The condensation product of 1-hexyne with crotonaldehyde and its anionotropic rearrangement', *J. Chem. Soc.*, pp. 264–5.
3. Committee on Penicillin Synthesis Reports, Nos 44, 62, 206, 378, 478 and 674.
4. Williams, D. H. (1986), 'Ralph Alexander Raphael – an acetylene chemist at 65', *Aldrichimica Acta* **19**, pp. 3–9.
5. Franz Sondheimer and Ralph met in 1946 when Franz was a Ph.D. student and Ralph an ICI Research Fellow at Imperial College. They collaborated on the synthesis of the essential fatty acids until 1949 at Imperial, and at a distance thereafter. Sondheimer went on to synthesise (*inter alia*) the steroids christmasterone, testosterone, cortisone and cholesterol in collaboration with David Taub as a post-doctoral member of R. B. Woodward's research group at Harvard, and then by way of Syntex and the seminal synthesis of the annulenes at the Weizmann Institute to a Royal Society chair, held first at Cambridge and then at University College, London (Jones, E. and Garratt, P. (1982), 'Franz Sondheimer, 1926–1981', *Biographical Memoirs of the Royal Society*, **28**, pp. 505–36). On 27 April 1981, Ralph gave the memorial appreciation at the thanksgiving meeting for Sondheimer's life and work at Chancellor's Hall, Senate House, University of London (Prudence Raphael, personal communication, 2003).
6. Raphael, R. A. and Sondheimer, F. (1950), 'The synthesis of long-chain aliphatic acids from acetylenic compounds. Part III. The synthesis of linoleic acid', *J. Chem. Soc.*, pp. 2100–3.

7. Raphael, R. A. (1947), 'Synthesis of the antibiotic, penicillic acid', *Nature*, **160**, pp. 261–2.

8. Crombie, L. (2000), 'Ralph Alexander Raphael, CBE (1 January 1921–27 April 1998), *Biographical Memoirs of Fellows of the Royal Society*, **46**, pp. 463–81; many of Ralph's remarks quoted in this Memoir were recounted by Prudence Raphael to Leslie Crombie.

9. Raphael, R. A. (1953), chapters 1–9, in Rodd, E. H. (ed.), *Chemistry of Carbon Compounds*, vol. IIA.

10. Cook, J. W., Gibb, A. R., Raphael, R. A. and Somerville, A. R. (1951), 'Tropolones. Part I. The preparation and general characteristics of tropolone', *J. Chem. Soc.*, pp. 503–11.

11. Cook, J. W., Raphael, R. A. and Scott, A. I. (1951), 'Tropolones. Part II. The synthesis of α-, β-, and γ-thujaplicins', *J. Chem. Soc.*, pp. 695–8.

12. The thujaplicins, produced in abundance by the Western Red Cedar, protect the tree from fungal disease, and also show activity in several disease states of mammalian organisms.

13. Raphael, R. A. (1955), *Acetylenic Compounds in Organic Synthesis*, London: Butterworths Scientific Publications.

14. Raphael, R. A., handwritten notes about his Glasgow days (in the collection of Prudence Raphael).

15. Dr Erich Klar was a refugee from the Sudetanland who was given space in the basement of the Glasgow department by J. W. Cook, where he produced some brilliant work. Ralph respected Klar highly and persuaded the university to give him a personal chair. He had a large collection of French antique furniture, bought and restored by himself over many years, with which he eventually retired to Spain (Prudence Raphael, personal communication, 2003).

16. Parker, W., Raphael, R. A. and Wilkinson, D. I. (1958), 'The structure and synthesis of bullatenone', *J. Chem. Soc.*, pp. 3871–5.

17. McCrindle, R., Overton, K. H. and Raphael, R. A. (1960), 'A stereospecific total synthesis of D-(–)-shikimic acid', *J. Chem. Soc.*, pp. 1560–5.

18. Mills, R. W., Murray, R. D. H. and Raphael, R. A. (1973), 'A new stereo-selective synthesis of *trans*-chrysanthemic acid [2,2-dimethyl-3-(2-methylprop-1-enyl)cyclopropanecarboxylic acid]', *J. Chem. Soc., Perkin Trans. I*, pp. 133–7.

19. Colvin, E. W., Malchenko, S., Raphael, R. A. and Roberts, J. S. (1973), 'Total synthesis of (±)-trichodermin', *J. Chem. Soc., Perkin Trans. I*, pp. 1989–97.

20. Colvin, E. W., Purcell, T. A. and Raphael, R. A. (1976), 'Total synthesis of the macrocyclic antibiotic (±)-pyrenophorin', *J. Chem. Soc., Perkin Trans. I*, pp. 1718–22.

21. Sanders, J. K. M. (2004), 'Ralph Alexander Raphael', *New Dictionary of National Biography,* Oxford: Oxford University Press (in press).

22. MacAlpine, G. A., Raphael, R. A., Shaw, A., Taylor, A. W. and Wild, H.-J. (1976), 'Synthesis of the germination stimulant (±)-strigol', *J. Chem. Soc., Perkin Trans. I*, pp. 410–16.

23. Becker, D., Hughes, L. R. and Raphael, R. A. (1977), 'Total synthesis of the antileukaemic lignan (±)-steganacin', *J. Chem. Soc., Perkin Trans. I*, pp. 1674–81.

24. Jackson, R. F. W., Raphael, R. A., Stibbard, J. H. A. and Tidbury, R. C. (1984), 'Formal total synthesis of (±)-pseudomonic acids from dihydropyran', *J. Chem. Soc., Perkin Trans. I*, pp. 2159–64.

25. Hill, M. L. and Raphael, R. A. (1990), 'Total synthesis of the antiviral (±)-virantmycin', *Tetrahedron* **46**, pp. 4587–94.

26. Andrew, R. G. and Raphael, R. A. (1987), 'A new total synthesis of aaptamine', *Tetrahedron* **43**, pp. 4803–16.
27. Hughes, I., Nolan, W. P. and Raphael, R. A. (1990), 'Synthesis of the indolo[2,3-*a*]carbazole natural products staurosporinone and arcyriaflavin B', *J. Chem. Soc., Perkin Trans. I*, pp. 2475–80.
28. On his arrival in Cambridge, Ralph found the undergraduates curiously polite – they did not interrupt his lectures as his Glasgow students had done (Prudence Raphael, personal communication, 2003).
29. Barlow, C. Y., Edwards, P. P., Millward, G. R., Raphael, R. A. and Rubio, D. J. (1988), 'Wood treatments used in Cremonese instruments', *Nature* **332**, p. 313.
30. Ralph also assisted Nest Rubio, the wife of David Rubio, with her work on the recreation of lost carpets, by giving her technical advice on traditional dyestuffs.
31. For further accounts of the life and work of Ralph Raphael, see Williams, D. (1999), 'Ralph Alexander Raphael', *J. Chem. Soc., Perkin Trans. I*, pp. 835–7; Crombie (2000); Sanders (2004).

11

Discovering the wonders of how Nature builds its molecules

Alan Battersby

Department of Chemistry, University of Cambridge

> *Oats, peas, beans and barley grow,*
> *Oats, peas, beans and barley grow.*
> *Do you or I or anyone know*
> *How oats, peas, beans and barley grow?*

The ability of living things to construct the molecules they need in order to grow and exist seems almost magical. As a plant develops, it builds chlorophyll, the green pigment in its leaves essential for photosynthesis; gorgeous colours are produced for the flowers and scents are generated, attractive to insects and to ourselves. How do they do it? The same question can be asked about micro-organisms or animals; for example, how do we build the red pigment in our blood that carries oxygen to our muscles? Even as an undergraduate, I was fascinated by these problems, but the means were not then available to allow them to be studied experimentally. Nevertheless, I hoped eventually to be able to carry out research on what became known as biosynthesis and what follows in this chapter will show that research on biosynthesis was at the centre of the work of my team in Cambridge. It is interesting for me (Figure 11.1) to look back at the sequence of events that led to that final outcome. All parts of this account, including those where molecular structures, spectroscopy and synthesis are to the forefront, are written for the non-specialist audience, so explanations of the less obvious terms, concepts and techniques will be given to help.

The 1702 Chair of Chemistry at Cambridge: Transformation and Change, ed. Mary D. Archer and Christopher D. Haley. Published by Cambridge University Press. © Cambridge University Press 2005.

Alan Battersby

Figure 11.1. Alan Battersby in 1995.

Early days to the 1950s: Manchester, St Andrews and Bristol

I knew from an early age that I wanted to be a chemist. As a boy of around 10 years of age, I started to read the books in my father's library. He was a builder by trade but he was widely read and remarkably his collection of books included some about chemistry. These were the ones that most caught my interest because they were full of directions for doing experiments such as the generation of hydrogen from zinc and acid. It was not long before my pocket money was being saved up to buy test tubes, flasks, a precious Bunsen burner, a tripod and similar treasures. Also, since at that time regulations concerning safety had not gone completely over the top, it was possible to buy zinc, sulphur, copper sulphate, hydrochloric and nitric acids. Everything was assembled to allow many different experiments and I was hooked.

Fortunately, my chemistry teacher, Mr Evans, at Leigh Grammar School that I attended was quite superb. So the strong interest in chemistry arising from my 'laboratory' at home was encouraged and nurtured. By the time entrance to a university was being considered, my clear goal was to become a professional chemist. Fortunately, I won a scholarship from my county of Lancashire, an essential step for otherwise my family could not have provided the necessary funds. Financial practicalities led to my entering the local university, the University of Manchester. This proved to be an excellent choice because the professor of organic chemistry there was then the brilliant Alexander Todd, later to become the twelfth 1702 Professor at Cambridge. The department at Manchester was an exciting place to be. It was buzzing with activity and, as an undergraduate, I could watch Alex Todd engaged with his large research team in the pioneering work on the building blocks of nucleic acids and his syntheses of the nicotinamide and related enzymic cofactors.[1] It was inspirational. Also, as part of the undergraduate teaching, I attended lectures by Alex on heterocyclic chemistry and the co-enzymes; so the fourteenth 1702 Professor was taught at one stage by the twelfth! Indeed, I was still in the midst of my undergraduate course when Alex Todd was invited to the 1702 Chair and I remained in Manchester to graduate.

My research supervisor in Manchester was Hal Openshaw, an outstanding young chemist whose interests included the chemistry of emetine, a basic substance extracted from a plant. A member of the group of natural organic bases called alkaloids, emetine is used for the treatment of amoebic infections. Its molecular structure was unknown and my task was to solve it. None of the spectroscopic methods such as infrared or nuclear magnetic resonance spectroscopy or mass spectrometry were available at that time, only a manual ultraviolet spectrometer. The classical approach of chemical degradation was the only one we could use. Organic chemists in the twenty-first century find it hard to imagine how a relatively complex structure could be worked out by these methods without spectroscopic tools. The classical approach involved the substance of interest being chemically manipulated or, more graphically, being tortured by degradative methods to squeeze out structural information. So the substance would be oxidised, reduced, pyrolysed with various additions, treated with acids and bases both mildly and vigorously and the products were purified and examined. In the case of emetine, some more controlled methods to open up the molecule were used before oxidative degradation. As a result, various fragments representing different parts of the original emetine molecule were isolated and when the structures of these simpler molecules had been worked out, they had to be fitted together like pieces of a jigsaw puzzle. Many other experiments were needed to cross check the information gained from the fragments and, eventually, after three years of full-time research, the structure of emetine was solved.[2] Roughly 100 grammes of emetine had been consumed in this work. The contrast with the situation today is impressive; modern tools would allow the structure of emetine to be determined in three days at most using about 10 milligrammes of recoverable material (365 times faster using 10 000 times less material).

During the work on emetine, Openshaw was appointed to a readership at the University of St Andrews and I moved with him from Manchester. So the structural research on emetine was completed in Scotland and shortly thereafter (1948) I gained my first academic post of assistant lecturer in chemistry at St Andrews. For my first teaching assignment, Professor John Read (author of *Humour and Humanism in Chemistry*) asked me to give the course on *inorganic* chemistry to the medical students! It turned out to be fun, and having been dropped in at the deep end and survived, subsequent teaching assignments seemed straightforward to me. My eight years in St Andrews gave me a great fondness for that university and for Scotland as a whole, especially for its mountains and wild places, and that fondness holds to this day.

Part way through the period at St Andrews, I had the marvellous opportunity to study in the United States as a Commonwealth Fund Fellow (1950–2). The first year was spent with Lyman Craig at what was then the Rockefeller Institute for Medical Research in New York City. Here my research focused on the peptide antibiotics tyrocidine[3] and gramicidin S,[4] projects that gave me experience of separating mixtures of closely related water-soluble materials. This experience was later to prove invaluable to me. But just as important was the large lunchroom at Rockefeller where all the research scientists mingled in a random way; the great names and the beginners (like me) were thoroughly mixed together. So one rubbed shoulders with Moore and Stein (Nobel Laureates for structural work on proteins), Kunitz (a pioneer in the isolation of pure enzymes) and Granick (another pioneer working on the biosynthesis of chlorophyll). What struck me with particular force was that many of the biologists and medical research workers were aiming to solve complex chemical problems. Surely a chemist should have some advantage, so I resolved for the future to use my training and experience as an organic chemist to tackle problems in the biological area. It was one of a small number of crucial decisions that moved my work forward in a highly beneficial direction. This decision was reinforced as a result of my joining Herbert Carter to work on pyruvate oxidation factor in the biochemistry department of the University of Illinois during the second period of my stay in the USA.

Research on alkaloids at Bristol and Liverpool

On returning from the USA in 1952, my first task was to start building a research team and soon three students were working with me on my first independent research projects. We were not to be much longer in St Andrews for in 1954 we moved to the University of Bristol, where I joined Professor Wilson Baker's department as lecturer in chemistry. This was an excellent move to a department fizzing with activity and ideas; moreover, my group rapidly doubled in size so we packed some punch.

Now my earlier fascination with the mysteries of biosynthesis, the decision to move into biological areas together with my experience of the plant alkaloids all blended together. We would take up the challenge of elucidating the biosynthetic pathways to the major families of alkaloids; then nothing was known of this. The timing was perfect since the radioactive isotope of carbon, carbon-14, had recently become commercially available, initially as simple starting materials such as labelled carbonate, cyanide and acetate. This provided a tool to allow the fate of molecules carrying a ^{14}C-atom (^{14}C-labelled molecules) to be determined after they had been introduced in some way into a living system. Carbon-14 is a low energy β-emitter and it was to be introduced in very small amounts (tracer levels) so with proper care about containment, it was perfectly safe to use. At last it was possible to initiate direct experiments on plants and we started these in 1955. Our first target was morphine (the molecule labelled as 3 in Figure 11.2), produced by the opium poppy and of long-standing medical value as a powerful analgesic. The experimental planning was strongly influenced by the remarkable speculations of Sir Robert Robinson, based on his recognition of structural relations among natural products. He suggested that

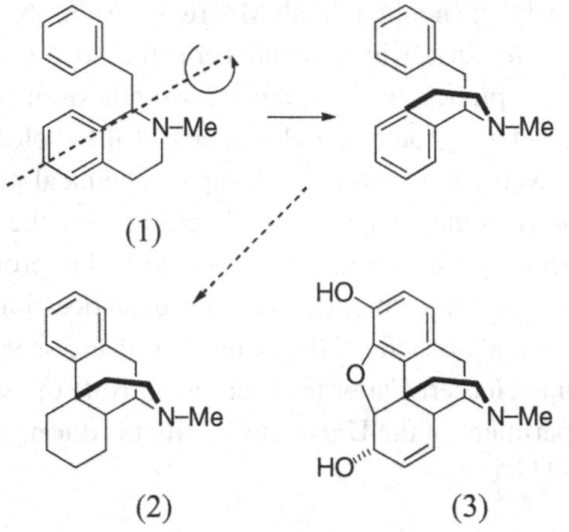

Figure 11.2. Possible scheme for derivation of morphine.

the morphine skeleton is derived from some unknown 1-benzylisoquinoline system (the molecule labelled as 1) by rotation as indicated in Figure 11.2, followed by oxidative coupling of the aromatic rings in some way to generate the morphine skeleton (2). (This system is illustrated in its simplest form without any substituents to show how the carbon skeleton might be generated.)

It was a brilliant, simple idea. Our plan was to synthesise a 1-benzylisoquinoline, building in one ^{14}C-atom at a single specific site (so-called, singly labelled) and the aromatic rings were to carry what on mechanistic grounds should be the appropriate phenolic hydroxyl groups. Note the importance of *synthesis*. The precursors for the work on morphine and for the many subsequent researches had to be built by methods that allowed the specific introduction of one ^{14}C-atom (or later, more than one), usually starting from ^{14}C-cyanide, ^{14}C-carbonate or ^{14}C-methyl iodide. The labelled 1-benzylisoquinoline had then to be introduced into the biosynthetic machinery of the living plant. It is fortunate that at the outset we were not over-aware of the difficulties of working with living plants; our biological colleagues were not optimistic about our plans! But sometimes a little ignorance of all the hurdles is an advantage so we took the simple approach and injected a solution of the ^{14}C-labelled 1-benzylisoquinoline into the seed capsules of opium poppies.

Gaining permission from the Home Office to hold supplies of morphine and especially to grow substantial numbers of opium poppies required careful negotiation. Once obtained, however, the injected plants were allowed to grow for some time before the morphine was isolated. To our delight it carried ample ^{14}C-radioactivity to take forward to the next experiments. Now it was

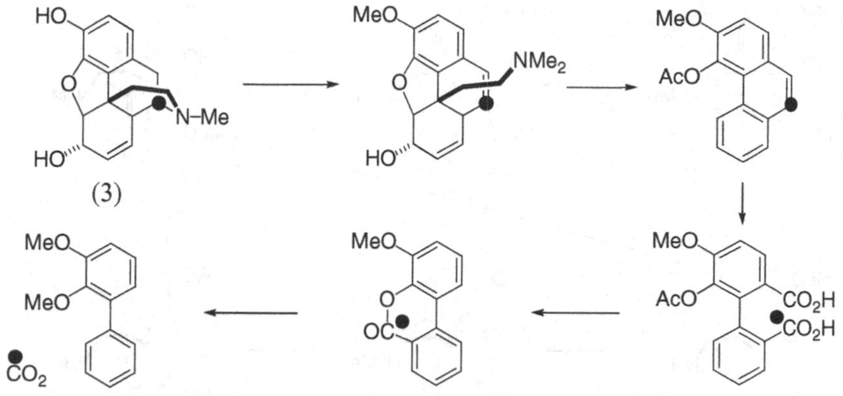

Figure 11.3. Degradative sequence for checking labelled morphine.

essential to prove that the isolated morphine carried its ^{14}C-label solely at the one carbon matching the corresponding site in the 1-benzylisoquinoline used as precursor. Without this evidence, it could be argued that the original labelled precursor had been degraded in the plant and the resulting small fragments then built into morphine in an uninformative way to scatter ^{14}C-atoms around the molecule. So specific degradative methods had to be designed to pluck out unambiguously that one carbon for radio-assay. The same problem of specific degradation also faced us for all the other alkaloids we studied; this work was often as time-consuming as the original synthesis of the ^{14}C-labelled precursors. Our degradative sequence for labelled morphine (molecule 3) is shown in Figure 11.3. The black dot shows where the ^{14}C-label should have been had the labelled 1-benzylisoquinoline been specifically incorporated into the morphine molecule. The results from the controlled degradation proved that this was indeed so.

This success was the crucial first step that opened the way to many other similar studies probing further along the pathway. These eventually led to reticuline (4) being pin-pointed[5] as the actual 1-benzylisoquinoline system that undergoes the oxidative coupling step, shown in Figure 11.4. The product was proved to be salutaridine (5), which is reduced to the dienol (6) ready for the illustrated ring closure to yield thebaine (7); thebaine occurs alongside morphine in the opium poppy. The final steps were shown to involve the formation of codeinone (8), then codeine (9) (familiar in codeine-aspirin) before the end product, morphine (3), is reached.[6]

The elucidation of the biosynthetic pathway to the opium alkaloids had a dramatic effect on our confidence. Beforehand, we dreamed that such a discovery might be possible and afterwards we knew it could be done. The

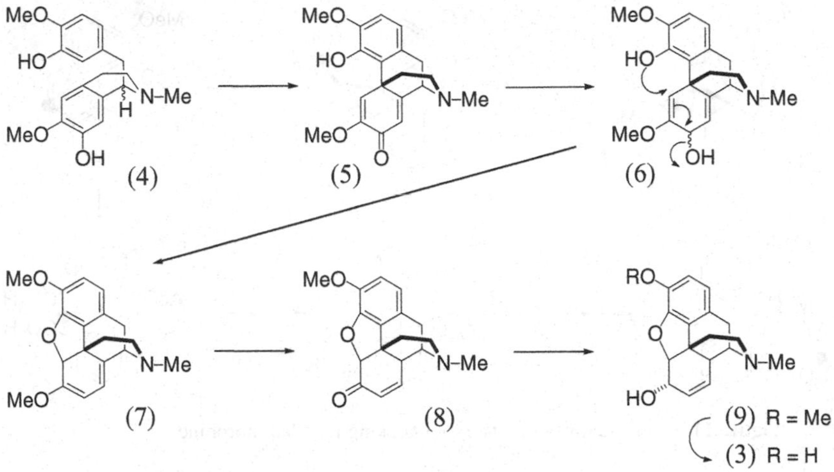

Figure 11.4. Precursors in morphine biosynthesis.

stage was set for adventures with the other major families of plant alkaloids and as a result of many years of research, we were able to show how all the important ones are biosynthesised. Representatives of some of these families are shown in Figure 11.5, including colchicine (10), the antimitotic agent which is biosynthesised in a most surprising way,[7] narcotine,[8] valuable for its antitussive activity, and berberine,[9] familiar to any gardener as the brilliant yellow pigment visible on damaging the roots of a *Berberis* plant. Then there was emetine[10] (11), discussed at the outset of this chapter, lycorine,[11] which is partly responsible for the poisonous nature of daffodil bulbs, and quinine[12] (12), the well-known antimalarial drug. There are over a thousand alkaloids based on indole, the most famous being the deadly poison strychnine. Three members of this huge family that we studied – corynantheal (13), vindoline (14) and catharanthine (15) – are shown in Figure 11.5.[13] The skeletons of the latter two alkaloids (14) and (15) are found joined together in the alkaloids vincristine and vinblastine from *Vinca rosea*. Their interest and value stems from being effective agents for the treatment of childhood leukaemia and our understanding of how the systems (14) and (15) are built reveals, in general terms, how the anti-cancer drugs are biosynthesised.

Of the many fascinating features of the natural pathways to the alkaloids in Figure 11.5, there is only space to refer to two. Firstly, the ten-carbon part of the indole alkaloids (13), (14) and (15), and also of emetine (11), that is picked out in bold black was unexpectedly found to be derived from geraniol (16) or nerol (17), members of the terpene family. These C_{10}-alcohols interconvert in

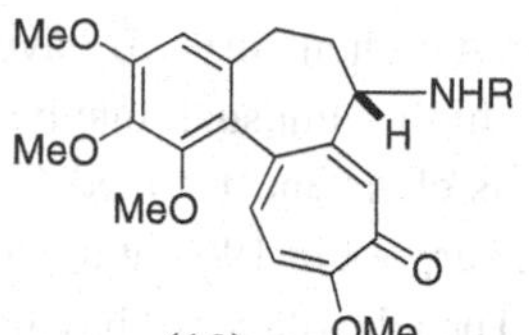

Figure 11.5. Some of the alkaloids investigated by Battersby's group.

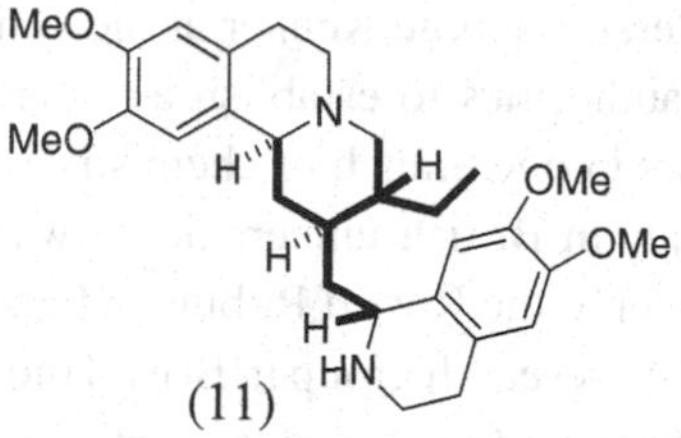

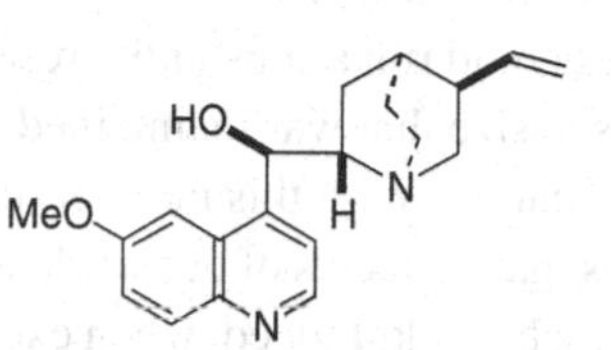

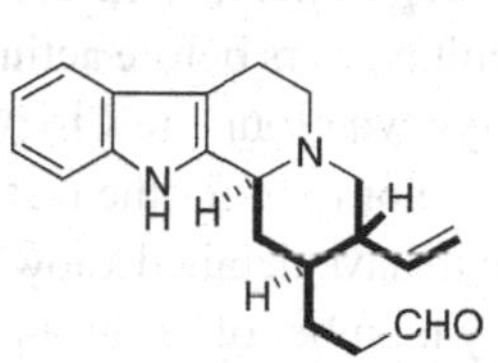

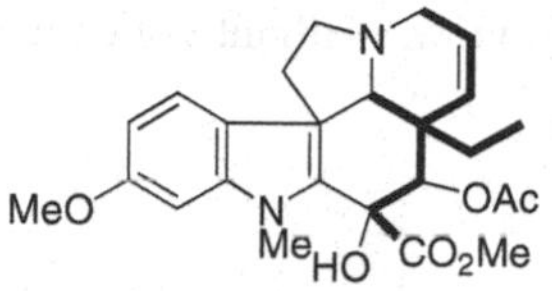

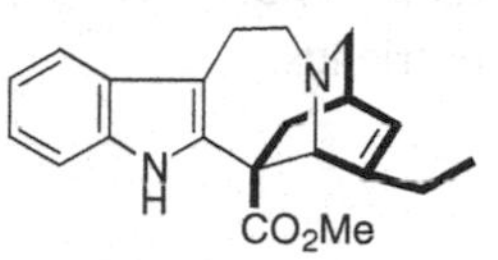

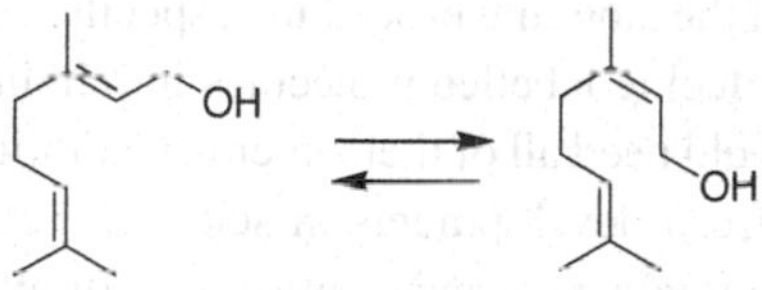

the plant and they are built by a quite separate pathway from the indole portion of the alkaloids. This terpene pathway leads to a huge variety of plant products including, for example, camphor and lavender oil, together with the brilliant colours of tomato and peppers. Nature moulds the available clay and can use intermediates on separate pathways to make totally different final products. Secondly and surprisingly, quinine (12) is formed by extensive modification of corynantheal (13), even though at first sight they appear to be unrelated.

The research on morphine (3) and on some of the other alkaloids collected in Figure 11.5 was started in Bristol but the major effort required to solve these biosynthetic puzzles was carried out by us in another university.

Professor George Kenner at the University of Liverpool was able to convince the authorities to establish a second chair of organic chemistry. To have two chairs in one branch of chemistry was at the time highly unusual, maybe even unique in British universities. I was invited to this chair and I joined George Kenner at the Robert Robinson Laboratories in Liverpool in 1962 at the age of 37. My research group in Bristol moved with me. The advantages of having two professors of equal status rapidly became clear. We worked as a team in looking after the department, so spreading the load, and we could talk over confidential or difficult matters before acting and each of us could run a substantial research team. My own group grew to twenty members, a size that was maintained, give or take one or two, for the rest of my research career. With this number, it was possible to have detailed knowledge of all the separate facets of our work and to see every member of the group several times each week. Indeed, when exciting results were popping out, we would be in discussion at the bench several times a day. It was a highly productive period. Much depended on my having George Kenner as a wonderfully supportive and, in my view, brilliant colleague; we were firm friends.

A new direction and new location: Cambridge

By 1967–8, many of the labelling procedures, specific degradative methods and techniques for introducing labelled molecules into living systems had all been finely honed; we would need all of them later in Cambridge. I say 'many' rather than 'all' because fresh developments in science, then just appearing on the horizon, would offer totally new and exciting opportunities, as I will describe in the next section. Other advances, especially in genetics and molecular biology, were still to burst onto the scene and were to have a major impact on our work in the 1980s.

It was clear during 1967 that the natural routes to all the major families of plant alkaloids had been mapped out by work in our laboratory and several others. The time had come to select another major problem; that choice is one of the most crucial a scientist makes. Surveying the literature describing the then available knowledge of the biosynthesis of the major classes of natural materials, it was clear that the class known as the tetrapyrroles (explained later) should be selected as our next target. This was because they are of central importance in metabolism and also little was known about large sections of the pathways by which they are built. So there were great opportunities for breaking new ground. Having made the firm decision to move into the tetrapyrrole field, I made a start with one postdoctoral colleague in 1968.

Another crucial decision had soon to be taken. In 1968, I was approached by Lord Todd (known to his friends as Alex Todd or Alec Todd), then the holder of the 1702 Chair. He asked me whether I would be interested in joining him in Cambridge as the holder of a second chair of organic chemistry that he was keen to make available. Alex was the driving force behind gaining the necessary funding for this second chair from Roche Products, the British arm of the Swiss pharmaceutical giant Hoffmann–La Roche. During the earlier part of Alex's career, he had been a consultant for Roche and it was his brilliant and highly practical route for the synthesis of the vitamin thiamine that Roche subsequently used commercially. As a result, the company coffers were appreciably enriched and it seems probable that Alex was able to call on this indebtedness. Certainly the funds were provided by Roche, the second chair of organic chemistry was set up in Cambridge and I was invited to it. I suppose deep down I knew that my answer would finally be 'yes' but the decision to accept was not as straightforward as many would imagine. The work of my research team was rolling rapidly forward in the splendid new Robert Robinson Laboratories in Liverpool, where the facilities and spectroscopic equipment could not be matched by any other chemistry department in the UK. James Baddiley has described in Chapter 9 the gulf that existed at his time between the laboratories and equipment in Manchester and those in Cambridge. The change that would face my research team in moving from Liverpool to Cambridge was nowhere near as stark; however, there is no doubt that the Cambridge scientific facilities and equipment were substantially inferior to those in Liverpool. But this was a challenge to be overcome and the opportunity to work in Cambridge greatly excited me. I accepted the second chair in 1969 and it is a decision I have never regretted. This chance also attracted my research team so that essentially everyone moved with me.

Alex Todd managed, in what seemed to me an almost magical way, to fund a lectureship for my colleague in Liverpool, Dr (now Professor) Jim Staunton. I very much hoped that Jim would move with me and was delighted when he agreed to do so. He and I had collaborated in our research for seven years in Liverpool and, in parallel, Jim built a vibrant independent group studying the biosynthesis of polyketides. This family of natural products includes the antibiotics erythromycin and the tetracyclines, and Jim's work in this field was to come to fruition in a spectacular way in Cambridge; but that is another story.

One must picture the arrival in Cambridge in 1969 of around twenty chemists from Liverpool all eager to get into action. Some of the more laid-back Cambridge research students had not seen anything quite like the intensity or hours of work of the new arrivals, who were given the sobriquet 'The Northern Hordes'.

An essential early step was to improve the equipment for the whole laboratory including such basic items as rotary evaporators, reliable vacuum pumps, and more ultraviolet and infrared spectrometers. Alex had tucked away some funds to allow this equipment to be bought, more funding came from the University with my chair and my research was supported by generous grants that allowed other items to be acquired. Within about one year, all was in place to allow us to work as efficiently and effectively as previously.

It is worth comparing my move as a Ph.D. student, from Manchester to St Andrews, in the late 1940s with the later one from Liverpool to Cambridge. For the latter move it was critically important, as has just been outlined, to assemble a wide range of equipment – not only the simpler items mentioned above, but also the seriously expensive gear such as mass spectrometers and nuclear magnetic resonance instruments, of which some were already in Cambridge, but more were needed. Only by having the full kit of modern equipment was it possible to solve the problems being studied at that time. The contrast with the former move is striking. Then, we simply carried with us our collection of research samples. One only needed simple glassware such as flasks, condensers and separating funnels and the like to work quite happily. Chemistry had changed out of all recognition, over the 20-year period, from a relatively inexpensive research area to one where a large investment was necessary.

The foregoing paragraphs have given a feeling for how the move to Cambridge affected our research team. How was it for me? I was arriving as a relatively young man to join the most powerful and distinguished professor in the country. It was, I admit, a somewhat daunting prospect. But there need not have been any concern; Alex was a marvellous colleague and friend. He gave me massive support and wise advice. During our discussions in 1968 about my possible move to Cambridge, he put to me his characteristically clear vision of how we should go forward together. In his ringing Scottish accent he said, 'Look, Alan, I will run the Organic Department and its interactions with Inorganic and Theoretical Chemistry, deal with the General and Faculty Boards and carry out the financial negotiations with the Treasurer. Your job is to push ahead with the chemistry, build a powerful research team and aim high.' Again characteristically, he looked ahead to his own retirement and he hoped the electors would share his wish, when that time came, to move me to the 1702 Chair, upon which the second chair would lapse.

At that stage, Alex's retirement was roughly six years away and my feeling was that by then my research team would be firmly established, and running smoothly. Further, I would have become familiar with the intricacies of the working of the university committees and administration in order to get things done. So, as in Liverpool, the tandem arrangement with two professors, at any rate for a good number of years, was a very practical and attractive one. It strongly affected my thinking about how to organise the transition of our research effort from the plant alkaloids to the tetrapyrroles. The choice was either to make the change as quickly as possible or do it gradually over many years. The problem can be appreciated when one realises that we knew the chemistry of the alkaloids inside out but almost nothing of the new area of chemistry that we were thinking of jumping into rather rapidly. In these circumstances, the team leader is initially in the same boat as his Ph.D. and postdoctoral students; everyone has to learn together. It is a humbling experience! But with around six years of unencumbered research before us, the decision was to live dangerously and go for the speedy transition. The effort and numbers of research workers studying tetrapyrroles built up rapidly until this became our major focus.

Though the final statement in the previous paragraph remained true, the original plan was frustrated by a very sad and unfortunate development that could not have been foreseen. In 1971, Alex Todd suffered a serious heart attack that caused him to resign from the 1702 Chair on medical advice. So only a little over one year in Cambridge, we were face-to-face with the possibility of having only one professor of organic chemistry, a situation that had not been expected to arise for six years. Alex was keen that I should move to the 1702 Chair to succeed him. However, if I did this, my original chair would then lapse since it had been established for only one tenure. I found myself in great difficulty – pulled on the one hand to follow Alex's wishes, especially as he had been so supportive, and on the other by my conviction that reverting to having just the one chair at that stage would be the wrong decision. To go that way would have involved my taking on the whole administration, as Alex had done, at a very early stage of my experience in Cambridge, while simultaneously leading a large research group still not fully equipped and striving to enter a new field of chemistry where I had much to learn. To try to retain two chairs seemed by far the best option, especially after my experience in Liverpool had shown the huge advantages of having two professors. It was that way forward that I proposed to the board of electors for the 1702 Chair, of which I was a member. The proposal was not well received and I was pressed hard by the other electors to accept the 1702 Chair despite my arguments against it. The meeting was adjourned for thought and discussion. When it reconvened a week or two later, my proposal was accepted: Ralph Raphael was elected to the 1702 Chair from October 1972 while I continued to hold the other chair. It is the mark of a great man that, with the matter settled, Alex accepted what had been done and backed us both very strongly.

As had happened at Liverpool with George Kenner and myself, Ralph and I pulled together as a wonderful team. All important decisions were made after joint discussion and I cannot recall anything of real significance having been put in place that left one or the other of us unhappy. We apportioned many tasks; for example, I carried responsibility for all matters concerning the technical staff, for some postgraduate teaching and for seeking joint research funding from the granting bodies, while Ralph handled the general running of the department – certainly a greater administrative load than mine. Many jobs were done jointly, such as allocating research studentships to different sections of the whole department, conducting many oral examinations (especially of research students), and negotiating finance with the University Treasurer. It all worked extremely well. Looking back now to the period from the 1970s through to my retirement in 1992, seeing how the department flourished and how the research output set Cambridge organic chemistry at the forefront, I am wholly confident that the decision to fight for the maintenance of two chairs was absolutely right. When Ralph retired in 1988, I was elected to the 1702 Chair and so became the fourteenth holder. Happily a very generous endowment was provided at this point by Herchel Smith to maintain the second chair in perpetuity from 1988. It was named the Herchel Smith Chair of Organic Chemistry, and Cambridge was fortunate to be able to elect Alan Fersht as the first holder. In 1990, the 1702 Chair was re-endowed by a contribution of £1.5 million from British Petroleum, with the agreement that its title be changed to the BP 1702 Professorship of Organic Chemistry. The highly satisfactory outcome of preserving two chairs in 1972 followed by these two endowments has been that two established chairs are available in organic chemistry in Cambridge now and for the future.

Biosynthesis of the Pigments of Life

I was hugely excited about the new field of research that my group was taking up as we moved to the Cambridge laboratories (Figure 11.6 comes from a somewhat later period, but the excitement remained with us). Haem (18), chlorophyll (19) and vitamin B_{12} (20), shown in Figure 11.7, are of fundamental importance in living creatures and plants and this has led to their being known as the *Pigments of Life*. Other important members are the cytochromes, responsible for electron transport, and phytochromes, vital in controlling the growth of plants. Though these substances carry out completely different functions, they have closely related structures; the remarkable reason for this is that they are all biosynthesised from the *same parent molecule*, uroporphyrinogen III (23),

Figure 11.6. Alan Battersby discussing with his student, Simon Bartholomew, their synthesis of a molecule that mimics one oxygen-binding site of haemoglobin (1982).

generally shortened to uro'gen III, the structure of which is shown in Figure 11.8. As can be seen, uro'gen III (23) is a macrocycle constructed from four pyrrole rings connected together by one-carbon bridges.

Thus, we have another example of Nature building very different end products from one starting material in a highly economical way. For the three Pigments of Life being considered here, the building process goes forward step by step and then the first branching occurs, leading to vitamin B_{12} (20), with the other track proceeding ahead until a second splitting occurs to give chlorophyll (19) from one branch and haem (18) from the other. One can see the exact points of the chemical evolutionary process.

The central role for uro'gen III (23) in the biosynthesis of the Pigments of Life picked it out as the key first target for our studies. The knowledge that was available[14] at the outset is shown in Figure 11.8. Briefly, two molecules of 5-aminolaevulinic acid (21) are enzymically condensed to form porphobilinogen (22) (PBG), then two enzymes, hydroxymethylbilane synthase (often called deaminase) and uro'gen III synthase (shortened to cosynthetase) cooperate in some way, then unknown, to convert four molecules of PBG (22) into uro'gen III (23). These final stages present an intriguing problem because the uro'gen III

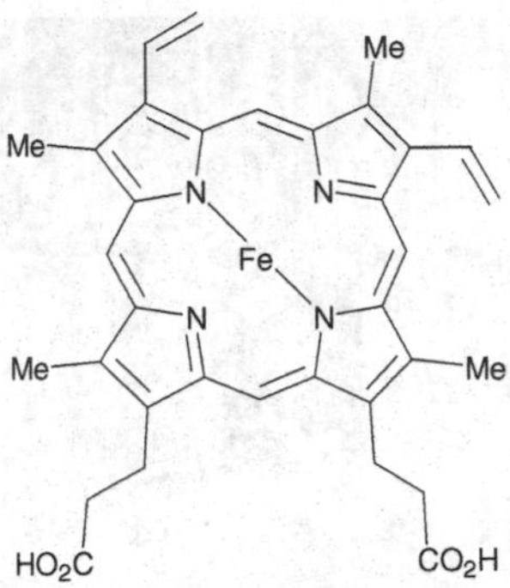

Figure 11.7. 'Pigments of Life' haem (18), chlorophyll (19) and vitamin B₁₂ (20).

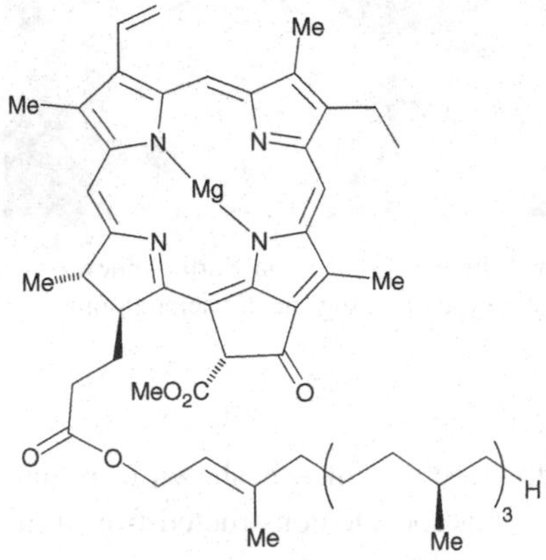

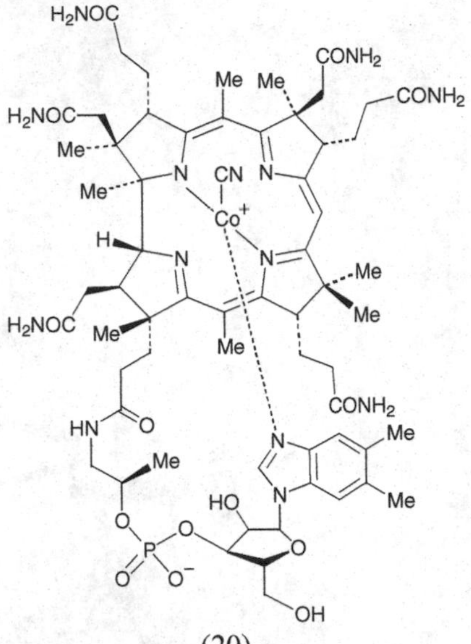

structure (23) is unexpected. The acetate and propionate groups on ring D are reversed relative to these substituents on the other rings. The expected product from straightforward joining of four PBG molecules (22) would have these groups running in sequence around the macrocycle. How is this rearrangement carried out and what roles do deaminase and cosynthetase play in this whole process? We recognised, some decades after the work collected in Figure 11.8, that the key to further progress was to use labelling with the stable isotope of carbon, carbon-13, combined with detection by nuclear magnetic resonance (NMR). Carbon-13 yields a signal in an appropriate NMR spectrometer and the position of that signal on the scale is governed by the environment of the ^{13}C-atom in the molecule of interest. So it is possible to determine from the

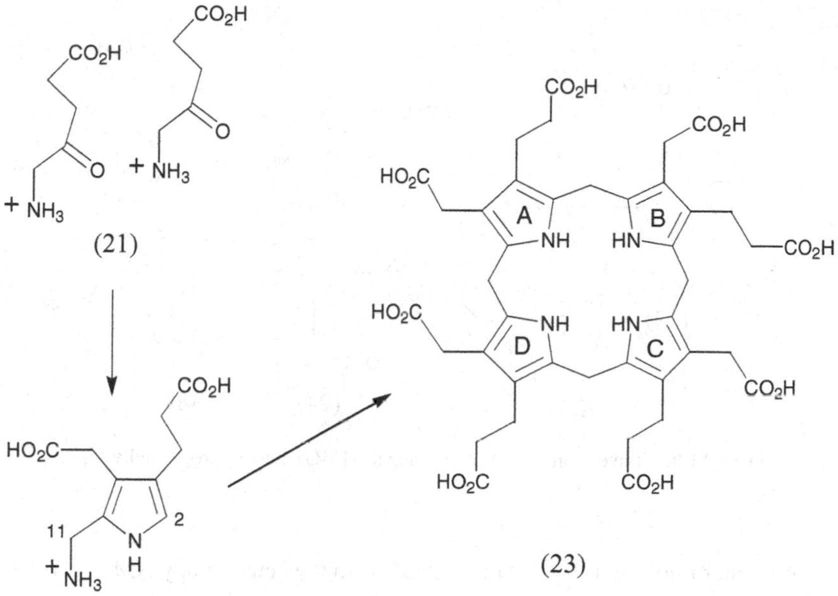

Figure 11.8. Uroporphyrinogen III (23) and its precursors.

NMR spectrum where the ^{13}C-atom is sited. Moreover, multiple ^{13}C-labels can be used (up to 11 labels being used for the work on B_{12}) and their NMR signals, observed in a single spectrum, give valuable information about the environment of all of them.

What a marvellous contrast from the ^{14}C-labelling used as described earlier for the alkaloids, where the labelled sites had to be located by controlled degradation. We were so convinced that the ^{13}C-approach was the only way to tackle the uro'gen III problem, and later also the B_{12} problem, that we initiated the synthetic work to produce the necessary ^{13}C-labelled precursors before we even had a ^{13}C-NMR spectrometer; fortunately, the funds to buy one were found in time for these precursors to be used without delay.

It is outside the scope of this chapter to give a full account of the many labelling studies that we carried out; in any event, they have been reviewed elsewhere.[14,15] But the results from our studies of uro'gen III (23) were a joy to see. They showed several things: firstly, that only one rearrangement step occurs; secondly, that this rearrangement occurs *after* the assembly of an unrearranged linear open-chain tetrapyrrole (called a bilane); and thirdly, that the rearrangement is *intramolecular*.

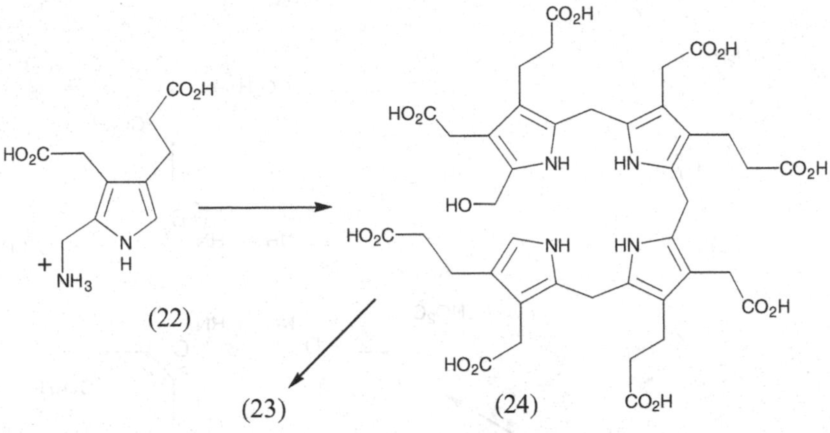

Figure 11.9. Conversion of porphobilinogen (PBG) into hydroxymethylbilane (HMB).

A further important development came from the detection by NMR in Texas A&M University[16] and by kinetic studies in Cambridge[17] of a labile intermediate that was formed when deaminase acted alone on PBG (22). This was shown in Cambridge[18] to be the hydroxymethylbilane, HMB (24) which acted as the substrate for cosynthetase[16,18] generating uro'gen III (23), as shown in Figure 11.9. This settled the exact identity of the bilane referred to above. Deaminase was thus shown to be the assembly enzyme that builds HMB (24), while cosynthetase is the ring-closing and rearranging enzyme.

Another surprise followed. Deaminase was shown to use a novel dipyrrolic cofactor (25) bound to the protein on which to build HMB (24).[19] The building process starts from a dipyrrole (25) and progresses through tri-, tetra- and pentapyrroles until a linear unrearranged hexapyrrole (26) has been formed, as illustrated in Figure 11.10. Then HMB (24) is cleaved off, leaving the original dipyrrole (25) still attached to the protein ready for a further cycle.

We also made extensive studies of the rearrangement process required to generate uro'gen III (23) from HMB (24) and strong interlocking evidence was gained for the delightfully simple mechanism shown in Figure 11.11; this work is available in collected form.[20]

One contrast with the earlier work on alkaloids was mentioned above. There was another: rather than having to use whole living plants, we could now work with purified deaminase and cosynthetase isolated initially from duck's blood and later from the green alga *Euglena gracilis*. We were no longer dependent on growing seasons and, even more important, the isolated enzymes allowed very high conversions of precursor into product to be achieved.

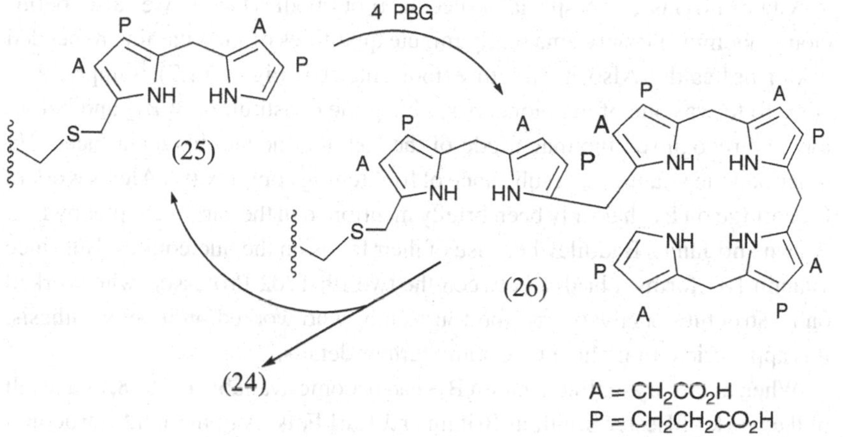

Figure 11.10. Mechanism of hydroxymethylbilane (HMB) formation.

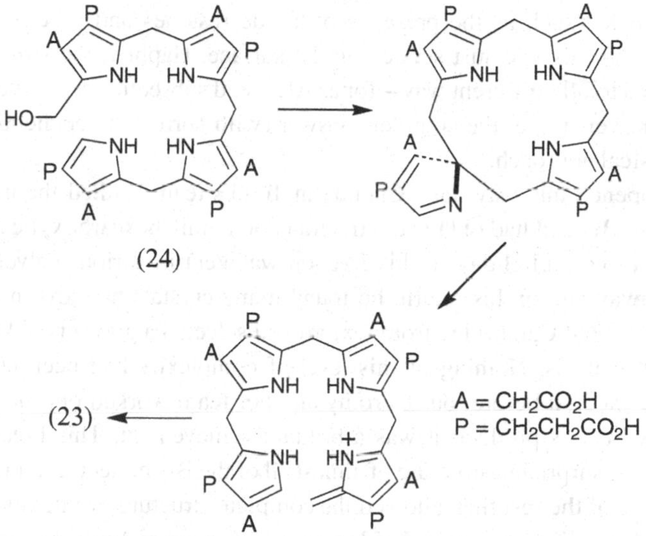

Figure 11.11. Spiro-mechanism for uro'gen III formation.

The foregoing outline of our research on uro'gen III (23) gives just a taste of the whole meal. Many more studies were carried out on the pathways going forward from uro'gen III (23) to haem (18) and chlorophyll (19). But the story so far will have given the reader a feeling for how we tackled those problems; that is sufficient and leads on to vitamin B_{12}. Separate literature references will not be given in this next section since all are available in two reviews.[14,21]

Vitamin B_{12} is a very special molecule. If our bodies lack it, we suffer pernicious anaemia and only amazingly minute quantities of the vitamin are needed to keep us healthy. Also, its structure (molecule 20 in Figure 11.7), is impressive. Alex Todd was one of the pioneers studying the constitution of B_{12} and before long, he recognised the magnitude of the task that he and his team faced. He described the vitamin as 'a substance of frightening complexity'. Alex's work in Cambridge on B_{12} has only been briefly mentioned in the earlier chapter by Dan Brown and James Baddiley because of their focus on the nucleotides. But since vitamin B_{12} forms a bridge between the twelfth 1702 Professor, who worked on its structural analysis, and the fourteenth, who worked on its biosynthesis, it is appropriate to include here some further details.

When at last crystalline vitamin B_{12} had become available in 1948, as a result of the work of Lester Smith in Britain and Karl Folkers in the USA, structural studies started. The classical degradative approach discussed earlier was again the only one available. Various fragments were isolated that indicated B_{12} was a member of the tetrapyrrole family. In addition, a few other structural features were revealed such as the presence of amide residues and a benzimidazole nucleus. But this was just scratching the surface. Happily, the structure was solved in a totally different way – for as Alex said subsequently, 'It would have taken for ever' to get the complete answer (with correct stereochemistry), by this classical approach.

It happened this way. Jack Cannon in Todd's team studied the hydrolysis of vitamin B_{12} and had obtained a fraction containing hexcarboxylic acids. He was due to go on holiday so this fraction was kept in various solvents while he was away and on his return, he found many crystals had grown. Dorothy Hodgkin visited Cambridge from Oxford and selected a good crystal for study by x-ray analysis. Nothing of this level of complexity had been tackled by the x-ray method before, but Dorothy and her team worked out the structure of the hydrolysis product; it was a brilliant achievement. This breakthrough revealed the surprising structure of almost all of the B_{12} molecule, and available knowledge of the rest then allowed the complete structure of vitamin B_{12} (20) to be deduced. The date was 1953, a mere seven years since the isolation of pure vitamin B_{12} had been announced.

The foregoing extremely brief outline covers only a fraction of the massive effort in many laboratories on the structure, chemistry and mechanism of action of vitamin B_{12} and the interested reader will find a recent review rewarding reading.[14]

It would not have been realistically possible to tackle the problem of B_{12} biosynthesis directly after the structure determination in 1953. Only when ^{13}C-labelling coupled with NMR came onto the scene in the sixties was it possible even to consider trying to solve such a dauntingly complex puzzle; I have often referred to it as the Everest of biosynthetic problems. It is not practicable to describe fully how the biosynthesis of B_{12} was eventually solved; this would take a whole chapter on its own. Rather, I would like to bring out how a broad interdisciplinary approach was needed, involving the synergistic combination of synthetic and structural chemistry, isotopic labelling, spectroscopy, enzymology, molecular biology and genetics and, of course, lots and lots of hard work.

Vitamin B_{12} differs from the other biosynthetic problems discussed so far. For the latter, helpful clues could be recognised from their structural relations to other natural products. For B_{12}, all one could see was the characteristic arrangement of the acidic side chains around the macrocycle suggesting derivation from uro'gen III (23), and this proved to be true. But the pathway leading from one to the other was like a 'black box'. At least seven methyl groups had been introduced into uro'gen III (23), the macrocycle had been contracted and the side chain at C–12 had been decarboxylated. These many steps, together with a few others, such as cobalt insertion, could be strung together in an enormous number of possible ways. The problem was to discover the one main pathway living systems use. Because of the nature of this book, the emphasis will be on the work in Cambridge together with that of the French team with whom we had the good fortune to collaborate. The important contributions of Duilio Arigoni, Vladimir Bykhovsky, Gerhard Müller, Ian Scott and David Shemin are covered in reviews that give leading references.[14,21]

Working with B_{12}-producing bacteria, *Propionibacterium shermanii*, we and others were able to isolate three early intermediates and determine their structures. One arose by monomethylation of uro'gen III (23), another by dimethylation and the third, having the structure (27) shown in Figure 11.12, by trimethylation. Of course, the chase started to find intermediates carrying four or more methyl groups. Despite enormous efforts over several years, none could be detected; my feeling at the time was that we had been sent to the Siberian salt mines. Something new was needed to renew forward progress and the exciting developments in genetics and molecular biology gave that fresh impetus. We had the great good fortune to be able to take part in the amazing surge forward that then resulted.

Our effort on genetics and molecular biology in Cambridge was relatively small, so when the opportunity came to make a scientific connection with a group of French biologists who were carrying out pioneering research on B_{12} using these two disciplines, I jumped at the chance. The main focus of the group in Paris was to use the power of genetics to locate the genes that were essential for the biosynthesis of vitamin B_{12} in their organism, *Pseudomonas denitrificans*.

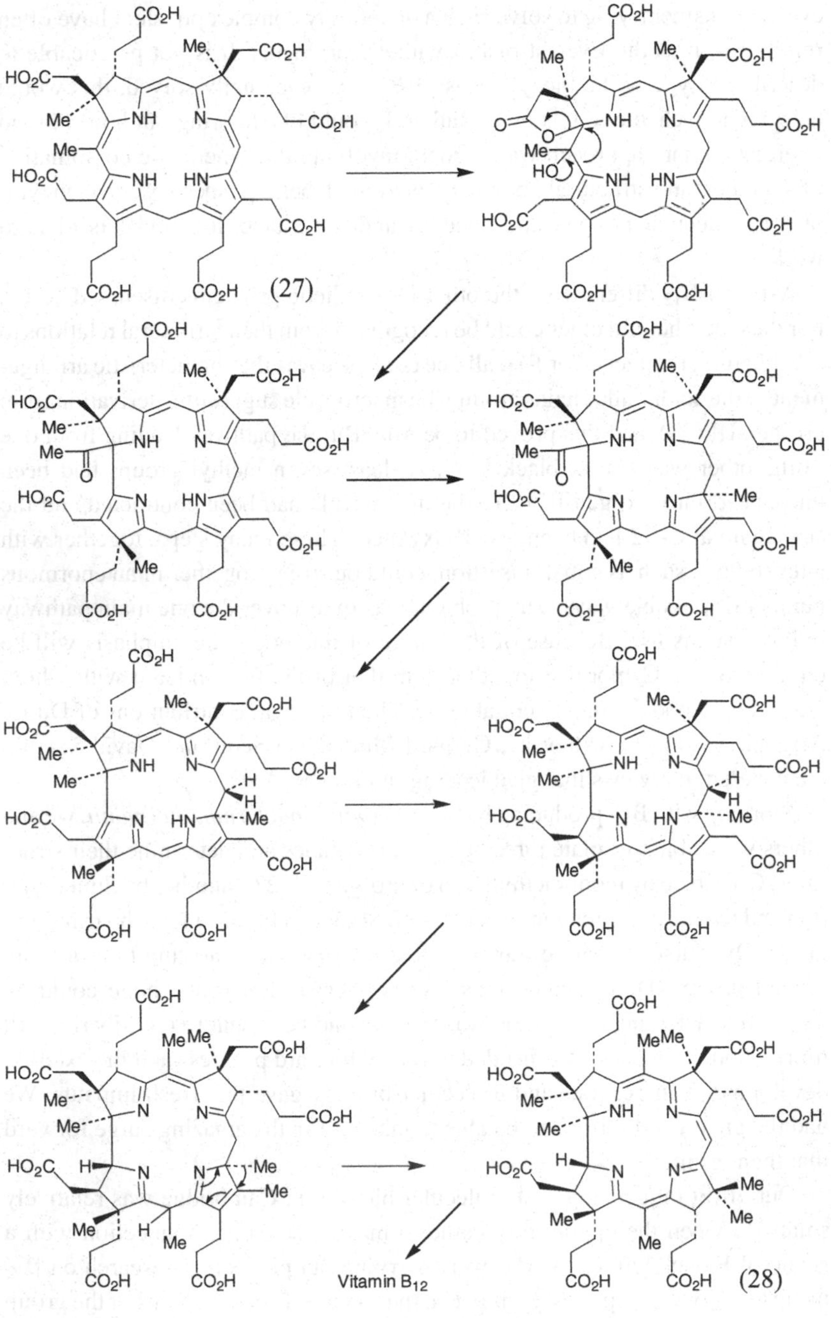

Figure 11.12. The biosynthetic pathway to vitamin B_{12}.

They had available a highly productive strain developed for the commercial production of vitamin B_{12} at the Rhône-Poulenc Rorer company. The senior scientists in Paris were Francis Blanche, Laurent Debussche and Denis Thibaut, who were responsible for the biochemistry, and Joel Crouzet, responsible for the genetics and molecular biology. Their names are given advisedly because their outstanding work was a decisive element in the solution of the B_{12} problem.

When the French group had pinpointed the necessary genes for the biosynthesis of B_{12}, these genes could be expressed to afford the corresponding enzymes, which were then used to try to find new intermediates. This approach turned the one everyone had previously used upside down. Now the enzymes were sought out first and the intermediates afterwards, rather than the other way round. It turned out that seven enzymes were needed to convert the trimethylated intermediate (27) mentioned earlier into the complete fully methylated macrocycle (28) for building vitamin B_{12}, as shown in Figure 11.12. This intermediate (28) only needed less mysterious modifications of the side chains to produce the vitamin itself (20). The elucidation of the critical and fascinating part of the biosynthetic pathway to vitamin B_{12} by which (27) is converted finally into (28) – the 'black box' – thus depended on discovering what these enzymes do. Further, the structures of all their products had to be determined; only then could the various enzymes be assigned to specific biosynthetic conversions. Initially, one did not know the function of any of the seven enzymes. It was this structural chemistry that our Cambridge group contributed to the collaborative effort. We had to design new ways for structure determination involving multiple ^{13}C-labelling combined with NMR to solve the structures of complex substances that were available in only tiny amounts, normally less than one milligramme. It is fair to say that this work was also an essential part of the joint research effort, in that it set the full structures of the intermediates on the biosynthetic pathway and allowed the function of the various enzymes to be understood. Each research group needed the other and this outstanding collaboration led to dramatic advances.

It is difficult in cold print to convey the intense excitement of this period of research. Labelled precursors, labelled products and ^{13}C-NMR spectra were shuttling between Paris and Cambridge and the latest results were being exchanged by fax or phone. This was a wonderfully satisfying time.

The outcome was that the logical approach described in simplified form above led to the entire biosynthetic pathway to vitamin B_{12} (20) being established.[14,21] This is shown in Figure 11.12, which acts as a good finale to this chapter. No organic chemist can fail to be impressed by the route Nature uses to fashion the complex B_{12} molecule (20) drawing on relatively simple chemistry. Moreover, each step sets up the required reactivity for the next one. It is fantastic. For those who are not organic chemists, there is also I believe real pleasure to be had from Figure 11.12 by seeing how the methyl (Me) groups gradually accumulate and how the molecules change in shape and complexity before finally the summit of 'Everest' (vitamin B_{12}) is reached. The approach used to solve the problem of B_{12} biosynthesis stands as an exemplar for any future biosynthetic research on a complex molecule.

To have been involved in this great adventure with our French colleagues, part of it during my tenure of the 1702 Chair, gave me much satisfaction. But at least as great has been the pleasure of working with outstanding young Ph.D. and post-doctoral students in Cambridge and then watching them develop as independent scientists ready for their own achievements, not a few of these being quite outstanding. By 1992, the time had come for the 1702 Chair to pass to Steve Ley so that he could continue the forward thrust for organic chemistry in Cambridge as the first holder of the BP 1702 Professorship of Organic Chemistry.

Notes and References

1. Enzymes are proteins of substantial molecular weight that act as catalysts, so speeding up and controlling chemical transformations in living systems. The mechanism of action of some enzymes requires that they work in partnership with a specific smaller molecule called a co-enzyme or cofactor.
2. Battersby, A. R. and Openshaw, H. T. (1949), 'Studies of the structure of emetine. Part IV. Elucidation of the structure of emetine', *J. Chem. Soc.*, pp. 3207–13.
3. Battersby, A. R. and Craig, L. C. (1952a), 'The chemistry of tyrocidine. Part I. Isolation and characterisation of a single peptide', *J. Amer. Chem. Soc.* **74**, pp. 4019–23; Battersby, A. R. and Craig, L. C. (1952b), 'The chemistry of tyrocidine. Part II. Molecular weight studies', *J. Amer. Chem. Soc.* **74**, pp. 4023–7.
4. Battersby, A. R. and Craig, L. C. (1951), 'The molecular weight determination of polypeptides', *J. Amer. Chem. Soc.* **73**, pp. 1887–8.
5. Battersby, A. R., Binks, R., Francis, R. J., McCaldin, D. J. and Ramuz, H. (1964), 'Alkaloid biosynthesis. Part IV. 1-Benzylisoquinolines as precursors of thebaine, codeine and morphine', *J. Chem. Soc.*, pp. 3600–10.
6. Battersby, A. R., Foulkes, D. M. and Binks, R. (1965), 'Alkaloid biosynthesis. Part VIII. Use of optically active precursors for investigations on the biosynthesis of morphine alkaloids', *J. Chem. Soc.*, pp. 3323–32; Barton, D. H. R., Battersby, A. R., Kirby, G. W., Steglich, W., Thomas, G. M., Dobson, T. A. and Ramuz, H. (1965), 'Alkaloid biosynthesis. Part VII. Investigations on the biosynthesis of morphine', *J. Chem. Soc.*, pp. 2423–38; Battersby, A. R., Brochmann-Hanssen, E. and Martin, J. A. (1967), 'Alkaloid biosynthesis. Part X. Terminal steps in biosynthesis of the morphine alkaloids', *J. Chem. Soc. C*, pp. 1785–8; Battersby, A. R., Foulkes, D. M., Hirst, M., Parry, G. V. and Staunton, J. (1968), 'Alkaloid biosynthesis. Part XI. Studies related to the formation and oxidation of reticuline in morphine biosynthesis', *J. Chem. Soc. C*, pp. 210–16.

7. Battersby, A. R., Herbert, R. B., Pijewska, L., Santavy, F. and Sedmera, P. (1972), 'Alkaloid biosynthesis. Part XVII. The structure and chemistry of androcymbine', *J. Chem. Soc., Perkin Trans. II*, pp. 1736–40; Battersby, A. R., Herbert, R. B., McDonald, E., Ramage, R. and (the late) Clements, J. H. (1972), 'Alkaloid biosynthesis. Part XVIII. Biosynthesis of colchicine from the 1-phenethylisoquinoline system', *J. Chem. Soc., Perkin Trans. I*, pp. 1741–6.
8. Battersby, A. R., Hirst, M., McCaldin D. J., Southgate, R. and Staunton, J. (1968), 'Alkaloid biosynthesis. Part XII. The biosynthesis of narcotine', *J. Chem. Soc. C*, pp. 2163–72.
9. Battersby, A. R., Francis, R. J., Hirst, M. and Staunton, J. (1963), 'Biosynthesis of the "berberine bridge"', *Proc. Chem. Soc.*, p. 268.
10. Battersby, A. R. and Gregory, B. (1968), 'Biosynthesis of the ipecac alkaloids and of ipecoside: a cleaved cyclopentane monoterpene', *J. Chem. Soc., Chem. Commun.*, pp. 134–5.
11. Battersby, A. R., Binks, R., Breuer, S. W., Fales, H. M., Wildman, W. C. and Highet, R. J. (1964), 'Alkaloid biosynthesis. Part III. *Amaryllidaceae* alkaloids: the biosynthesis of lycorine and its relatives', *J. Chem. Soc.*, pp. 1595–609.
12. Battersby, A. R. and Parry, R. J. (1971), 'Biosynthesis of the *Cinchona* alkaloids: middle stages of the pathway', *J. Chem. Soc., Chem. Commun.*, pp. 31–2.
13. Battersby, A. R. (1967), 'Biosynthesis of the indole and colchicum alkaloids', *Pure Appl. Chem.* **14**, p. 117; Battersby, A. R., Burnett, A. R. and Parsons, P. G. (1969), 'Alkaloid biosynthesis. Part XIV. Secologanin, its conversion into ipecoside and its role as a biological precursor of the indole alkaloid', *J. Chem. Soc. C*, pp. 1187–92.
14. Battersby, A. R. (2000), 'Tetrapyrroles: the Pigments of Life. A millennium review', *Nat. Prod. Rep.* **17**, pp. 507–26.
15. Battersby, A. R. and McDonald, E. (1979), 'Origin of the Pigments of Life: the type-III problem in porphyrin biosynthesis', *Accounts Chem. Res.* **12**, pp. 14–22.
16. Burton, G., Fagerness, P. E., Hosozawa, S., Jordan, P. M. and Scott, A. I. (1979), '[13]C N.M.R. evidence for a new intermediate, pre-uroporphyrinogen, in the enzymic transformation of porphobilinogen into uroporphyrinogens I and III', *J. Chem. Soc., Chem. Commun.*, pp. 202–4; Scott, A. I., Burton, G., Jordan, P. M., Matsumoto, H., Fagerness, P. E. and Pryde, L. M. (1980), 'N.M.R. spectroscopy as a probe for the study of enzyme-catalysed reactions. Further observation of pre-uroporphyrinogen, a substrate for uroporphyrinogen III cosynthetase', *J. Chem. Soc., Chem. Commun.*, 1980, pp. 384–7.
17. Battersby, A. R., Fookes, C. J. R., Matcham, G. W. J., McDonald, E. and Gustafson-Potter, K. E. (1979), 'Biosynthesis of natural porphyrins: experiments on the ring-closure steps and with the hydroxy-analogue of porphobilinogen', *J. Chem. Soc., Chem. Commun.*, pp. 316–19; Battersby, A. R., Fookes, C. J. R., Gustafson-Potter, K. E., McDonald, E. and Matcham, G. W. J. (1982a), 'Biosynthesis of porphyrins and related macrocycles. Part 17', *J. Chem. Soc., Perkin Trans. I*, pp. 2413–26.
18. Battersby, A. R., Fookes, C. J. R., Gustafson-Potter, K. E., McDonald, E. and Matcham, G. W. J. (1982b), 'Biosynthesis of porphyrins and related macrocycles. Part 18', *J. Chem. Soc., Perkin Trans. I*, pp. 2427–44.
19. Battersby, A. R. and Leeper, F. J. (1990), 'Biosynthesis of the Pigments of Life: mechanistic studies on the conversion of porphobilinogen to uroporphyrinogen III', *Chem. Rev.* **90**, pp. 1261–74; Battersby, A. R., Hart, G. J., Miller, A. D. and Leeper, F. J. (1987), 'Biosynthesis of the natural porphyrins: proof that hydroxymethylbilane synthase (porphobilinogen deaminase) uses a novel binding group in its catalytic action', *J. Chem. Soc., Chem. Commun.*, pp. 1762–5; Jordan, P. M. and Warren,

M. J. (1987), 'Evidence for a dipyrromethane cofactor at the catalytic site of *E. coli* porphobilinogen deaminase', *FEBS Lett.* **225**, pp. 87–92.

20. Battersby, A. R., Spivey, A. C., Capretta, A., Frampton, C. S. and Leeper, F. J. (1996), 'Biosynthesis of porphyrins and related macrocycles. Part 45', *J. Chem. Soc., Perkin Trans. 1*, pp. 2091–102.

21. Blanche, F., Cameron, B., Crouzet, J., Debussche, D., Thibaut, D., Vuilhorgne M., Leeper, F. J. and Battersby, A. R. (1995), 'Vitamin B_{12}: how the problem of its biosynthesis was solved', *Angew. Chem.* **107**, pp. 421–52; reprinted as *Angew. Chem. Int. Ed. Engl.*, 1995, **34**, pp. 383–411.

12

Chemistry in a changing world: new tools for the modern molecule maker

Steven Ley

Department of Chemistry, University of Cambridge

The Chair of Chemistry at Cambridge was established in 1702 and has been occupied without a break since then, save for a brief interregnum in World War II. I was appointed in 1992 as the fifteenth professor on which this honour has been bestowed (Figures 12.1 and 12.2), and the first to hold it under the title of BP 1702 Professor, and it falls to me to celebrate the tercentenary of the chair. This is an appropriate time to reflect on the rich history of the past three hundred years of chemistry at Cambridge, to take stock of where we now stand, and to speculate about the future.

Although chemistry has changed remarkably during the last three hundred years, there are some interesting coincidences. For example, John Francis Vigani, the first 1702 professor, was a fellow of Trinity College and I too am a fellow of Trinity. Vigani was interested in medical substances that are similarly part of our interests too, and, like Vigani, I have currently served nearly eleven years in post – but there the similarities end.

The dissimilarities are more obvious, no more so than when we make a comparison between laboratories then and now. Yet, while the chemical laboratory of three hundred years ago must have been a very unsafe environment in which to work, certainly when compared with modern standards, great discoveries were to be made. What came out of them over the centuries enriches our lives today. Indeed, the chemical sciences owe so much to the early pioneers, whose inquisitive minds and quest for knowledge of mechanisms at atomic and molecular levels, underpin everything we do.

The world is changing at a remarkable rate and it is wise to look at the impact that chemical synthesis – that is, the making of new molecules – has had on our daily lives. Everything around us involves chemistry, and many of the things

The 1702 Chair of Chemistry at Cambridge: Transformation and Change, ed. Mary D. Archer and Christopher D. Haley. Published by Cambridge University Press. © Cambridge University Press 2005.

Figure 12.1. Steven Ley, current BP 1702 Professor of Organic Chemistry.

we use require the skills of the synthesis chemist. Figure 12.3, for example, illustrates some of these materials, devices and products that we often take for granted. Nevertheless, we have them because of our knowledge of chemistry. Although this is undoubtedly true, it is the interaction and collaboration of chemistry with the other sciences, and also with engineering, which really creates the value.

Figure 12.2. Steven Ley in the company of three previous holders of the 1702 Chair at St Catharine's College, *c.* 1991. From left to right: Ralph Raphael, Alexander Todd, Alan Battersby and Steven Ley.

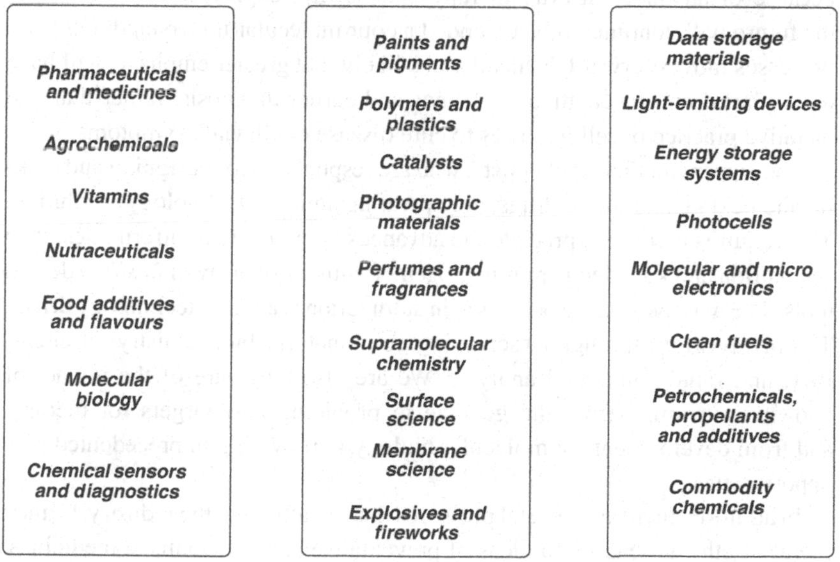

Pharmaceuticals and medicines	*Paints and pigments*	*Data storage materials*
Agrochemicals	*Polymers and plastics*	*Light-emitting devices*
Vitamins	*Catalysts*	*Energy storage systems*
Nutraceuticals	*Photographic materials*	*Photocells*
Food additives and flavours	*Perfumes and fragrances*	*Molecular and micro electronics*
Molecular biology	*Supramolecular chemistry*	*Clean fuels*
Chemical sensors and diagnostics	*Surface science*	*Petrochemicals, propellants and additives*
	Membrane science	*Commodity chemicals*
	Explosives and fireworks	

Figure 12.3. Impact of chemical synthesis on our daily lives.

In spite of the phenomenal achievements of the past, we have reached a point when chemistry, and particularly chemical synthesis, not only has to meet the ever-increasing demands, challenges and opportunities of a modern society, but we have also to recognise our responsibilities and legacy to future generations. We must discover much cleaner, and strategically important reaction processes that lead to the assembly of molecules free from waste products, and which proceed with high atom efficiencies. The environment, and our planetary resources, cannot sustain current practices. We need greater speed, more understanding of molecular diversity, and improved process versatility, more than ever before. We must look beyond the simple molecule to supramolecular structures, and their assembly and applications. This must all be done with an awareness not just of the benefits of modern products, but also of the risks. A better reputation and image of our subject will have to be earned, and justified, if it is to be fully sustainable.

Rather than attempt to comment on the issues facing us from too wide a perspective of the chemical sciences and of the associated chemical industry, I will restrict my remarks to the methods of synthesis of the healing drugs of today. I will show how the methods and technologies we discover and invent in our research laboratories may impact on the future of synthesis in general, and especially on the pharmaceutical industry.

It is a sobering thought that nearly 40% of us in this country are alive today because of advances in drug therapy. Indeed, life expectancy now and into the future will continue to be extended as our molecular understanding of life processes moves forward. Ethically, it is likely that greater emphasis will have to be placed on preventative medicines and earlier diagnosis, rather than the lucrative practice of selling drugs to cure disease or alleviate symptoms.

The current medicinal chemists, who are responsible for designing and making the next generation of drugs, are operating amidst a technology revolution. This requires them to appreciate the advances being made in informatics, data acquisition and knowledge mining, together with a vast array of *in silico* design tools. The massive change occurring in automation is another technology driver. This involves high throughput screening, combinatorial biochemistry and chemistry, and rapid compound analysis. We are also all aware of the impact of knowledge coming from the genome in providing new targets for therapy, and from developments in molecular biology in providing unprecedented new opportunities.

In addition, there are societal pressures that are affecting the industry, be they moral or ethical, leading to ideas of preventative and personalised medicines. Life-style changes, and healthcare and work practices, are also impacting significantly. Yet, if we look at the pharmaceutical industry, it seems somewhat contradictory that despite its obvious success, we find it is spending more and more on research and development, with fewer and fewer drugs finding their way on to the market. The attrition rate of potentially useful drugs is far too high: typically only one in ten promising chemical entities gets through clinical trials. It also seems remarkable, given the tremendous advances in chemical synthesis, that for every kilogramme of drug made today we still generate 20–25 kg of waste products. Perhaps this is not too surprising when one realises that all of the top 30 drugs were complicated to prepare, involving an average of about twelve separate steps when they were first produced in medicinal chemistry research laboratories. It is noteworthy that 80% of these drugs are assembled by only fifteen different types of reaction process. There is therefore a huge demand for improvement, both in terms of speed and quality, for the production of new drugs.

At this point, I should explain what constitutes the synthesis process and the complex decision-making that is required even in the assembly of a single new molecule. First, one has to appreciate that every intermediate molecule is new, and not likely to have existed previously on the planet. In this respect, synthesis is therefore a very creative and inventive science. Even deciding what to make is a non-trivial problem, given the almost infinite number of ways that molecules and atoms can be brought together to provide new molecular architectures. Remember too, that they must eventually do the job for which they were designed; that is to say they must be formulated and remain stable as a pill or liquid that can be safely consumed. Then they must go to the correct site of action within the body, there to cure the specific illness – and with no side effects. Clearly to do all this is a phenomenally complex task, and one needs a large and talented group, which fortunately I have (Figure 12.4), to make headway.

The synthesis or the assembly of this very specific compound is equally demanding. First, one needs a plan of how to bring together commercially available building blocks, with control of molecular shape and function, using all one's accumulated knowledge of mechanism and chemical reactivity. Here one uses logic, but also it is at this point that innovation and creativity can have their most powerful influence on the outcome. Next, one has to decide on the reaction conditions that will have to be used to bring about the coupling of the chemical fragments. Then, before any one step is undertaken, there are many, many thousands of reagents, temperatures and solvents to choose between. Based upon one's experience of scale versus cost, and of the safety factors concerned, one has to make a judgement as to the best set of conditions to bring

Figure 12.4. Steve with his research group, October 2003.

about the specific transformation. This can be a daunting task: get it wrong and not only would you have no product, but you might cause a disaster such as an explosion!

Once a reaction is underway for the first time, we have to work out how correctly to monitor the process so as to be aware of reaction exotherms and times for the process to go to completion. This makes use of a vast array of analytical tools, knowledge of which is crucial for success. But even assuming all has progressed satisfactorily so far, the process is by no means over. One still has to isolate, safely and in pure form, the product of the synthesis process. This can be the most frustrating and time-consuming part of the whole process, and of course this is the stage at which all the waste and by-products need to be removed. This requires the synthesis chemist to have considerable experimental knowledge of all the tools, such as filtration, distillation, crystallisation, liquid extraction and chromatography, needed to achieve chemical separation. This is where the subject is not just a craft, but an art.

When one appreciates all these aspects, it is possible to understand why we still have such a long way to go to make molecules more efficiently. In

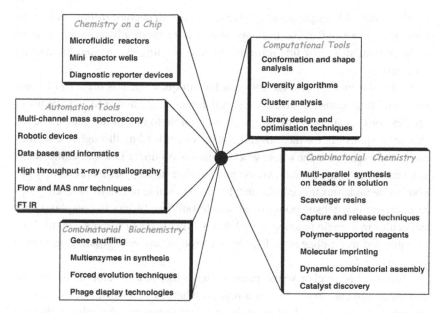

Figure 12.5. The molecule maker's new tools.

fact, we have only just scratched the surface of what might be possible in the future.[1] It is clear to see that for every carefully designed and prepared molecule, the processes are complicated using these conventional methods. If we are to increase the rate of successful discovery to meet future needs, we will require new thinking and new tools.

Fortunately some new tools, shown in Figure 12.5, have recently become available to the molecule maker, and these are changing the paradigm in which we work. For example, computational tools now help us predict the shape and properties of molecules. Nevertheless, it is still difficult to determine *a priori* the potential toxicity of a drug substance, or its solubility and absorption in a biological system. The computational tools do, however, give us a better handle to design compounds with wider structural diversity and can help in the more rapid optimisation of compounds as biological lead structures. We are seeing considerable developments in the miniaturisation of chemical processes sometimes referred to as lab-on-a-chip devices. Here we have the prospect of using microfluidics, or thin channel devices, for fast serial processing, in order to investigate a greater variety of reagents in a much shorter space of time.

We have also seen rapid development of automation tools to aid the synthesis chemist. In particular, robotic devices, multi-channel mass spectrometry, and advances in other areas of spectroscopy and microscopy are creating countless opportunities. Although we now have the ability to capture information more rapidly than we did in the past, we still require more advanced methods to make maximum use of this information, and to mine the created knowledge base better.

Molecular biology and its associated techniques, such as the area of combinatorial biochemistry, have also helped the molecule maker to provide novel species for evaluation. For example, it is possible to obtain the gene cassettes that are responsible for the biosynthetic pathways leading through to enzymes and proteins and on to the whole world of natural products. One can shuffle these gene cassettes to have them express and produce novel systems previously not obtainable under normal conditions. It is also possible to use isolated enzymes in multi-step synthesis, going from simple building blocks to complex target molecules in a single reaction vessel. Many other ways of forcing evolutionary changes and using phages to display new proteins are now possible, as shown in Figure 12.5.

Over the last twelve years, chemists have likewise developed the area of combinatorial chemistry. This is a process whereby new chemical entities are prepared in greater numbers by using parallel or fast serial methods, thereby overcoming some of the labour-intensive practices of the past. Our research group has been very active in developing new tools for this type of work. The chemical and pharmaceutical industries have devoted much of their effort in this area to using polymer beads to support substrates; after these have been subjected to chemical change, they are cleaved from the support to give a clean product. We have challenged this dogma, preferring instead to support the reagents (rather than the substrates) used in the chemical processes, which allows the spent reagents or catalysts to be removed simply by filtration, again leading to clean products.[2]

We believe these concepts offer a better practical solution to the problem of product isolation and, moreover, are more amenable to scale-up and more suited to multistep processes. The procedures can be readily automated and therefore easily run in parallel. Figures 12.6 and 12.7 show some of the procedures that use these immobilised reagents (●-Reagent) to provide clean products.

A number of special opportunities arise from using these systems. In the simplest case, one can consider the conversion of a substrate in solution, reacting with a polymer-supported reagent and leading directly to a clean product, obtained by just filtering away the spent reagent at the end of the process. What is important here is that, using modern analytical methods, formation of the product can be monitored in real time to provide information that can be used as an intelligent feedback mechanism to make the process self-optimising. Not only could this save considerable time for the synthesis chemist, but it offers

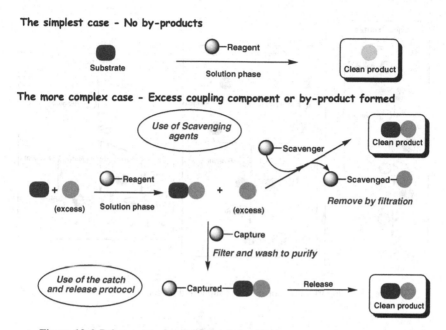

Figure 12.6. Polymer-supported reagents in synthesis.

new and exciting opportunities, given that the process can be automated with the robot itself evaluating and interrogating the informatics databases.

Chemistry is not always as straightforward as in this simple case, and more complex tasks may need to be performed. For example, it is more usual that two (or more) compounds need to be brought together and reacted by use of a coupling reagent. Also in order to guarantee that the reaction proceeds well (that is, gives a high yield), it is not uncommon to use one of these coupling components in excess, as shown in Figure 12.6. This causes a problem. Even though the coupled product is formed nicely, there is inevitably some excess component left over which must be removed. Normally this can be done using sophisticated, but operator-intensive procedures, such as distillation, crystallisation, or even expensive chromatography. These procedures are not easily carried out using robots or automation methods, especially under multi-parallel conditions.[3]

There are two techniques which can be applied to solve this problem. Either one can add to the mixture a polymer-supported scavenger or quenching agent that will combine selectively with only the excess component (or by-product) so that a clean coupled product can be recovered by the simple process of filtration. Alternatively one can add a polymer material that is designed to capture the coupled product, and leave behind the excess component or other by-products. The captured material is then filtered and simply washed free of impurities, and

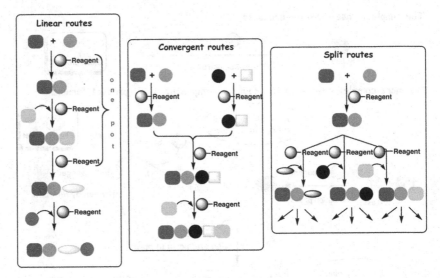

Figure 12.7. Opportunities for polymer-supported reagents.

can then be released to give a clean product. This is known as a catch-and-release protocol, and is also shown in Figure 12.6.

In Figure 12.7 we can see additional ways of using these systems for synthesis. First, long, linear, multi-step, syntheses can be performed, leading to complex products. Again, in solution, real-time analysis of product formation leads to rapid optimisation of the parameters necessary to complete the reaction scheme.

However, another important concept can come into play. It should be recognised that reagents that are immobilised on a polymeric support matrix are effectively site-isolated, that is to say, they cannot react with one another but only with the substrates. This means that we can consider combining steps (and reagents) into a single one-pot operation. Compare this with conventional chemical synthesis in which each step is conducted in a separate reaction vessel, inevitably leading to extensive water washes, work-ups and solvent changes. The new concept minimises all this, with obvious cost and environmental savings.

Next, one could use convergent synthesis routes, which are always considered to be more efficient than linear sequences. Here components are prepared in separate reaction channels and then combined to give chemically more elaborate structures.

Finally, polymer supports can be used to prepare even more diverse fragment molecule sets, by splitting the reaction products and diverting them to different

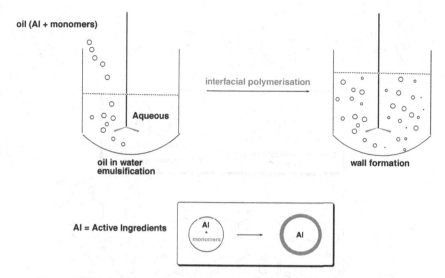

Figure 12.8. Microcapsule manufacture by interfacial polymerisation.

coupling steps again using immobilised reagents to effect the chemical elaboration. This process is especially important to industry because it rapidly affords new, proprietary, building blocks for synthesis that are not currently available from the catalogues of chemical suppliers.

Other benefits of immobilisation include the ability to affix toxic, or malodorous, species to the support, rendering them much safer, or more pleasant, to handle. They can also be used profitably when product properties, such as volatility or diastereoisomeric structures, cause problems. Because the work-up of the spent reagent is so simple, the opportunity to recycle the reagent makes commercial sense. Before illustrating how all these concepts can be operated with real chemical examples, I should mention one further technique we have used and developed, which employs a process of reagent microencapsulation using interfacial polymerisation,[4] as shown in Figure 12.8.

Here an oil containing an active ingredient (AI) reagent, together with precursor monomers, is added to a rapidly stirred aqueous medium to give an emulsion. Interfacial polymerisation of the monomers occurs to give a solid wall formation so as to entrap the active ingredient inside the polymer bead, effectively making a micro-reactor well. For this process we chose to use isocyanate monomers since these are known to react at a water interface to give polyureas, as shown in Figure 12.9.

Furthermore, the urea was expected to act as a chemical barrier to entrap species such as commonly used metal catalysts, but still be porous enough

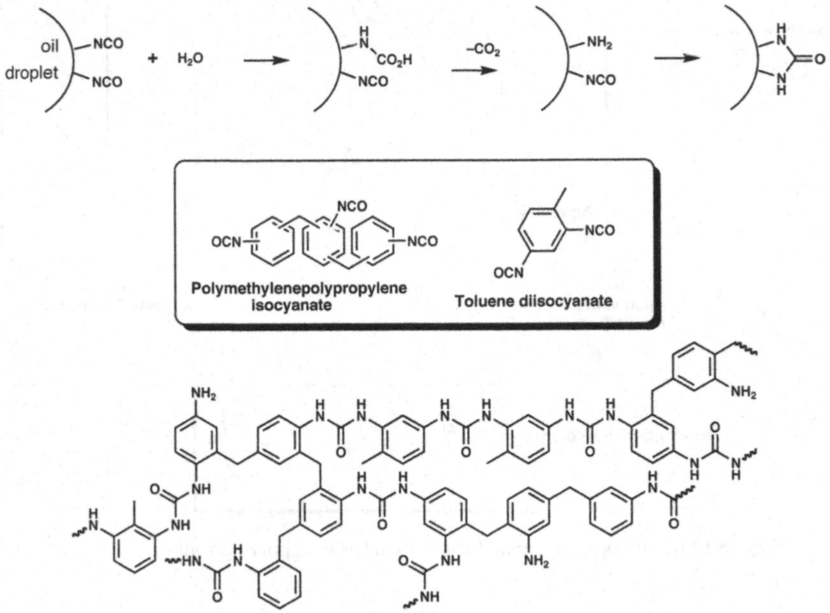

Figure 12.9. Polyurea microcapsules made by *in situ* interfacial polymerisation.

to allow chemical substrates to enter the particle, react, then exit as products. In this way one could expect to prepare compounds cleanly, with all the by-products and spent and expensive catalysts being recovered by simple filtration. These systems work, they are easy and cheap to prepare, and can be used to trap important metallic species such as palladium and osmium. The polymer beads that are produced are robust, flocculent in nature, and can be used in a variety of important chemistries.

The future of these systems is also exciting, in that they could be used in flow reactor systems, or in readily manipulated plugs. Moreover, encapsulation of other reagent species or enzymes, and even whole cells, can be envisaged. So far we have only investigated their use in palladium-catalysed cross-coupling reactions to make products of interest to the agrochemical and pharmaceutical industries. However, these methods could also be employed when high-purity products are demanded, such as in novel materials like those used in light-emitting devices. The encapsulated palladium catalysts are also especially useful in hydrogenation and transfer hydrogenation processes.[5]

Osmium tetroxide, when encapsulated with polyurea, also becomes an attractive and stable oxidant for alkenes. All these new catalysts can be readily recovered and reactivated, and can therefore be reused. In hydrogenation reactions,

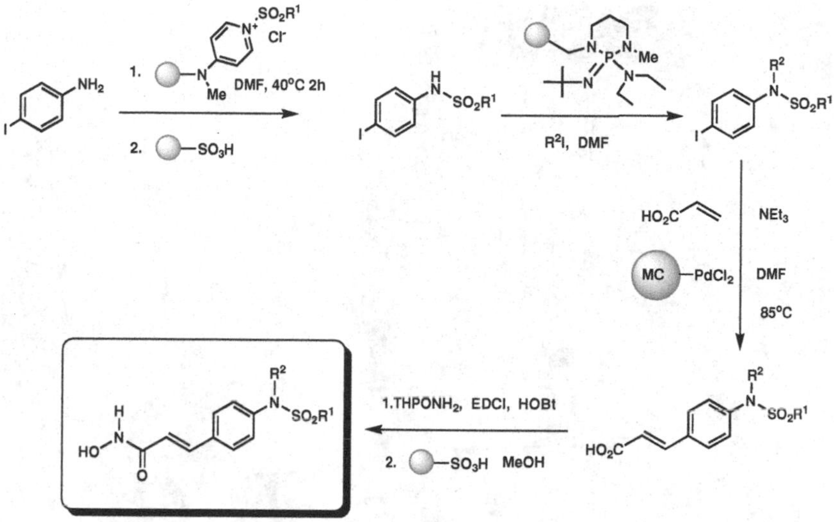

Figure 12.10. Synthesis of histone deacetylase inhibitors.

for example, we have recycled these species up to thirty times without serious degradation. In the future this will become especially important, as we need to have cleaner (greener) processes for the production of chemical entities.[6]

Having established this new set of immobilised reagents and scavengers as new tools for synthesis, we now need to apply these to real chemical examples to evaluate their effectiveness and potential advantages properly. In order to show how these supported reagents and scavenging agents can be applied in synthesis programmes, I want to discuss a few specific research examples. In the first of these, shown in Figure 12.10, a short route has been developed leading to histone deacetylase inhibitors for potential application as anti-tumour agents.[7]

Here the route allows the chemist to make variations at the R^1 and the R^2 functional positions of the molecule to create a collection of compounds, known as a chemical library. These can be biologically screened to determine the most active compound. In this scheme, six steps are used to make the final products, which are hydroxamic acids (containing the HO.NH.COR functional group), which are also known to be strong binders of metal atoms. For this reason, when metals such as palladium are used in their synthesis under traditional methods of compound preparation, the metal often contaminates the final product. Extensive purification is then necessary to obtain the clean product. In this new synthesis, all the key reagents and scavengers are bound to polymeric supports and thus cannot contaminate the resulting products. Especially important is the use of the microencapsulated palladium chloride reagent in step four of the

Figure 12.11. The Sophas Synthesiser: the new laboratory workbench.

process, to couple the fragments together in what is known as a Heck reaction. This reaction proceeds readily at 85 °C in dimethylformamide as the solvent and simple filtration gives a product substantially free from contaminating palladium metal. In fact the whole synthesis process was conducted using a Zinsser Sophas Synthesiser shown in Figure 12.11 as the robot laboratory workbench. By being able to carry out syntheses in this automated way, where reactions with different substituents are introduced by parallel reactions, the chemist is free to spend more time designing experiments and devising reactions, having relegated the more routine tasks to an instrument.

Figure 12.12 shows a more elaborate process, showing that even quite complex natural products, such as carpanone, which is not readily available from natural sources, can be prepared in just five stages from the commercially available building block sesamol.[8]

There are a number of subtle transformations taking place during this reaction scheme that are worth commenting upon. The first of these uses a highly hindered amine-supported base to conduct an alkylation reaction on the precursor phenolic oxygen atom. Next, this is heated to high temperature to effect a rearrangement reaction in which a new carbon–carbon bond is made. When chemists perform reactions under conventional heating, especially at 220 °C,

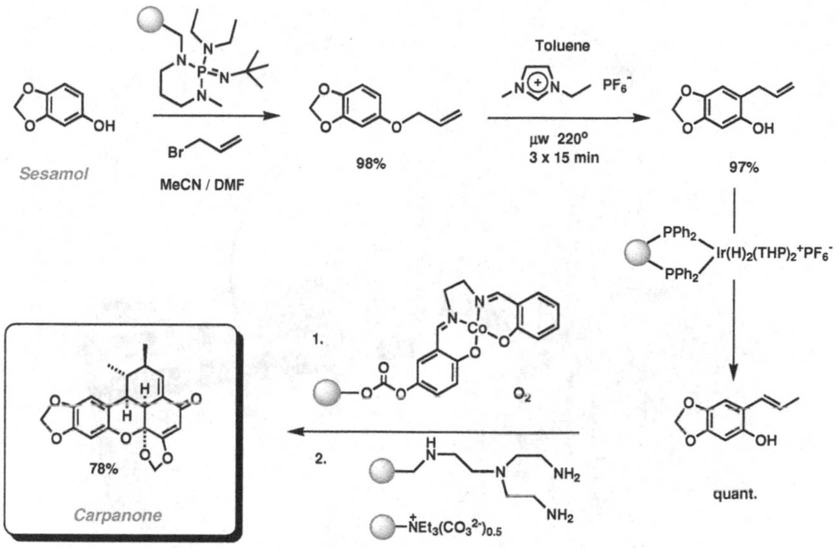

Figure 12.12. Synthesis of carpanone.

many things can go wrong and many products are formed, leading to black, tarry, polymeric materials. Here, however, we are using a new technique, first reported by our group, to effect heating by using focused microwaves to heat an ionic liquid that has been added to the mixture. This process affords cleaner products and allows much easier work-up of the reaction. The microwaves are provided by a rather special and expensive piece of apparatus and not the conventional kind of microwave oven we have in our kitchens. These reactors, shown in Figure 12.13, are today's equivalent of the traditional Bunsen burner.

The last steps of the carpanone synthesis required us to develop special reagents, first an immobilised iridium complex to effect movement of double bonds and then a new cobalt complex to achieve an oxidative dimerisation. Finally, two scavenging reagents are needed to clean up the whole process, and the result is carpanone in a yield of 78%, and as a very pure product.

The next scheme, shown in Figure 12.14, illustrates how very complicated natural products, such as the alkaloid plicamine, can be made. This scheme uses eleven immobilised systems, in an orchestrated fashion, and we were able to make the product in only six weeks, which is far quicker than if we had used a conventional approach. Also, no chromatographies, solvent washes or complex work-up of the products were necessary, even though sophisticated transformations were carried out under carefully controlled conditions. This procedure also makes use of the convergent synthesis protocol discussed earlier.

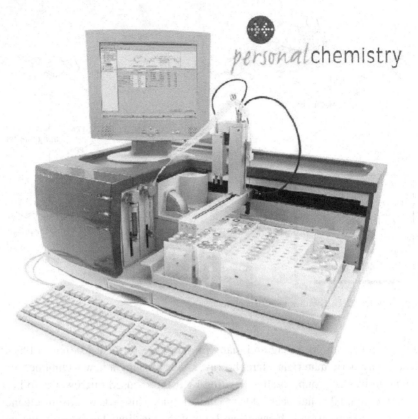

Figure 12.13. The microwave reactor: today's Bunsen burner.

Several of the reagents were specially designed to achieve these important transformations, and the reactions were readily optimised using the new systems but by conducting experiments in parallel. This synthesis represents a significant achievement in the field.[9]

Finally, as a state-of-the-art synthesis using supported reagent systems, scavengers, and a catch-and-release protocol, reported in the earlier discussion, we have been able to carry out a total synthesis of the complex natural product, epothilone C. (The epothilones are currently in late-stage clinical trials as anti-tumour drugs.) Molecules of this type are important, and contain many functional groups such as thiazoles, lactones, ketones, alcohols, epoxides and double bonds, that make even their conventional synthesis difficult. Also, traditional synthesis of the compounds inevitably uses excessive chromatography to separate minor diastereoisomers and other by-products formed during the multi-step process. Here we are challenged to progress the synthesis to the end

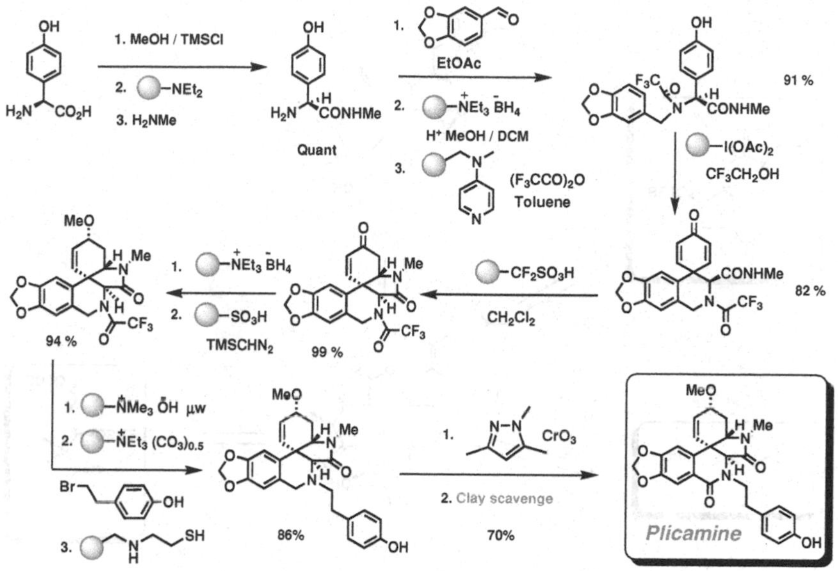

Figure 12.14. Synthesis of plicamine.

product, essentially without use of conventional separation tools. Figure 12.15 shows the plan of the synthesis, leading to potential fragments that we anticipate may be coupled, in a designed way, to lead to the natural product itself. I am happy to say that the process, detailed in Figures 12.16–12.19, was successful and made use of no less than 27 supported reagents and scavengers.[10]

The final purification was also achieved via a catch-and-release protocol. This synthesis example clearly demonstrates the power of these new tools and concepts for the preparation of molecules that display important biological properties. However, we are only at the beginning and one can conceive of a very exciting future whereby synthesis today will be changed forever. These new tools can be expanded to study reactions in a flow-mode, rather than the conventional single batch-mode currently employed. We can imagine the use of reagent chips capable of expressing a vast amount of reagent space, and envisage that process optimisation will become much facilitated. These systems are capable of discovering new chemistry, especially when used in the one-pot concept. They could in the future lead us to new catalysts, cleaner reactions, and even discover molecular transformations that are not currently possible.

When new supported materials and presentational formats are developed, whole new areas of science are likely to arise, and they will be limited only by our imagination.

This really is a great time to be a chemist!

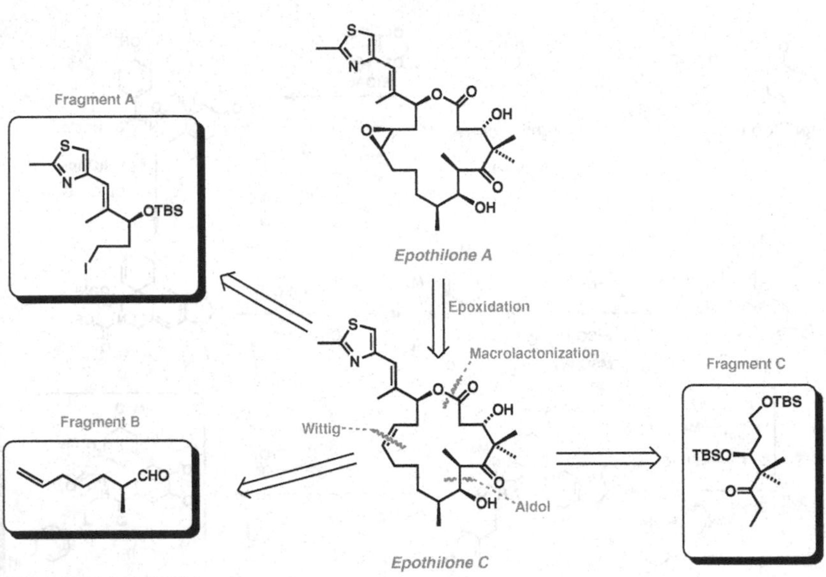

Figure 12.15. Synthesis of epothilones C and A.

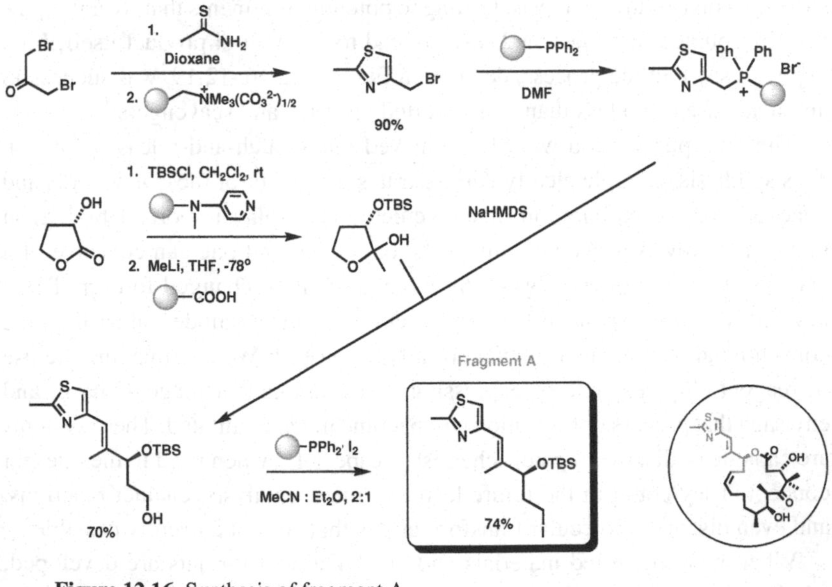

Figure 12.16. Synthesis of fragment A.

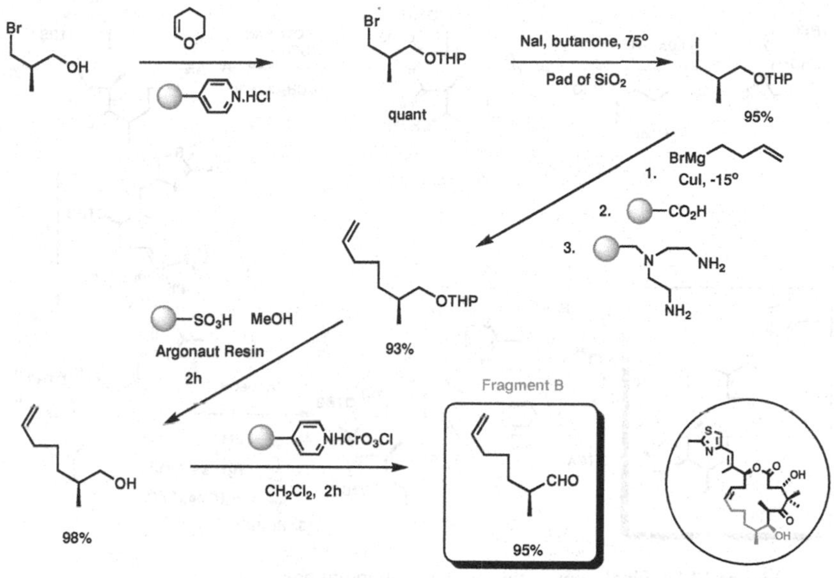

Figure 12.17. Synthesis of fragment B.

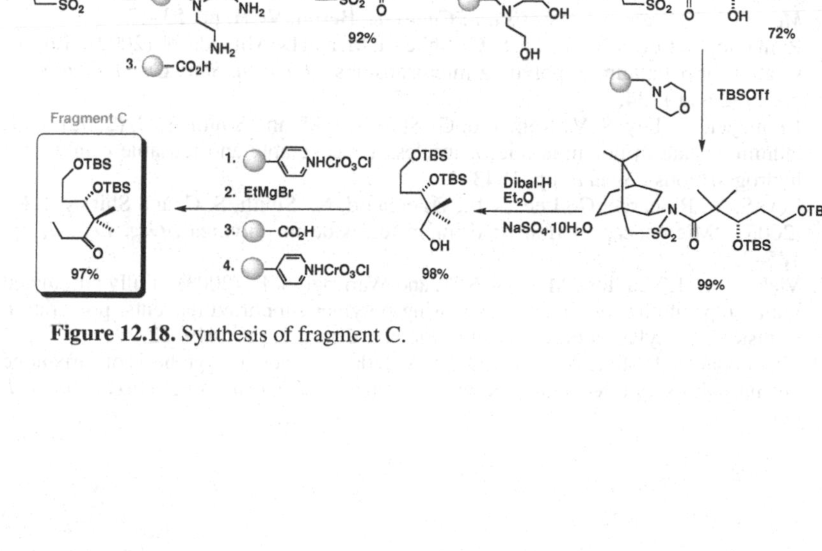

Figure 12.18. Synthesis of fragment C.

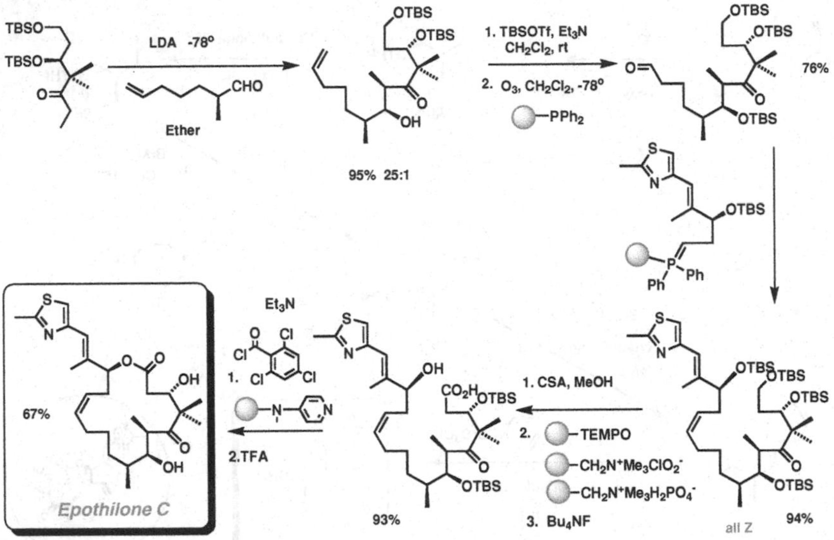

Figure 12.19. Final steps in the synthesis of epothilone C.

Notes and References

1. Ley, S. V. and Baxendale, I. R. (2002), 'New tools and concepts in modern organic synthesis', *Nature Reviews* **1**, pp. 573–86.
2. Ley, S. V., Baxendale, I. R., Bream, R. N., Jackson, P. S., Leach, A. G., Longbottom, D. A., Nesi, M., Scott, J. S., Storer, R. I. and Taylor, S. J. (2000), 'Multistep organic synthesis using solid supported reagents and scavengers: a new paradigm in chemical library generation', *J. Chem. Soc., Perkin Trans. I*, pp. 3815–4195.
3. Baxendale, I. R., Storer, R. I. and Ley, S. V. (2003), 'Supported reagents and scavengers in multi-step organic synthesis', in Buchmeiser, M. R. (ed.), *Polymeric Materials in Organic Synthesis and Catalysis*, Berlin: VCH, pp. 53–136.
4. Ramarao, C., Ley, S. V., Smith, S. C., Shiley, I. M. and DeAlmeida, N. (2002), 'Encapsulation of palladium in polyurea microcapsules', *J. Chem. Soc., Chem. Commun.*, pp. 1132 and 1134.
5. Bremeyer, N., Ley, S. V., Ramarao, C., Shirley, I. M. and Smith, S. C. (2002), 'Palladium acetate in polyurea microcapsules: a recoverable and reusable catalyst for hydrogenations', *Synlett*, pp. 1843–4.
6. Ley, S. V., Ramarao, C., Lee, A-L., Østergaard, N., Smith, S. C. and Shirley, I. M. (2003), 'Microencapsulation of osmium tetroxide in polyurea', *Org. Lett.* **5**, pp. 185–7.
7. Vickerstaff, E., Ladlow, M., Ley, S. V. and Warrington, B. (2003), 'Fully automated multi-step solution phase synthesis using polymer supported reagents: preparation of histone deacylase inhibitors', *Org. Biomol. Chem.* **1**, pp. 2419–22.
8. Baxendale, I. R., Lee, A.-L. and Ley, S. V. (2002), 'A concise synthesis of carpanone using solid-supported reagents and scavengers', *J. Chem. Soc., Perkin Trans. I*, pp. 1850–7.

9. Baxendale, I. R., Ley, S. V., Nesi, M. and Piutti, C. (2002), 'Total synthesis of the amaryllidaceae alkaloid (+)-plicamine using solid-supported reagents', *Angew. Chem. Int. Ed.* **41**, pp. 2194–7.
10. Storer, R. I., Takemoto, T., Jackson, P. S. and Ley, S. V. (2003), 'A total synthesis of epothilones using solid-supported reagents and scavengers', *Angew. Chem. Int. Ed.* **42**, pp. 2521–5.

Index

1702 Chair
  appointment of Alan Battersby, 270
  appointment of Alexander Robertus Todd, 200, 213
  appointment of George Downing Liveing, 168
  appointment of Isaac Pennington, 85–86
  appointment of James Cumming, 142
  appointment of John Francis Vigani, 31
  appointment of John Hadley, 45
  appointment of John Mickleburgh, 45
  appointment of John Waller, 45
  appointment of Ralph Alexander Raphael, 247, 269
  appointment of Richard Watson, 60
  appointment of Smithson Tennant, 107
  appointment of Steven Victor Ley, 280, 283
  appointment of William Farish, 103
  appointment of William Jackson Pope, 190
  death in office of James Cumming, 156
  death in office of John Francis Vigani, 37
  death in office of John Hadley, 60
  death in office of John Mickleburgh, 45
  death in office of John Waller, 10
  death in office of Smithson Tennant, 134
  death in office of William Jackson Pope, 199
  election procedure in 1771–73, 85
  election procedure in 1813–15, 1, 142, 205
  endowed by BP, 270
  endowment, 61
  possible appointment of Alan Battersby, 269
  renamed BP 1702 Chair, 280
  renamed Chair of Organic Chemistry, 216
  resignation of Alexander Robertus Todd, 233, 247, 269
  resignation of George Downing Liveing, 184
  resignation of Isaac Pennington, 103
  resignation of Richard Watson, 66
  resignation of William Farish, 105, 142
  retirement of Alan Battersby, 280
  retirement of Ralph Alexander Raphael, 253

aaptamine, 249, 250
actomyosin, 224
Addenbrooke's Hospital, Cambridge, 85, 87, 253
adenosine triphosphate, 222, 223
Agricola, Georg, 44
Alchemists' Club, Glasgow, 245
Alexander, Albert Ernest, 204
alkaloids, 259, 263, 265, 266, 269, 273, 274
  and chemotherapy, 264
  biosynthesis of, 261, 263, 264
  indole family, 264
Allan Glen's School, Glasgow, 210
Amati, Nicola, 253
amatol, 204
Ampère, André-Marie, 157, 159
Anderson's College, Glasgow, 179
Andrew, Raymond, 204
Anglicanism, 2, 3, 15
anthocyanins, 210, 217, 232
anti-gas respirators, 181
anti-leukaemic compounds, 248
anti-malarials, 204
Antiphlogistians, 85, 93
anti-tumour agents, 295
aphids, 232

apothecaries, 32, 37, 38, 39, 45, 53
Apothecaries Act, 148
arachidonic acid, 241
Arigoni, Duilio, 277
Armstrong, Henry Edward, 184, 190
Ashby, Eric, 244
Askesian Society, 131
AstraZeneca, 244
Astronomer Royal, 153
asymmetric quaternary ammonium salts, 183, 184
Atherton, Alice, 138
atomic theory, 141, 143, 144
ATP. *See* adenosine triphosphate
Avery, Oswald T., 219
Avogadro, Amadeo, 152
Axford, Douglas, 204

B₁₂. *See* vitamin B₁₂
Babbage, Charles, 24, 134, 142, 146
Bacchus, Henry, 155
Bacon, Francis, 65, 67
Baddiley, James, 221, 223, 224
  research student of Todd, 234
  synthesis of ATP, 222–223
Bader, Alfred, 189
Bailey, Kenneth, 224
Baker, Wilson, 261
Bamford, Harold Firth, 204
Banks, Joseph, 113, 114, 127
Banks, Robert, 45
Barger, George, 210
Barlow, William, 190
Barrow, Isaac, 5
Bartoli, Daniello, 35
Barton, Derek, 244
Battersby, Alan Rushton, 217, 257–282
  and Alexander Todd, 259, 267, 268
  and Ralph Raphael, 270
  appointed to second chair of organic chemistry at Cambridge, 267
  appointed to 1702 Chair, 60
  biosynthesis of morphine, 261–263
  biosynthesis of plant alkaloids, 264–266
  biosynthesis of the Pigments of Life, 266, 270–275
  biosynthesis of vitamin B₁₂, 276–280
  early life, 259
  portrait, 258
  possible move to 1702 Chair, 269
  University of Bristol, 261
  University of Liverpool, 266

University of Manchester, 259
University of St Andrews, 260
Baumé, Antoine, 61
Bayle, Pierre, 32
Beale, John, 5
Becher, Johann Joachim, 17, 84
Beddoes, Thomas, 20, 21, 22, 49, 50
Bedford College, London, 202
Belasyse, Thomas. *See* Fauconberg
Bell, Henry, 154
Bene't Hall. *See* Corpus Christi College, Cambridge
Bentham, Jeremy, 23
Bentley, Richard, 3, 8, 9, 10, 13, 24, 37, 38, 41
Benwell, John, 45
benzylisoquinoline, 262, 263
Bergman, Torbern Olof, 48
Berington, Joseph, 21
Bernal, John Desmond, 198, 206
Berthollet, Claude Louis, 84, 119
Berzelius, Jöns Jacob, 130, 132, 144, 146
Bible Society, 90
  formation at University, 90
Birkbeck Laboratory, University College, London, 148
Bishop of Lincoln. *See* Tomline, George
bismuth compounds, 179
Black, Joseph, 20, 23, 45, 48, 65, 92
Blackburn, G. Michael, 231
Blackett, Patrick Maynard Stuart, 198
Blagden, Charles, 63
Blanche, Francis, 279
Blanes, Lord, 15
Bletchley Park, 204
Board of Longitude, 24
Boerhaave, Hermann, 8, 17, 31, 45, 92
Bond, Henry, 148
Borsche, Walther, 210
Botanic Garden, Cambridge, 149, 168
  1784 building, 90
  establishment, 90
  head gardener, 91
  lecture room, 168
Botany, professor of, 91, 168
Boyle, Robert, 5, 6, 8, 10, 13, 35, 36, 37, 41, 43, 44
BP 1702 Chair, 280
  endowment, 270
Bradley, Richard, 90
Brady, Robert, 35
Bragg, (William) Lawrence, 228
brewing, 176

Brewster, David, 3
Brigham, Samuel, 45
Bristol, University of
    and Alan Battersby, 261
British Association for the Advancement of
    Science, 159, 160, 174
British Dyestuffs Corporation, 201
British Mineralogical Society, 131
British Museum, 182
Brodie, Benjamin Collins Sr, 144
Brougham, Lord, 130
Brown, Daniel M., 215, 226
    reaction to double helix, 228
    research with Todd, 234
Brown, Roger F. Challis, 214
Brownrigg, William, 96
Buchanan, George, 245
Buchanan, J. Grant, 232
Bun Shop (public house), 226
Burke, Edmund, 21
Bykhovsky, Vladimir, 277
Byron, Lord, 144

Cain, Tubal, 31
Calgarth Park, Westmorland, 71, 73, 76, 77
calico-printing, 176
California Institute of Technology, 231
Calzolari, Francesco, 32
Cambridge Chemical Society, 203
Cambridge Mathematics Laboratory, 203
Cambridge Philosophical Society, 96, 105,
    156, 157, 158, 159, 160, 183
    and Farish, 156
Cambridge School of Agriculture, 174
Cambridge Training College for Women, 181
Cambridge University Act, 150
Cambridge University Rifle Volunteers, 174
Cameron, Alastair, 204
Campbell, Hugh, 204
camphor, 265
camphor derivatives, 184
camphorsulphonic acids, 184
cannabis, 219
Cannizzaro, Stanislao, 152
Cannon, Jack, 232, 276
carbon-13, 272, 273, 276, 279
carbon-14, 261, 262, 263, 273
Carlisle Grammar School, 96
Carlisle, Anthony, 113
carpanone, 296
Carter, Herbert, 261

Catharine Hall, Cambridge. *See* St Catharine's
    College, Cambridge
Cavendish family, 138
Cavendish Laboratory, Cambridge, 215, 228
Cavendish, Henry, 95
    determination of density of the Earth, 97
Central Institution, London, 184
Chambers, Ephraim, 58
Chaptal, Jean Antoine, 85
    Elements of Chemistry, 99
Chapter Coffee House Society, 49
charcoal cylinders, 75
Charles II, 31
Charterhouse School, 17, 46, 95
chemical elements
    year of discovery, 114
chemical engineering, 205
Chemical Society, 153, 183
    James Cumming as a founder member,
        159
Chesterfield, Philip Stanhope, Earl of,
        33
chlorophyll, 257, 261, 270, 271, 275
Christ's College, Cambridge, 35, 178, 232,
        245, 250, 251
    and Alexander Todd, 232
chromatography, 214, 232
chrysanthemic acid, 246
Church Missionary Society, 90
Churchill College, Cambridge
    and Alexander Todd, 233
Churchill, Winston Spencer, 205
    as Minister of Munitions, 202
City and Guilds College, London, 190
Clapham Sect, 89
Clare College, Cambridge, 182
Clark, William, 146
Clarke, Edward Daniel, 21, 22, 142, 143, 147,
        156
Clarke, Samuel, 69
Classical Tripos, 147, 153
Clausius, Rudolf, 177
Clothworkers' Company, 178
coal gas, 176
cobalt complexes, 297
codeine, 263
codeinone, 263
Cohn, Waldo E., 225
Colchester, Henry, 41
colchicine, 264
college laboratories, 168, 169, 174, 195

combinatorial chemistry, 286
Committee on Scientific and Industrial
    Research, 201, 202
Commonwealth Fund Fellowship, 260
Congreve, William, 75
contraceptive pill, 214
convergent synthesis, 292, 297
Cook, James W., 243
copperas, 65
cordite, 204
Corpus Christi College, Cambridge, 14, 45,
    180
    and Hales, 37, 41
    and Mickleburgh, 14
    and Stukeley, 37, 40
    and Waller, 45
corynantheal, 264, 265
Cotes, Roger, 9, 37
Cottenham, Charles Christopher Pepys, 1st
    Earl of, 99
Coulson, Charles, 160
Covel, John, 35
Craig, Lyman, 260
Crane, John, 38
Cremonese violins, 252
Crick, Francis, 215, 228
Croucher Foundation, 233
Crouzet, Joel, 279
Crum Brown, Alexander, 171
Cudworth, Ralph, 5
Cullen, William, 17, 23, 48
Cumming, George William, 140
Cumming, Hannah Sophia, 155
Cumming, Isabella, 155
Cumming, James, 21, 138–165, 168
    admitted to Trinity College, Cambridge, 140
    appointed to 1702 Chair, 142–143
    course structure, 143
    death, 156
    friendship with F. J. H. Wollaston, 142
    portrait, 139
    religion, 141
    withdraws from 1702 Chair election, 132,
        142
Cumming, James John, 155
Cumming, James Sr, 138
Cumming, Joseph Notzel, 140
Cumming, Revd Joseph George, 140, 154
Cumming, William George, 140
Cusano, Maffeo, 32
Cuthbert, George, 45

cyanine dyes, 202
cytochromes, 270

D'Oyly, George, 145
Dainton, Frederick Sydney, 203, 204
Dale, Henry Hallett, 199, 212
Dalton, John, 74, 113, 141, 215
Darwin, Charles, 141, 146, 168
Darwin, Erasmus, 49, 145, 146, 160
Daubeny, Charles Giles Bridle, 151, 152, 153
David, George E., 205
Davy, Humphry, 19, 22, 24, 62, 66, 102, 114,
    141, 146, 152
    isolation of Na, K, Ba, Sr, Ca and Mg, 114
Dawson, John, 74
Day, Thomas, 38
DDT, 205
de Kok, Johan Egbert Frederik ('Frits'), 201
De La Pryme. *See* Pryme
de Morveau, Louis Bernard Guyton, Baron,
    84
de Quincey, Thomas, 73
Dealtry, William, 145
deaminase, 274
Debussche, Laurent, 279
Dee, John, 4
Dekker, Charles (Chuck) A., 226, 228
Delamethérie, Jean Claude, 119
Demonferrand, Jean Baptiste Firmin, 159
Dent, Peter, 38
Department of Scientific and Industrial
    Research, 199
    creation of, 201
Desaguliers, John Theophilus, 46
Descartes, René, 13, 35
Deville, Henri Sainte-Claire, 160
Devonshire, 5th Duke of, 138
Devonshire, 7th Duke of, 150
Devonshire, William Cavendish, Earl of, 33
Dewar, James, 150, 170, 171, 172, 177, 182,
    183, 195, 199
    book on spectroscopy, 183
    spectroscopic work with Liveing, 183
Diderot, Denis, 47
dienol, 263
Divinity, Regius chair of
    appointment of Richard Watson, 66
DNA, 219, 220, 228
    discovery by Friedrich Miescher, 219
    work of Oswald Avery, 219
Dollar Academy, Scotland, 171

double helix, 228
Downing College, Cambridge
    laboratory, 169
Downing, Catherine, 166
Downing, George, 166
Doyle, Peter, 244
Drake, Richard, 4
duck's blood, 274
Dulong, Pierre Louis, 146
Dundonald, Earl of, 64
Dunlop, John, 200
Dunthorne, Richard, 87
dyestuffs, 176

Eachard, John, 35
Eagle (public house), 215
East Anglia, University of, 251
East India Company
    saltpetre, 76
Edinburgh Review, 58
Edinburgh, University of, 146, 171, 210
Edleston, Revd James, 141, 154, 155, 156
Eglinton, Geoffrey, 244
Eirich, Frederick Roland, 203
Eley, Daniel Douglas, 204
Elliston, William, 103
Eméleus, Harry Julius, 203, 216
emetine, 259, 260, 264
Emmanuel College, Cambridge, 5
epothilone C, 298
Euglena gracilis, 274

Fage, John, 38
Faraday, Michael, 144, 157
Farish, William, 22, 24, 66, 103–107, 121,
    142, 146, 147
    accident with hydrogen, 146
    appointed to 1702 Chair, 103
    appointed to Jacksonian chair, 105, 132
    appointed to Observatory Syndicate, 97
    appointed vicar of St Giles' Church, 105
    awarded Smith's Prize, 96
    competes for Jacksonian chair, 95
    death, 106
    education by his father, 96
    engineering ingenuity, 106
    enters Magdalene College, 96
    examination marking, 106
    first president of Cambridge Philosophical
        Society, 156
    lectures, 103, 142–144

Plan of a Course of Lectures, 105
    portrait, 104
    residence at Merton Cottage, 106
    resignation from 1702 Chair, 142
    University offices, 96
Fauconberg, Thomas Belasye Earl of, 33, 35
Fenton, Henry John Horstman, 178
    books on chemistry, 179
Fenton's reagent, 178
Fersht, Alan, 270
Fisons, 233
flavin adenine dinucleotide, 224
Folkers, Karl, 276
Forbes, James David, 171
forestry, 76, 78
Fourcroy, Antoine François de, 84, 85, 99
Fownes, George, 151, 153
Frank, Charles, 204
Franklin, Benjamin, 17, 96
Franklin, Rosalind Elsie, 203, 206
Freind, John, 14, 31, 44
Frend, William
    expulsion from University, 95
Freund, Ida, 181
Fry, Elizabeth Gurney, 155
Fuller, Watson, 229
Fullerian Chair, Royal Institution, 171

Gahn, Johann, 119
galvanoscope, 144, 157, 158
Gassendi, Pierre, 35
Gatebeck, Cumbria, 75
Gatty, Oliver, 204
Gay-Lussac, Joseph Louis, 143
gene cassettes, 290
genome, 286
Geoffroy, Etienne-François, 47, 61
Geological Society of London, 131, 145,
    159
George I, 8
George III, 72
    and Richard Watson, 58
George, Philip, 204
geraniol, 264
Gibbon, Edward, 70, 121
Gibbs, Josiah Willard, 177
Gilson, Ralph, 213
Girton College, Cambridge, 180, 181
    laboratory, 169, 181
Glasgow, University of, 179, 210
    Ralph Raphael appointed lecturer, 242

Glaxo laboratories, 232
Glisson, Francis, 4
Glynn, Robert, 1, 3, 87
Goldsmiths' Company, 196
Goldsmiths' Institute, London, 184, 190
Goldsmiths' Metallurgy Laboratory,
    Cambridge, 216
Goldsmiths' Professorship, 198
Goldsmiths' Reader in Metallurgy, 180, 186
Gonville and Caius College, Cambridge, 4, 14,
    96, 155, 179
    laboratory, 169, 174, 179, 182
Gough, John, 74
gout, 90
Gowland Hopkins, Frederick, 198, 200,
    206
Grafton, Duke of, 70
Graham Commission, 148, 151
gramicidin S, 260
Granick, Sam, 261
Grantchester Meadow, 204
Gray, Stephen, 37
Greene, Robert, 3, 13, 14
Gregory, Duncan Farquharson, 145
Gregory, William, 153
Gren, Friedrich Albrecht Carl, 99
Gresham College, 3, 5, 8
Griffith, Fred, 219
Grossmith brothers, 138
Gunning, Henry, 92
gunpowder, 57, 65, 74, 75, 78, 102
    British inferiority to French, 74
    Cumbrian manufactories, 75
    improvement by Watson, 74
    manufacture, 94
Gurney, Daniel, 153, 154
Gurney, Revd William Hay, 154
Guthrie, Frederick, 202
Guy's Hospital, London, 45
Gyles, Henry, 35, 36

Hadley, John, 15, 17, 18, 23, 24, 45–48, 50,
    60, 87
    and Charterhouse School, 46
    and practical chemistry, 47
    and St Thomas's Hospital, London, 46
    appointed to 1702 Chair, 45, 85
    death, 59
    description of chemistry, 47
    lectures, 46
    portrait, 16

haem, 270, 271
Hales, Stephen, 9, 14, 37, 40, 44, 45, 46, 47,
    63, 64
Hamer, Neil Kenneth, 231
Harris, John, 46
Harrison, Thomas, 145
Hartley, David, 21
Hartley, Harold, 201
Hartley, W. N., 180
Harvard University, 231
Harvey, William, 4
Harwood, Busick, 96, 119, 134
hashish, 219
Hatchett, Charles, 114
Hauksbee, Francis, 46
Haviland, John, 147, 148
Heck reaction, 296
Heilbron, Isidor (Ian) Morris, 238
Helm Crag, Cumbria, 63
Helmont, Johann van, 37
Henckel, Johann, 17
Henry, Thomas, 65
Henslow, John Stevens, 141, 145, 146, 147,
    148, 154, 156, 160
Herschel, John, 142, 143, 145, 147
Heversham, 58, 62
Heycock, Charles Thomas, 172, 179
    election to Royal Society, 180
Hinchliffe, John, 85
histone deacetylase inhibitors, 295
Hodgkin, Dorothy Crowfoot, 232, 276
Hodson, William, 85
Hoffmann (Richard Watson's assistant), 61,
    108
Hoffmann-La Roche, 222, 267
Hofmann Prize
    won by Ralph Raphael, 238
Hofmann, August Wilhelm, 167
Holland, Lord and Lady, 130
Hooker, Joseph Dalton, 168
Hopkins, Frederick Gowland, 212
Horne, George, 21
Horner, Francis, 131
Horner, Leonard and Francis, 130
Hughes Hall, Cambridge, 181
Humfrey, Sarah, 153
Hutchinson, Arthur, 181
Hutchinson, Eric, 204
Hutchinsonianism, 14
Hyde, West, 123
hydroxamic acids, 295

Hyson Club, 88
Hythe firing range, 74

Illinois, University of, 261
immobilised reagents, 290, 293
Imperial College, London, 238, 240
indole, 264
Ingram, Rowland, 168
Institute of Chemistry, London, 176
International Union of Pure and Applied
    Chemistry, 196
iridium, 114, 137
iridium complexes, 297
isocyanate monomers, 293

Jackson, Richard, 170
    endowment of Jacksonian chair, 90
Jacksonian chair, 19, 20, 88, 142, 147, 148,
    156, 158, 168, 170, 171, 172, 182, 185,
    186, 195
    appointment of Francis Wollaston, 97
    appointment of Isaac Milner, 90
    appointment of William Farish, 105, 142
    death in office of William Farish, 106
    endowment, 90
    first accommodation for, 90
    increased stipend, 91
    resignation of Francis Wollaston, 142
    resignation of Isaac Milner, 95
Jacksonian demonstrators, 172, 182
James, Keith, 244
Jesuits' bark, 38, 42
Jesus College, Cambridge, 95
John Humphrey Plummer bequest, 198
Johnson, Paley, 204
Johnson, Samuel, 62, 70
Jones oxidation, 240
Jones, Ewart Ray Herbert, 238
Jones, Humphrey Owen, 172, 181, 182
Jones, Muriel, 182
Jurin, James, 15, 44, 45

Kekulé, August, 171
Kelvin, William Thomson, Baron, 177
Kemp, George, 146
Kenner, George Wallace, 221, 231, 266, 270
Khorana, H. Gobind, 224, 230
King of Clubs, 130
King's College, Cambridge, 90, 179
King's College, London, 178
Kingsley, Revd Charles, 141

Kipping, F. Stanley, 184
Kirby, Anthony John, 231
Kirwan, Richard, 49
Königliche Gewerbeschule, Berlin, 167
Kunitz, Moses, 261

Laboratory of Molecular Biology, Cambridge,
    198
Lapworth, Arthur, 231
latitudinarianism, 67, 141
laughing gas, 102, 146
lavender oil, 265
Lavoisier, Antoine Laurent, 23, 47, 50, 65, 84,
    99, 113, 141, 143
    adoption of his chemical nomenclature in
        Cambridge, 97
    chemical nomenclature, 84
    oxygen as the principle of acidity, 98
    Traité Élémentaire de Chimie, 84, 85, 99
Lawrence, Arthur Stuart Clark, 204
Le Bel, Joseph, 190
Le Clerc, Jean, 13
lead smelting, 63, 64
Lefèvre, Nicolas, 31, 37, 39
Leigh Grammar School, Lancashire, 259
Lennard-Jones, John Edward, 198, 199, 206,
    210, 216
    Chief Superintendent of Armament in
        World War II, 203
Lensfield Road laboratory, 217
    departmental structure, 216, 247, 268, 270
    departmental unification, 217
Levene, Phoebus A., 220, 225
Ley, Steven Victor, 283–303
    appointed to BP 1702 Chair, 280, 283
    portrait, 284
    research group, 288
    synthesis of carpanone, 296–297
    synthesis of histone deacetylase inhibitors,
        295
    synthesis of plicamine, 297–302
    use of polymer beads in synthesis,
        290–293
Leyden, 31, 37, 45
liberal education, 150, 151, 152, 153, 160,
    176, 190
Lincoln's Inn, London, 166
linoleic acid, 240
linolenic acid, 241
liquid air, 184
liquid oxygen, 184

Lister Institute of Preventive Medicine, London, 210, 231
Little Go examination, 147, 152
Little Stonham Rectory, 106
Liveing *née* Ingram, Catherine, 168, 175
Liveing, Edward, 166
Liveing, George Downing, 152, 153, 160, 166–188, 190
    appointed to 1702 Chair, 168
    as magistrate, 174
    assumes duties of James Cumming, 150, 156
    book on chemical equilibrium, 177
    book on spectroscopy, 183
    death, 174, 175
    lectures on agricultural chemistry, 174
    lecturing style, 177
    national scientific offices, 174
    performance in Natural Sciences Tripos, 150
    portrait, 167
    resignation from 1702 Chair, 174, 184
    spectroscopic work with Dewar, 183
    visits to European laboratories, 173
Liverpool, University of
    and Alan Battersby, 266
    and George Kenner, 266
    second chair of organic chemistry, 266
Llandaff, 18, 73
    See also Watson, Richard
    dilapidation of cathedral, 71
    poverty of See, 71
Locke, John, 58
London Association, 58
Lowry, Thomas Martin, 196, 199
Lowwood gunpowder works, Cumbria, 75
Lucasian chair of Mathematics, 24
Lucretius, 35
Lunar Society, Birmingham, 49
Luther, John, 59, 60
lycorine, 264
Lythgoe, Basil, 221, 231

Macquer, Pierre-Joseph, 17, 92
    Dictionary of Chemistry, 99
Macrae, Thomas F., 232
Maddock, Alfred Gavin, 204
Magdalene College, Cambridge, 96
Maier, Michael, 5
Malthus, Thomas, 130

Manchester, University of, 184, 210, 212, 213, 231
    arrival of Alan Battersby, 259
Manhattan Project, 205
Mann, Frederick George, 204, 206
Margaret, H.R.H. Princess, 218
Marggraf, Christiaan, 17, 28
Markham, Roy, 226
Marlborough School, 140
Martin, Archer, 206, 232
Martin, Benjamin, 18, 47
Martineau, Harriet, 76
Martyn, Henry, 141
Martyn, Thomas, 91, 97
Mary, Queen of Scots, 138
Maskelyne, Nevil, 153
Mason, Charles, 45
mass spectrometry, 259
Mathematical Tripos, 141, 145, 147, 151, 166
Maxwell, James Clerk, 159, 177
Maxwell, James Rankin, 244
May & Baker Ltd, 239
McKillop, Alexander (Sandy), 244
McKillop, Tom, 244
Meldola Medal
    receipt by Ralph Raphael, 240
Merck laboratories, 232
Merton Cottage, 106
metallurgy, 180, 186
Méthode de Nomenclature Chimique, 84
Michell, John
    torsion balance, 97
Michelson, A. Michael (Mick), 228
Mickleborough. *See* Mickleburgh
Mickleburgh, John, 14, 15, 17, 23, 24, 45–47, 50, 60
    appointed to 1702 Chair, 45
    death, 45
    students of, 45
    trade as dispensing chemist, 45
microanalysis, 214
microwave reactors, 297
Miescher, Johann Friedrich, 219
Miller, William Hallowes, 147
Mills, William Hobson, 183, 199, 202
Milner, Isaac, 18, 19, 20, 21, 23, 24, 76, 88–95, 96, 121
    appointed Dean of Carlisle, 94
    appointed to Jacksonian chair, 90
    appointed to Lucasian chair, 106
    collection of scientific instruments, 97

Milner, Isaac (*cont.*)
  death, 106
  early years and education, 88
  elected President of Queens' College, 94
  elected Vice-Chancellor, 95
  election to Royal Society, 88
  illness, 94
  improvement in health, 106
  inhales noxious gas, 88
  lecturing style, 88, 92
  low-church evangelism, 89
  oxidation of ammonia by manganese
    dioxide, 93
  portrait, 89
  Rector of St Botolph's Church, 88
  resignation from Jacksonian chair, 95
  University offices, 89
Ministry of Aircraft Production, 204
Ministry of Fuel, 203
Ministry of Munitions, 201
Ministry of Supply, 203, 204
Moelwyn-Hughes, Emyr Alun, 203
Molteno Institute, Cambridge, 215, 226
Money, Tom, 244
Moore, Stanford, 261
More, Henry, 5
Morgan, John, 45
morphine, 261, 262, 263, 265
Moses, 31
Muir, Matthew Moncrieff Pattison, 179
  textbooks by, 179
Müller, Gerhard, 277
mustard gas, 196, 201

Napoleonic Wars, 74
narcotine, 264
National Gallery, London, 253
Natural Sciences Tripos, 149, 150, 151, 152,
  153, 161, 166, 168, 175, 176, 178, 179,
  180, 181
Needham, Joseph, 206
nerol, 264
nerve gases, 204
Neville, Francis Henry, 180
New Museums Site, 91, 168, 173
Newark-on-Trent, 32, 35, 37
Newborough, Thomas, 36
Newcastle, 1st Duke of, 60
Newnham College, Cambridge, 180, 181, 182
  laboratory, 169, 181
Newton, Humphrey, 6

Newton, Isaac, 2, 5, 6, 8, 13, 14, 15, 21, 32, 36,
  39, 40, 43, 44, 45, 46, 48, 64, 68, 140, 150
  Opticks, 14, 44
Newtonianisation, 9, 14
Newtonianism, 3, 15
Nicholson, William, 113
  Dictionary of Chemistry, 84
nicotinamide, 259
nicotinamide adenine dinucleotide, 224
Nidd, Joseph, 5
nitrocellulose, 204
nitrous oxide, 102, 146
Noad, Henry, 159
Nobel Prize, 212, 219, 231
Norrish, Ronald George Wreyford, 199, 200,
  206, 210, 216
North, Christopher, 74
Norton, Samuel, 4
Norton, Thomas, 4
Nottingham, University of, 240
Nuclear Magnetic Resonance spectroscopy.
  See spectroscopy:NMR
nucleic acids, 259
nucleosides and nucleotides, 212, 219, 220
Nuffield Foundation, 233

Oak Ridge National Laboratory, 225
Oersted, Hans, 157, 159
Ohm, Georg, 159
onchalum, 64
Openshaw, Hal, 259, 260
opium, 261, 262
Ordnance Department, 74, 75
organophosphorus compounds, 204
Orgel, Leslie Eleazer, 215
osmium, 114, 127
osmium tetroxide, 294
Overton, Karl, 244
Owens College, Manchester, 179, 180
Oxford, University of, 20, 21, 23, 31, 40, 48,
  210, 231

Paley, William, 18, 69, 88, 95
  influence on Isaac Milner, 89
palladium, 126
palladium catalysts, 294, 295
Parker, William, 244
Parkes, Samuel, 48, 50, 62
Parkinson, James, 62
Partington, James Riddick, 203
Partridge, S. Miles, 232

Parys, William, 4
Patterson, Thomas Stewart, 210
Pauling, Linus Carl, 231
Peacock, George, 156
Pembroke Street laboratory, 172, 184, 213, 217
    1908 extension, 186
    addition of physical chemistry, 196
penicillic acid, 241
penicillin, 205, 212, 239, 246
Pennington, Isaac, 85–88
    appointed to 1702 Chair, 86
    appointed to Regius chair of Physic, 103
    death, 106
    early years and education, 87
    knighthood, 106
    elected President of St John's College, 106
    portrait, 86
    resignation from 1702 Chair, 103
Pepper, David Charles, 204
Perkin, William Henry, Jr, 184, 190, 193
Perkin, William Henry, Sr, 160
pernicious anaemia, 232
Perry, Sam, 224
Perse School, Cambridge, 181, 196
Perutz, Max Ferdinand, 198
Peterhouse, Cambridge, 4
pharmaceutical industry, 287
Ph.D. degree
    introduction in Cambridge, 198
phlogiston, 84, 92, 93, 98, 99, 100
phosgene, 201
phosphate esters, 220
phosphorus pentoxide, 222
phosphoryl chloride, 224
phosphorylation, 215, 220, 221, 222
photographic sensitisers
    development in Cambridge, 202
Physic, Regius Professor of, 4, 14, 167, 175
phytochromes, 270
Pigments of Life, 270
Pitt, William (the younger), 21, 71, 93
Place, John, 35
plastic explosives, 204
platinum, 123–126
Playfair, Lyon, 171
plicamine, 297
Plumian chair of Astronomy (and
    Experimental Philosophy), 24, 36, 98
Plumptre, Russell, 15, 85, 87
    death, 103
Poggendorff, Johann, 157

Polanyi, Michael, 213, 231
poll men, 147
polyketides, 267
polymer beads
    use in chemical synthesis, 290–293
polyurea, 294
Pope, William Jackson, 184, 189–209, 210
    appointed to 1702 Chair, 190
    appointed to Manchester chair, 190
    as a collector of alchemical paintings, 189
    death, 199
    experiments with alkyl selenides, 192
    financial support from oil companies, 196
    first president of IUPAC, 196
    first president of the Solvay Chemistry
        Conferences, 196
    knighthood, 196
    portrait, 191
    preparation of mustard gas, 196, 201
    preparation of phosgene, 201
    Presidency of the Chemical Society, 196,
        205
    trip to Australia, 196
    work on optically active non-carbon
        compounds, 190
Porter, Francis, 41
Porter, George, 206
Porter, Henry, 140
potassium nitrate, 94
Pott, Johann, 17
'practising chemists', 176
Preston, William, 58
Price, William Charles, 203
Priestley, Joseph, 21, 24, 49, 62, 65, 84, 100
Propionibacterium shermanii, 277
prostitution, 6
Prout, William, 144
Pryme, Abraham de la, 36, 39
Pseudomonas denitrificans, 279
pseudomonic acids, 247, 248, 249
public health, 180
Purvis, John Edward, 180, 186
pyrenophorin, 246
pyridine derivatives, 178
pyrolysis of coal, 64

Quarterly Review, 57, 58
Queen Mary College, London, 202, 203, 240
Queen's University, Belfast, 244, 251
Queens' College, Cambridge, 6, 9, 15, 20, 36,
    41, 43, 88, 90

Queens' Lane laboratory, 45, 60
Quinan, Kenneth Bingham, 205
quinine, 264, 265

radioactive labelling, 261, 273, 279
  See also carbon-14
radium, 174
Ramage, Robert, 244
Rammelsberg, Karl, 167
Raphael *née* Gaffikin, Prudence Marguerite
    Anne, 240, 250, 251, 252, 253
Raphael, Ralph Alexander, 237–256
  acetylene chemistry, 238, 239, 240
  administrative role, 251
  and Alan Battersby, 270
  appointed to 1702 Chair, 247, 269
  appointed CBE, 251
  chair at Belfast, 244
  chair at Glasgow, 244
  childhood, 237
  Ciba–Geigy Award of the Chemical Society,
    251
  Davy Medal of the Royal Society, 251
  death, 253
  Fellowship of the Royal Society, 246
  ICI Research Fellowship, 240
  Imperial College, 238
  marriage, 240
  May & Baker Ltd, 239
  Pedler Lecturer, 251
  retirement from 1702 Chair, 253, 270
  sense of humour, 245, 249, 250
  synthesis of natural products, 240–243,
    246–250
  Tilden Lecturer, 246
  violin manufacture, 252
  war gases, 239
  war work on Penicillin, 239
Ray, John, 5, 8, 32, 36
Read, John, 186, 199, 260
Regius Professor of Divinity, 66
Regius Professor of Physic, 14, 167, 175
reticuline, 263
Reynolds, Joshua, 77
rhodium, 126
Rhône–Poulenc Rorer company, 279
Ricardo, David, 130
Richmond, 3rd Duke of, 74
Rideal, Eric Keightly, 198, 199, 200, 203, 204,
    206, 216
  war work of his group, 204
Ripley, George, 5

RNA, 220, 222, 225, 226, 227, 228
  linkage points, 226
Robertson, Andrew John Blackford, 204
Robinson, Robert, 210, 231, 232, 261
Robert Robinson Laboratories, University of
    Liverpool, 266, 267
Robinson, Tancred, 5, 8
Robison, John, 21
Roche Products, 233, 267
  and chair of organic chemistry at
    Cambridge, 267
Rockefeller Foundation, New York, 198
Rockefeller Institute for Medical Research,
    New York, 219, 260
Rockingham, 2nd Marquis of, 60, 61
Romilly, Joseph, 141, 154
Romilly, Samuel, 130
Romney, George, 59
Roper, John, 45
Roscoe, Henry, 179
Rouelle, Guillaume-François, 47
Rousseau, Jean-Jacques, 47
Royal (Dick) Veterinary College, Edinburgh,
    171
Royal College of Chemistry, London, 148,
    167
Royal College of Music, London, 240
Royal College of Physicians, London, 17
Royal College of Science, Dublin, 180
Royal Commission for the Exhibition of 1851,
    212, 253
Royal Dutch Shell Company, 201
Royal Holloway College, London, 181
Royal Institute of Chemistry, London, 240
Royal Institution, London, 171, 172, 182, 183,
    195, 200
Royal Navy, 74
Royal Society, 1, 5, 8, 9, 17, 20, 33, 36, 45,
    48, 144, 159, 160, 174, 180, 183, 246,
    251
  election of Battersby, vii
  election of Cumming, 144
  election of Hadley, 17
  election of Ley, ix
  election of Liveing, 174
  election of Milner, 88
  election of Pope, 190
  election of Raphael, 246
  election of Tennant, 119
  election of Watson, 63
  election of Wollaston, Francis, 96
  election of Wollaston, William, 103, 123

presidency of Dale, 199
presidency of Todd, 233
Wollaston family connections, 95
Rubio, David, 252
Ruhemann, Siegfried, 182
Rumford, Benjamin Thompson, Count, 99
Runcton, Parish of North, 150, 153, 154, 156
Rupe rearrangement, 248
Rutherforth, Thomas, 66
Rutland, Duke of, 18, 62, 70

salmon sperm, 219
Salò, Mario, 32
salutaridine, 263
Sanders, Jeremy Keith Morris, 253
Sandhurst, Royal Military College, 168
Sandys, Francis, 45
Sanger, Frederick, 232
Saunders, Bernard Charles, 204
Saunders, John Tennant, 198
Scheele, Karl Wilhelm, 48, 119
  experiments with Vital Air (oxygen), 98
  On Air and Fire, 97
  theory of heat and light, 97
Schulman, James Herbert, 204
Schweigger, Johann Salomo Christoph, 157
Science Research Council, 246
Scott, (Alastair) Ian, 243, 245, 277
Scott, Alexander, 182
Scott, Walter, 74
Sedburgh School, 87
Sedgwick, Adam, 60, 146, 151, 156, 160
Sedgwick Gunpowder Works, Cumbria, 75
Seebeck, Thomas, 157, 159
Sell, William James, 178
Senate House examinations, 106
Sendivogius, Michael, 5
sesamol, 296
Shaw, Peter, 46
Shelburne, Lord, 70
Shemin, David, 277
Shepherd, Anthony, 98
shikimic acid, 246
Shrewsbury, Earl of, 138
Sidgwick, Henry, 141
Sidney Sussex College, Cambridge, 95, 180
  and Francis Wollaston, 95, 103
  laboratory, 169, 174, 180
  Mastership regulations, 103
Simeon, Charles, 73, 90, 141

Sir William Dunn Institute of Biochemistry, Cambridge, 198
sizars, 14, 58, 87, 88, 96, 140
Slaughterhouse Lane laboratory, Cambridge, 167
Sloane, Hans, 8, 17
Smart, Christopher, 14
Smith, Adam, 17
Smith, Herchel, 206, 214
  endowment of chair of organic chemistry, 270
Smith, John, 226
Smith, Lester, 276
Smith, Samuel, 36
Smith, Thomas, 130
Snow, Charles Percy, 198, 206
soap making, 22, 176
Society for the Abolition of Ethyl Esters, 249
Society for the Promotion of Philosophy and General Literature, 96
Society for the Propagation of the Gospel, 69
Socinians, 95
Solvay Chemistry Conferences, 196
Sondheimer, Franz, 240
Southey, Robert, 73, 146
Special [Gas] Service, World War I, 201
spectroscopy, 172, 177, 180, 183
  infrared, 259, 268
  NMR, 217, 249, 259, 272, 273, 274, 276, 279
  ultraviolet, 183, 238, 260, 268
Spivey, William Thomas Newton, 182
Sprat, Thomas, 5
St Andrews, University of, 259, 260, 261, 268
  and Alan Battersby, 260
St Bartholomew's Hospital, London, 202
St Catharine's College, Cambridge, 35, 88
St Giles' Church, Cambridge, 105
St John's College, Cambridge, 145, 150, 166, 175, 180, 181
  baths, 175
  contested 1702 Chair, 85
  laboratory, 168, 169, 175, 178
  observatory, 87
St Thomas's Hospital, London, 17
staganacin, 248
Stahl, Georg Ernst, 17, 84
Staunton, Jim, 267
staurosporinone, 232, 249
steganacin, 247, 248

steganone, 248
Stein, William H., 261
stereochemistry, 183, 184
steroids, 210, 217
Stevenson, J. J., 173, 184
Stirling, University of, 251
Stokes, George Gabriel, 146, 160
Stradivarius violins, 252
Strathclyde, University of, 233
Streptococcus pneumoniae, 219
strigol, 247, 248
strychnine, 264
Stukeley, William, 8, 9, 36, 37, 39, 40,
    41
Sugden, Morris, 204
sulphur wells, 64
sun-spots, 183
Sydenham, Thomas, 36
Synge, Richard, 206
synthetic rubber, 205

Tate, Thomas, 159
Technische Hochschule, Berlin, 167
Telford, Thomas, 62
Temple, William, 10, 13
Tennant, Smithson, 20, 107, 113–137, 142,
    143, 145
    analysis of emery, 127
    and Jacksonian chair, 107
    antiphlogistic doctrine, 119
    appointed to 1702 Chair, 132
    award of Copley Medal, 114
    becomes Vice-President of Geological
        Society, 132
    character, 115, 122
    childhood and schooling, 116
    continental travels in 1784/5, 119, 204
    continental travels in 1792, 121, 204
    death in France, 134
    discovery of osmium and iridium, 114, 127
    early interaction with W. H. Wollaston, 121
    election to Geological Society of London,
        131
    enters Christ's College, Cambridge, 117
    friendship with Berzelius, 130, 132
    lectures as Professor of Chemistry, 134
    lectures on mineralogy, 131
    membership of 'King of Clubs', 130
    membership of Society for the Promotion of
        Philosophy and General Literature, 97
    obtains M.B., 121

paper on the action of nitric acid on gold
    and platina, 122
paper on the nature of diamond, 122
partnership with W. H. Wollaston, 122, 123,
    126, 128
platinum, 123
receives Cambridge M.D., 121
studies in Edinburgh, 117
studies on latent heat, 119
transfer to Emmanuel College, 119
withdraws candidacy for Jacksonian chair,
    121
terpenes, 264
Test and Corporation Acts, 58
tetrapyrroles, 266, 269, 273, 276
Texas A&M University, 274
thebaine, 263
thiamine. *See* vitamin B_1
Thibaut, Denis, 279
Thomas, Mary Beatrice, 181
Thomson, Thomas, 143
Thorpe, Edward, 179
TNT
    production at Oldbury, 205
Todd née Dale, Alison, 199, 212
Todd, Alexander Robertus, 190, 210–236, 247
    advisory roles, 232
    and Alan Battersby, 259, 267, 268
    and Christ's College, Cambridge, 232
    and deoxynucleoside structure, 228
    appointed to 1702 Chair, 200
    arrival at Cambridge, 213
    death, 233
    description of Pembroke Street Laboratory
        in 1944, 189, 224
    early work, 217
    honours received, 233
    job offer from Caltech, 206
    Nobel Prize, 219
    nucleosides and nucleotides, 219
    offered Cambridge chair of biochemistry,
        212
    portrait, 211
    Presidency of the Royal Society, 233
    relationship with Norrish, 216
    resignation from 1702 Chair, 233, 269
    role within department, 216
    style of management, 214, 215
    suffers heart attack, 269
    views about synthesis, 221
'Toddlers', 213

toluene
  manufacture of explosives from, 201
Tomline, George, 141
Tottenham County School, London, 237
Trevelyan, Walter Blackett, 99
trichodermin, 246
Trinity College, Cambridge, 5, 9, 10, 23, 35,
    37, 38, 41, 44, 45, 58, 59, 61, 90, 140,
    141, 142, 144, 153
  and Mastership of Sidney Sussex, 103
  contested 1702 Chair, 85
  laboratory for Vigani, 37, 38
Trinity Hall, Cambridge, 96
tropolones, 243
Troutbeck, John, 33
trypanocides, 204
Tübingen University, 179
Tuckett, Ronald Francis, 204
Tutte, William, 204
tyrocidine, 260

UCNW Bangor, 182
ultraviolet spectroscopy. *See* spectroscopy
UMIST, 184
University Arms Hotel, Cambridge, 175
University of Liverpool, Robert Robinson
    Laboratories, 266, 267
uridine diphosphate glucose, 224
uroporphyrinogen III, 270, 271, 272, 273, 274,
    275, 277
  role in biosynthesis, 271
Utrecht, 37

Valentine, Basil, 5
van't Hoff, Jacobus, 190
Venice, 32, 33, 41
Verona, 31, 32
Vigani, Giovanni Francesco. *See* Vigani, John
    Francis
Vigani, John Francis, 5–9, 13–14, 24, 31–44
  and Henry Gyles, 35
  and John Covel, 35
  and Robert Boyle, 35
  and Roger Cotes, 37
  and Stephen Hales, 37
  and William Stukeley, 37
  appointed to 1702 Chair, 31
  cabinet of *materia medica*, 7, 38, 41, 42, 43
  death, 37
  laboratory in Trinity College, 37, 38
  links with printers, 36

Medulla Chymiae, 8, 24, 32
  patrons, 33
  reputation, 32
vinblastine, 264
Vinca rosea, 264
Vince, Samuel, 97, 98
vincristine, 264
violins, manufacture of, 252
virantmycin, 232, 249
vitamin A, 238
vitamin B_1, 217, 221, 231
vitamin B_{12}, 231, 243, 270, 271, 275, 276, 279
  crystallisation, 276
  French experiments on biosynthesis, 277
  production by bacteria, 277
  structure announced, 232
Vogel, Emanuel, 245
volcanic ash, 252
Volta, Alessandro, 113
voltaic piles, 113
vortex theory, 177

Wales, University of, 182
Waley Cohen, Robert, 201
Walker, Richard, 90
Waller, John, 9, 10, 13, 23, 45
  appointed to 1702 Chair, 45
Walsh, Arthur Donald, 204
Waltham Abbey, 75
war. *See* World War I and World War II
war gases, 196, 201, 204, 239
Warburton, Henry, 117, 118, 142, 144
Ware, Edgar, 237
Waring, Edward, 87, 88, 106
Warren, Stuart G., 231
water, spectrum of, 183
Watson, James, 215, 228
Watson, Richard, 1, 3, 18, 19, 24, 47, 48, 50,
    57–83, 87, 88, 142
  absenteeism, 70, 71, 72
  appeals for increased stipend for Jacksonian
    professor, 91
  appointed to 1702 Chair, 60
  appointed Bishop of Llandaff, 70
  appointed to Regius chair of Divinity, 66
  arrival at Cambridge, 58
  burning of books, 67
  character, 58, 62, 79
  Chemical Essays, 62, 63, 65, 66
  chemical philosophy, 65
  communication of chemistry, 66

Watson, Richard (*cont.*)
  complaints of examiners' bias, 106
  dress, 58
  election to 1702 Chair, 60
  examiner of Isaac Pennington, 87
  friendship with George Wollaston, 95
  gunpowder manufacture, 94
  industrial practice, 66, 76
  influence on Isaac Milner, 89
  Institutiones Metallurgicae, 62
  landscaping, 76, 77
  metal smelting, 63
  political views, 57, 58, 69, 72
  portrait by George Romney, 59
  portrait by Joshua Reynolds, 78
  preferments, 71
  residence, 60, 71, 73
  resignation from 1702 Chair, 85
  theology, 66, 67, 70
Watt, James, 49
Weedon, Basil Charles Leicester, 240
Wesley College, Dublin, 237
Wesley, John, 73
Westheimer, Frank H., 231
Whewell, William, 23, 24, 147, 150, 151, 153,
    156, 159, 160
Whigs, 57, 58, 59, 60, 62, 67, 70, 72
Whishaw, John (biographer of Tennant), 115,
    116, 117, 122, 130, 131, 133, 134,
    145
Whiston, William, 3, 8, 13
Whitefield, George, 73
Wilberforce, William, 94, 106, 110
Wilkinson, Geoffrey, 238
Williams, Dudley Howard, 217
Willis, Robert, 146, 147, 158, 160, 170
Wilson, Dorothy, 62
Wilson, Woodrow, 205
witchweed, 247
Wollaston, Charlton, 17, 95
Wollaston, Francis (1694–1774), 95
Wollaston, Francis (1731–1815), 95
Wollaston, Francis John Hyde, 21, 121, 142,
    143, 144
  acts as examination moderator, 96
  adoption of Lavoisier's system of chemistry,
    97, 98, 100
  appointed to Observatory Syndicate, 97
  archdeacon of Essex, 106
  appointed to Jacksonian chair, 95

  fellowship of Trinity Hall, 96
  Mastership of Sidney Sussex College,
    103
  on the manufacture of charcoal, 102
  problems with phlogiston theory, 98
  proof of the composition of water, 101
  publication of *A Plan of a Chemical Course
    of Lectures*, 99
  resignation from Jacksonian chair, 103,
    132
  synthesis of ammonia, 101
  synthesis of laughing gas, 102
  view on the nature of heat, 99
Wollaston, George (1738–1826), 95
Wollaston, George Hyde, 123
Wollaston, William (1659–1724), 95
Wollaston, William Hyde, 19, 95, 114, 142,
    143, 145
  declines to stand for 1702 Chair, 103
  discovery of palladium, 126
  discovery of rhodium, 126
  early interaction with Smithson Tennant,
    121
  medical student at Caius, 96
  partnership with Smithson Tennant, 122,
    123, 126, 128
  platinum, 123
women's suffrage, 181
Women's University Settlement, 181
Woodward, Robert Burns, 231
Woodwardian chair, 97, 151
Wordsworth, Christopher, 141, 147
Wordsworth, William, 73, 140
World War I, 181, 182
  impact on Cambridge chemistry, 200
  mustard gas, 196, 201
  phosgene, 201
  photographic sensitisers developed in
    Cambridge, 202
  Special [Gas] Service, 201
World War II, 238
  high-octane gasoline production, 205
  impact on Cambridge chemistry, 202
  Manhattan Project, 205
  synthetic rubber programme, 205

Young, Douglas Wilson, 245

Zerewitinoff determination, 239
Zinsser Sophas Synthesiser, 296